Natural Stone Resources for Historical Monuments

Geological Society books refereeing procedures

The Society makes every effort to ensure that the scientific and production quality of its books matches that of its journals. Since 1997, all book proposals have been refereed by specialist reviewers as well as by the Society's Books Editorial Committee. If the referees identify weaknesses in the proposal, these must be addressed before the proposal is accepted.

Once the book is accepted, the Society Book Editors ensure that the volume editors follow strict guidelines on refereeing and quality control. We insist that individual papers can only be accepted after satisfactory review by two independent referees. The questions on the review forms are similar to those for *Journal of the Geological Society*. The referees' forms and comments must be available to the Society's Book Editors on request.

Although many of the books result from meetings, the editors are expected to commission papers that were not presented at the meeting to ensure that the book provides a balanced coverage of the subject. Being accepted for presentation at the meeting does not guarantee inclusion in the book.

More information about submitting a proposal and producing a book for the Society can be found on its web site: www.geolsoc.org.uk.

It is recommended that reference to all or part of this book should be made in one of the following ways:

PŘIKRYL, R. & TÖRÖK, Á. (eds) 2010. *Natural Stone Resources for Historical Monuments*. Geological Society, London, Special Publications, **333**.

LAHO, M., FRANZEN, C., HOLZER, R. & MIRWALD, P. W. 2010. Pore and hygric properties of porous limestones: a case study from Bratislava, Slovakia. *In*: PŘIKRYL, R. & TÖRÖK, Á. (eds) *Natural Stone Resources for Historical Monuments*. Geological Society, London, Special Publications, **333**, 165–174.

GEOLOGICAL SOCIETY SPECIAL PUBLICATION NO. 333

Natural Stone Resources for Historical Monuments

EDITED BY

R. PŘIKRYL
Charles University in Prague, Czech Republic

and

Á. TÖRÖK
Budapest University of Technology and Economics, Hungary

2010
Published by
The Geological Society
London

THE GEOLOGICAL SOCIETY

The Geological Society of London (GSL) was founded in 1807. It is the oldest national geological society in the world and the largest in Europe. It was incorporated under Royal Charter in 1825 and is Registered Charity 210161.

The Society is the UK national learned and professional society for geology with a worldwide Fellowship (FGS) of over 9000. The Society has the power to confer Chartered status on suitably qualified Fellows, and about 2000 of the Fellowship carry the title (CGeol). Chartered Geologists may also obtain the equivalent European title, European Geologist (EurGeol). One fifth of the Society's fellowship resides outside the UK. To find out more about the Society, log on to www.geolsoc.org.uk.

The Geological Society Publishing House (Bath, UK) produces the Society's international journals and books, and acts as European distributor for selected publications of the American Association of Petroleum Geologists (AAPG), the Indonesian Petroleum Association (IPA), the Geological Society of America (GSA), the Society for Sedimentary Geology (SEPM) and the Geologists' Association (GA). Joint marketing agreements ensure that GSL Fellows may purchase these societies' publications at a discount. The Society's online bookshop (accessible from www.geolsoc.org.uk) offers secure book purchasing with your credit or debit card.

To find out about joining the Society and benefiting from substantial discounts on publications of GSL and other societies worldwide, consult www.geolsoc.org.uk, or contact the Fellowship Department at: The Geological Society, Burlington House, Piccadilly, London W1J 0BG: Tel. +44 (0)20 7434 9944; Fax +44 (0)20 7439 8975; E-mail: enquiries@geolsoc.org.uk.

For information about the Society's meetings, consult *Events* on www.geolsoc.org.uk. To find out more about the Society's Corporate Affiliates Scheme, write to enquiries@geolsoc.org.uk.

Published by The Geological Society from:
The Geological Society Publishing House, Unit 7, Brassmill Enterprise Centre, Brassmill Lane, Bath BA1 3JN, UK

(*Orders*: Tel. +44 (0)1225 445046, Fax +44 (0)1225 442836)
Online bookshop: www.geolsoc.org.uk/bookshop

British Library Cataloguing in Publication Data

A catalogue record for this book is available from the British Library.
ISBN 978-1-86239-291-5

Typeset by Techset Composition Ltd, Salisbury, UK
Printed by CPI Antony Rowe, Chippenham, UK

Distributors

North America
For trade and institutional orders:
The Geological Society, c/o AIDC, 82 Winter Sport Lane, Williston, VT 05495, USA
Orders: Tel. +1 800-972-9892
Fax +1 802-864-7626
E-mail: gsl.orders@aidcvt.com

For individual and corporate orders:
AAPG Bookstore, PO Box 979, Tulsa, OK 74101-0979, USA
Orders: Tel. +1 918-584-2555
Fax +1 918-560-2652
E-mail: bookstore@aapg.org
Website: http://bookstore.aapg.org

India
Affiliated East-West Press Private Ltd, Marketing Division, G-1/16 Ansari Road, Darya Ganj, New Delhi 110 002, India
Orders: Tel. +91 11 2327-9113/2326-4180
Fax +91 11 2326-0538
E-mail: affiliat@vsnl.com

Contents

Preface

The Earth's own building material – rock – has provided an excellent raw material for construction purposes from the very beginning of civilization. Rocks, termed natural stone in construction industry, are often incorrectly regarded as an everlasting material that can be extracted from any place and used for any purpose. If properly handled or dressed, natural stone can impart 'value' to the structure in a much broader way than any other building material. Natural stone, similar to other nature-derived materials, can be highly prone to deterioration, both due to improper use and/or deterioration in the quality of the environment to which it is exposed. This is followed by loss of its integrity and function. The knowledge of stone properties, their development through time and under weathering conditions therefore constitutes a crucial part of natural stone research. Replacement of once deteriorated parts of stone construction raises important issues concerning the compatibility of fresh versus weathered and/or original versus new (alien) stone varieties. The restoration of monuments also requires solid knowledge on the past resources that have ceased to be available over the past century. Although most of the papers within this volume were provided by geologists, the content of their contributions addresses a wider audience anchored in the field of cultural heritage care, monument conservation, civil engineering and architecture.

This volume brings together one general introductory and twenty original research papers grouped in four sections mirroring the major aims of the volume and dominant trends in current research in the field. These are: (1) decay processes, (2) performance and compatibility of natural stone, (3) properties of natural stone and (4) provenance studies and stone databases. Most of the papers were presented during the 'Natural stone resources for historical monuments' special session held under the framework of the 'Energy, Resources, Environment' programme session on the General Assemblies of the European Geosciences Union held in Vienna (Austria) annually during 2006–2008.

The preparation of this volume would not have been possible without help from numerous colleagues who kindly provided reviews: Mónica Alvarez de Buergo, Michael Auras, Thomas Bidner, JoAnn Cassar, Wim Dubelaar, Howel G.M. Edwards, Rafael Fort, Klaus Germann, Ciriaco Giampaolo, Andrew Goudie, Gabreile Grassegger, Ewan Hyslop, Jennifer M. McKinley, Vladimír Machovič, Radek Mikuláš, Derek Mottershead, Urs Müller, Dawn Nicholson, David Robinson, Carlos Rodriguez-Navarro, Jörg Rüdrich, Ricardo Sandrone, Oliver Sass, George W. Scherer, Barbora Schulamnnová, Heiner Siedel, Siegfried Siegesmund, Bernard Smith, Michael Steiger, Heather Viles, Patricia Warke, Tim Yates, Maureen Young, Konrad Zehnder and Fulvio Zezza. Their volunteer work significantly improved the quality of the papers.

The review process and editorial handling was financially assisted by the institutional research project of the Ministry of Education, Youth and Sports of the Czech Republic: MSM 0021620855 'Material flow mechanisms in the upper spheres of the Earth' and also benefited from the support of Hungarian Scientific Research Fund (OTKA, grant no. K63399). Both sources of funds are fully acknowledged.

Finally, we would like acknowledge help from the Geological Society staff during the production of this volume.

RICHARD PŘIKRYL &
ÁKOS TÖRÖK

Natural stones for monuments: their availability for restoration and evaluation

RICHARD PŘIKRYL[1]* & ÁKOS TÖRÖK[2]

[1]*Institute of Geochemistry, Mineralogy and Mineral Resources, Faculty of Science, Charles University in Prague, Albertov 6, 128 43 Prague 2, Czech Republic*

[2]*Budapest University of Technology and Economics, Department of Construction Materials and Engineering Geology, Sztoczek u. 2, H-1111 Budapest, Hungary*

**Corresponding author (e-mail: prikryl@natur.cuni.cz)*

Abstract: This paper represents the introduction to the volume focused on the various aspects of natural stone research. From the topic issues studied recently, four major aspects are covered: (1) availability of certain stone types for monuments, (2) strategies aiming to strengthen our knowledge of past resources through the establishment of natural stone databases and inventories, (3) evaluation of natural stone properties and assessment of the compatibility of repair stones, and (4) decay studies of natural stone on monuments and on natural exposures. This paper also aims to highlight the importance of the use of local stone resources, a practice that has seriously declined during the 20th century in most of the industrialized countries. The availability of natural stone from monuments is also discussed in terms of sandstones and travertine in the Czech Republic and Hungary as a typical example.

Natural stone has become the expression to describe the versatile, durable and aesthetically plausible building materials (Currier 1960). From the very beginning of civilization, important structures and monuments were built from or were principally based on natural stone. The use of local stone resources was mostly in balance with the local environment. Although highly durable when properly applied, no stone type can be considered immortal (Schaffer 1932) and most of the stone varieties are affected by a polluted atmosphere (Winkler 1997). Deteriorated stone on monuments should be preferably replaced by stone varieties of the same composition from the same quarries. However, this is often not possible, and so stone having similar properties and appearance must be sourced (Přikryl 2007). The main aim of this research is to find compatible stones in terms of appearance and geochemical-physical properties. This task requires a thorough understanding of both the rock properties and its response to external conditions and weathering agents.

Availability of natural stone

General

Availability of traditional building stone for the restoration of monuments is rarely discussed in the scientific literature although it represents one of the key points of the conservation of built heritage (Ashurst & Dimes 2004; Snethlage 2005). Exploitation and utilization of local resources of natural stone are a typical feature of world and European history until the turn of 19th/20th century (e.g. Frangipane 2004 and this volume). The tradition of natural stone utilization dramatically declined during the first half of the 20th century when new, artificial materials and previously un-used and alien stone varieties were introduced. Häfner (2007) provides an example of this decline from the sandstones of the Rhine area (Germany) where only 16 operations now exist when previously there were about a thousand quarries until relatively recent times. A similar situation can be found in the most European countries; the Czech Republic has access to about 5–10% of the originally available natural stone varieties (Přikryl *et al.* 2003; Přikryl 2004*a*). New materials such as concrete or imported stones are probably acceptable for new buildings in rapidly expanding suburbs but restored monuments and historic city centres require the use of traditional materials (Dreesen & Dusar 2004; Carta *et al.* 2005).

A common question that architects and restorers often ask geologists concerns the availability of the traditional stone types. The availability is limited not only by geological factors but is regulated by government requirements, local planning, natural and water resources protection (Häfner 2007). Even if the original stone types are still being quarried (which is not commonly the case in most of the

From: PŘIKRYL, R. & TÖRÖK, Á. (eds) *Natural Stone Resources for Historical Monuments.* Geological Society, London, Special Publications, **333**, 1–9.
DOI: 10.1144/SP333.1 0305-8719/10/$15.00

industrialized countries), the quality of recently available stone varieties can be different from those exploited in the past. The quantity of exploitable reserves is another key issue when considering the suitability of a certain site to supply material for monuments (Selonen *et al.* 2000). Small scale, non-continuous operation of local quarries, which has been practised in the past, does not fit modern, industrial-scale exploitation of internationally traded stone varieties. A similar contradiction applies to stone manufacturing for which diamond-saw cutting now prevails (Shadmon 1989). Hand-dressed stone elements are often required for most of the restored monuments, a fact hardly realized in practice.

Recultivation and/or renaturalization of ancient quarry sites and conversion of the land for other purposes is common. As well as agricultural or developer activities, numerous abandoned quarries have also become sites of 'natural protected' status. The protection of these sites is very often not due to the geological record but to other non-geological values (e.g. botanical, zoological) (Přikryl 2009). The question regularly arises as to whether or not small-scale operations aiming to exploit limited amounts of stone for a particular monument will be acceptable on those sites. The public responses, particularly in developed countries, are very aware of the potential environmental impact of any mining/quarrying activity.

In cases where the original stone is no longer available, interest commonly focuses on whether any alternative material can be supplied and used without changing the character of the monument. Even if the alternative stone types show similar petrographic macroscopic character, it can produce a significantly different appearance when exposed to weathering conditions. A typical example of such behaviour is provided by Blows *et al.* (2003) describing the currently unavailable Caen stone (cream-coloured fine-grained French limestone) that has often been replaced by Lepine limestone. The latter exhibits significantly different weathering features. Compatibility assessment is therefore an important research topic in recent natural stone studies as shown by several papers in this volume (Beck & Al-Mukhtar 2010; Nijland *et al.* 2010).

Availability of natural stone for the restoration of monuments: a case example of the Charles Bridge in Prague

The stone Charles Bridge, the oldest preserved bridge in Prague (Czech Republic), represents one of the iconic monuments of the city that shows long-term deterioration of its natural stone facing masonry. The bridge presents a mosaic of numerous types of local sandstones (quartz arenites, litharenites and arkosic arenites from Carboniferous and Upper Cretaceous sediments in the Prague neighbourhoods) that were utilized either during the original construction during 1357–1402 period or for later repairs (Fig. 1a). Based on the detailed petrographic research of stone samples from the bridge and geotechnical survey (Drozd & Přikryl 2003; Drozd *et al.* 2005), seven quarry areas that provided two major rock types, Carboniferous arkoses and Cretaceous sandstones (Table 1), can be traced. Categorization of these stones was facilitated by a lithotheque of historical dimension stones of the Czech Republic (Přikryl *et al.* 2001, 2004*a*), a fact-documenting exercise through study of local stone resources and their presentation on the form of stone databases, lithotheques and catalogues (see papers written by Cooke 2010; Frangipane 2010; Kampfová & Přikryl 2010 and/or Allocca *et al.* 2010). Until about the mid-19th century, the sources of stones for the Charles Bridge remained unchanged but, in the late 19th and 20th centuries, repairs did not respect the original types of stones and new stone

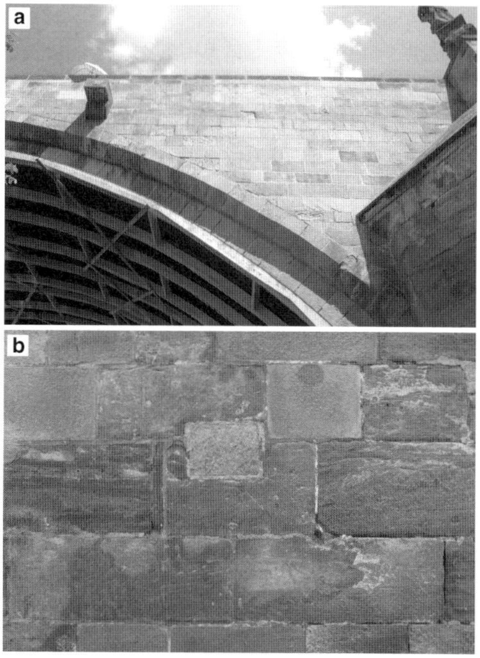

Fig. 1. The sandstone facing masonry of the Charles Bridge in Prague (Czech Republic) (**a**) has often been repaired using different types of local stone resources. The stone used during the latest repair in 1960–1970 (**b**) shows odd colours (beige to rusty yellow) and rock fabric which differentiate it from the original sandstones (which have a dark grey surface due to the polluted atmosphere in Prague).

Table 1. *Summary of natural stone types used for the construction and repairs of the Charles Bridge in Prague (modified after Přikryl 2006a)*

Rock type	Source area/locality	Period of application	Extent of use	Availability for restoration
Arkoses (medium to coarse grained, beige and yellow colour)	Carboniferous, limnic Kladno-Rakovník basin W and N from Prague, 4 major quarry areas at a distance of 20–50 km	Undated, probably 14th to early 19th centuries	Not measured on the whole bridge, 50–90% of test areas	Recently not available, great interest in using it (quarry site must be explored)
Quartz sandstone with clay matrix (generally fine-grained, whitish to grey colour)	Upper Cretaceous, Cenomanian, mostly abandoned quarry areas NW, N and E from Prague (distance up to 30 km)	Undated, probably 14th to early 19th centuries	Not measured on the whole bridge, 20–45% of test areas	Partly available, no interest in using it (low durability, unsure period of use)
Glauconitic quartz sandstone with clay matrix (extremely fine-grained, green-grey colour)	Upper Cretaceous, Cenomanian, abandoned quarry areas NW, N and E from Prague (distance up to 30 km)	Undated, probably 14th to early 19th centuries or repairs until the end of 18th century	Not measured on the whole bridge, up to 2% of test areas	Not available, no interest in using it (minor use, low durability)
Quartz sandstone with Fe-hydroxides cement (generally medium-grained, deep rusty brown colour)	Upper Cretaceous, basis of beds on Petřín hill (distance less than 1 km from the bridge)	Undated, probably during construction in 14th century as a recycled material from the antecedent Judita Bridge	Not measured on the whole bridge, up to 10% of test areas	Not available, potential interest in using it (minor use)
Quartz sandstone with clay matrix (fine-grained, yellow colour)	Upper Cretaceous, Cenomanian, active quarry area near Hořice E of Prague (distance 130 km)	After 1890 flood repair	Not measured on the whole bridge, broad use on three collapsed arches	Available, no interest in using it (extremely low durability)
Glauconitic quartz sandstone with clay matrix (generally fine-grained, yellowish green-grey colour)	Upper Cretaceous, Cenomanian, active quarry area near Libná NE of Prague (distance 170 km)	1960–1970 repair	Not measured on the whole bridge	Partly available, no interest in using it (low durability, new repair stone only)
Arkosic sandstone (generally medium-grained, light beige colour)	Upper Cretaceous, Turonian, active quarry area NE of Prague (distance 170 km)	1960–1970 repair, also for 2005 repair of pier nos 8 and 9	Not measured on the whole bridge	Available, no interest in using it (low durability)

varieties were introduced (Table 1). This was partly due to economic reasons but also to the fact that the 20th century repairs suffered from the closure of all sandstones quarries in the Prague area.

The Charles Bridge also suffered from incorrect maintenance during the 20th century. Major repairs were conducted after the damaging floods in 1890 (repairs continued till 1910) and again in 1960–1970. Between and after these repairs, there was no ordinary maintenance of the facing masonry partly caused by missing stoneworks of Charles Bridge and non-availability of the original stone. The most serious impact of improper modern repairs is linked to the fact that the original types of natural stone were not employed. As stated before, this was mainly caused by the closure of original quarries. As a consequence, other types of sandstones showing different quality (in terms of physical properties and of appearance – see Fig. 1b) were introduced which probably accelerated the deterioration of the facing masonry. These stones show pronounced granular disintegration by salt weathering followed by the surface retreat of 1 cm (rate 0.3 mm/a) for Božanov arkosic sandstone after 30 years of service and up to 3 cm retreat for Hořice sandstone after 100 years of service. Use of Portland cement-based concrete for fixing of newly inserted ashlars worsened the state of the bridge further.

Discussions for the new repair and maintenance plan opened the question of whether or not the dominant traditional natural stone (Carboniferous arkoses) could be used for the replacements (Přikryl 2004*b*, 2006*a*) in spite of the fact that its exploitation finished at the beginning of 20th century. In 2004–2008, an extensive desk study and field reconnaissance focused on several areas of the Czech Republic where this stone type was quarried in the past. From eleven abandoned quarries, two sites provided promising results; the possible re-opening of these sites is being discussed. Although the macroscopic similarity to the original stone was the main criterion, the physical properties, durability and amount of reserves were the main decision-making criteria for selection of these sites.

Property evaluation

Assessment of natural stone properties became routine practice throughout the 19th/20th centuries. The testing of physical properties, both index and mechanical, relies on procedures developed for other artificial inorganic silicate materials, namely concrete, bricks and glass. The specific features of natural stone concerning their genesis and wide range of composition makes 'blind' adoption of testing procedures often questionable and often requires test procedures that at least be modified. This also affects the evaluation of the test results. This broad issue is discussed below in the context of rock durability.

Durability of natural stone reflects its ability to withstand the external pressures leading to the deterioration of the physical properties, partial decomposition and physical breakdown. The durability is therefore proportional to the period during which the stone can preserve its properties, both physical and aesthetical. The resistance of natural stone to weathering action is a function of:

- internal parameters of the rock which encompass genetic conditions, mineralogical and chemical composition, rock fabric (understood as spatial arrangement of rock-forming minerals and pore space – see discussion provided by e.g. Přikryl 2006*b*) and isotropy or anisotropy of rock fabric;
- external factors that can generally be described as the environmental conditions to which the stone is exposed (climatic conditions, composition of atmosphere including presence of pollutants, presence of water, biospheric influence and/or interaction between stone piece and other materials present in the construction, Smith *et al.* 2001).

Some authors (e.g. Warke 1996) also distinguish between factors that influenced the rock before its emplacement in the construction (i.e. mode of the extraction, dressing of the stone) and factors that modify the stone during its service. The latter case has been found extremely significant for stone sculptures which have experienced numerous restoration/conservation treatments in the past (Přikryl *et al.* 2004*b*).

Weathering as a process leading to both functional and aesthetical loss of the original value of the stone (Smith & Přikryl 2007) is often manifested by numerous so-called weathering phenomena that have been studied in detail during the last few decades both from the point of view of the mechanisms of formation and the methodology of assessment (e.g. Warke *et al.* 2003). Monuments represent highly prized cultural objects that do not allow detailed study and sufficient sampling in most cases; the study of weathering processes on natural objects therefore presents a welcome alternative (see e.g. Siedel 2010).

The susceptibility of natural stone to weathering agents was tested by several approaches during past decades. Along with normalized procedures that focused on the salt crystallization action and freezing/thawing of water in pore systems of natural stone, widely adopted in practical assessment of recently produced stone types (EN 12370 2000; EN 12371 2002), numerous scientific experimental approaches have also been applied. The latter tests mostly rely on accelerated decay in laboratory conditions and employ just one type of weathering agent (e.g. variable types of salts) (see Angeli *et al.* 2010; Yu & Oguchi 2010; Oguchi & Yuasa 2010). Numerous studies published during the past decade document a significant increase in knowledge both regarding the processes and respective stone type response (Turkington & Paradise 2005).

Some of the tests are interpreted according to the old practices which, unfortunately, are linked to certain types of environment or climate and cannot be easily extrapolated and generalized. Assessment of the rock susceptibility to the frost action is a typical example (Thomachot & Jeannette 2002). When assessing this parameter, the rock specimens are tested for their uniaxial compressive strength in the dry state and after being subjected to 25 freeze-thaw cycles. The number of cycles – 25 – was originally suggested by Hirschwald (1911) based on an evaluation of the number of frost days that might have had an impact on natural stone in Germany (Berlin) during the period 1884–1892. The lowest number of frost days (Hirschwald considered the days during which the temperature was below 0 °C and were preceded by a rainy period which caused wetting of the rock) was 14 and the highest 25. Based on this observation, Hirschwald (1911) proposed to subject the rocks to 25 freeze-thaw cycles which should correspond to 1–3 years of real conditions. It is obvious that this can only be valid for regions having similar climatic conditions as Berlin in the late 19th century. The freeze-thaw impact of one winter in for example, New York area can be better deduced from 12–16 cycles as shown by Bortz & Wonneberger (1997).

The dynamics of the changes during the above-mentioned tests presents another important issue. It has been shown by numerous theoretical considerations and practical tests that the decrease of stone properties due to accelerating weathering is neither linear nor continuous (e.g. Smith *et al.* 1992). This is another weak point in the practical assessment of the quality of natural stone by standard procedures when only the values before and after the cycles are compared. In contrast, experimental studies often evaluate changes after each cycle (e.g. Goudie 1999).

The real weathering conditions involve several processes. The test design should therefore combine those processes that are known or likely expected from the site where the natural stone will be emplaced (Duffy & O'Brien 1996). The real conditions can be better modelled using large climatic chambers allowing control of several factors (temperature, humidity, concentration of salts etc.) (Taylor-Firth & Laycock 1999).

Performance in use and compatibility of different varieties of natural stone

From a great variety of stones, only one example is given in the wide use and compatibility of stones, namely travertine. Travertine is a common building and dimension stone that has been explored and used in many countries from prehistoric times (Pentecost 2005). Travertine was one of the favourite stones of the Roman Empire and historic monuments were made from this stone in the heart of the Empire in Rome (e.g. the Colosseum). Understanding the properties and workability of this stone allowed the Romans to search for similar deposits throughout the entire Roman Empire. This resulted in the widespread use of travertine during Roman times not only in Rome but far away from the '*locus typicus*' of travertine in Europe (e.g. present day London) and Asia (e.g. Hierapolis in Turkey). The fate of the Roman Empire brought a drastic decline in the use of travertine. Far fewer monuments and structures were built from this unique stone in the following periods. Nevertheless, numerous Renaissance and Baroque buildings made of travertine are known. The previous Roman travertine quarries (near Tivoli) were partly abandoned but later re-opened and new quarries were developed, especially near Rome. The existing deposit, which provides the source material, is therefore always available but its use depends on the traditions and understanding of workability of stones. Travertine is also gaining increasing popularity in modern architecture, especially when newly built emblematic structures are considered. From many possible examples just two are given: travertine slabs and blocks were used at the Metropolitan Opera House in New York and at J. Paul Getty Museum and Conservation Center in Los Angeles, to where large quantities of travertine were transported from Rome.

There is another aspect of the use of travertine, which is its long-term performance. Travertine is generally considered a durable stone, although air pollution-related weathering is commonly observed and black crust-covered façades are common in urban areas (Sidraba *et al.* 2004; Török 2006, 2008). In comparison with other porous limestones, travertine normally performs better than many of its porous carbonate counterparts in extreme stress conditions (Török 2004). A good example is from Budapest, where the façade of Parliament House (late 19th century building) was originally made of porous Miocene oolitic limestone. This showed rapid decay after only 20 years from completion of the building. Due to the deterioration (Török 2002) and structural damage (Török 2003*a*) of the porous limestone and unavailability of durable porous oolitic limestones, a decision was made

Fig. 2. Stone replacement at Parliament House, Budapest: the porous limestone façade is replaced by more durable travertine.

(after the Second World War) that a type of a Hungarian travertine would be used as a replacement stone. Consequently, during the restoration works the entire façade will be replaced by travertine (Török 2008) (Fig. 2). The improved long-term performance of the high-quality travertine has been demonstrated by checking other built examples. At Mathias Church (Budapest) the parts that are the most exposed were built from travertine (window frames, gargoyles, footing), while the rest of the church façade is covered by the less durable porous limestone. This understanding of the long-term durability of this stone resulted in use of travertine in monuments throughout the country for centuries. Long transport distances did not hamper the application of this workable but durable stone, as it is clearly seen on the map of the country where the two regions with travertine quarries and the localities where travertine in monuments are found are indicated (Fig. 3).

Although the Hungarian travertine is very similar to that of Italy and Turkey in terms of the origin (Török 2007), since it was also precipitated from lukewarm waters (Korpás 2003; Török 2003*b*) the selection of proper quality travertine must be based on quality checks and laboratory tests. These types of travertines are less porous and more durable than calcareous tufas of recent or relatively recent stream deposits. The use of travertine as building and/or replacement stone depends on its mechanical properties and on the long-term behaviour, which is primarily controlled by water content and its influence on rock strength, that is, freeze-thaw durability or thermal behaviour (Gomez-Heras *et al.* 2006).

Conclusions

The late 20th and 21st century brought a new era in the use of natural stones. Due to the availability and bulk mass of dimension stones, the use of imported stones has significantly increased in most industrialized countries. These almost unlimited sources of cheap stones and the strict environmental legislations resulted in the closing of many long-existing quarries in industrialized countries, which led to a shortage of available local stone resources.

Fig. 3. Map of Hungary showing travertine quarries and localities where travertine is used in the monuments. The wide distribution of monuments indicates that travertine was a popular dimension and building stone from the Roman period and that the raw material was also transported longer distances.

This lack of locally available stones is a significant drawback in monument restoration practice, since replacement stones are no longer available from the original source. Consequently, researchers and restorers are trying to solve this problem by various different approaches. One solution is the identification of old quarries and deposits that have similar stones to the monument. The re-opening of ancient quarries or beginning new quarrying activity at known deposits can provide material for use as replacement stones. Another approach is the *in situ* preservation of monumental stones by understanding the decay mechanism and using preventive conservation methods or techniques to slow down the decay processes. This approach is often not possible and the long-term performance of such interventions (e.g. consolidants) can only be judged after years. The role of scientists and practitioners in preserving our built stone heritage is therefore clear; it has to focus on: (i) providing sound scientific background and data based on the availability of possible replacements stones (ii) evaluating the properties of new stones and stones that are found in the structure; (iii) judging the risks of application of replacement stones or the long-term durability of historic stone structures; and (iv) modelling the long-term effect on any conservation intervention.

This paper benefited from financial support from the project of the Ministry of Education, Youth and Sports of the Czech Republic: MSM 0021620855 'Material flow mechanisms in the upper spheres of the Earth' and from Hungarian Scientific Research Fund (OTKA, grant no. K63399).

References

Allocca, F., Calcaterra, D., Calicchio, G., Cappelletti, P., Colella, A., Langella, A. & de'Gennaro, M. 2010. Ornamental stones in the cultural heritage of Campania region (southern Italy): the Vitulano marbles. *In*: Přikryl, R. & Török, Á. (eds) *Natural Stone Resources for Historical Monuments*. Geological Society, London, Special Publications, **333**, 219–231.

Angeli, M., Hébert, R., Menéndez, B., David, C. & Bigas, J.-P. 2010. Influence of temperature and salt concentration on the salt weathering of a sedimentary stone with sodium sulphate. *In*: Přikryl, R. & Török, Á. (eds) *Natural Stone Resources for Historical Monuments*. Geological Society, London, Special Publications, **333**, 35–42.

Ashurst, J. & Dimes, F. G. 2004. *Conservation of Building and Decorative Stone*. Elsevier, Amsterdam.

Beck, K. & Al-Mukhtar, M. 2010. Evaluation of the compatibility of building limestones from salt crystallization experiments. *In*: Přikryl, R. & Török, Á. (eds) *Natural Stone Resources for Historical Monuments*. Geological Society, London, Special Publications, **333**, 111–118.

Blows, J. F., Carey, P. J. & Poole, A. B. 2003. Preliminary investigations into Caen Stone in the U.K.; its use, weathering and comparison with repair stone. *Building and Environment*, **38**, 1143–1149.

Bortz, S. A. & Wonneberger, B. 1997. Laboratory evaluation of building stone weathering. *In*: Labuz, J. F. (ed.) *Degradation of Natural Building Stone*. Geotechnical Special Publications 72, ASCE, Reston (USA), 85–104.

Carta, L., Calcaterra, D., Cappelletti, P., Langella, A. & de'Gennaro, M. 2005. The stone materials in the historical architecture of the ancient center of Sassari: distribution and state of conservation. *Journal of Cultural Heritage*, **6**(3), 277–286.

Cooke, L. 2010. The 19th century Corsi collection of decorative stones – a resource for the 21st century? *In*: Přikryl, R. & Török, Á. (eds) *Natural Stone Resources for Historical Monuments*. Geological Society, London, Special Publications, **333**, 185–195.

Currier, L. W. 1960. Geological appraisal of dimension-stone deposits. *Geological Survey Bulletin*, **1109**, 1–78.

Dreesen, R. & Dusar, M. 2004. Historical building stones in the province of Limburg (NE Belgium): role of petrography in provenance and durability assessment. *Materials Characterisation*, **53**(2–4), 273–287.

Drozd, K. & Přikryl, R. 2003. Credibility of geotechnical exploration methods for the Charles Bridge. *Geotechnika*, **6**(4), 3–8 (in Czech).

Drozd, K., Přikryl, R. & Votoček, R. 2005. Charles Bridge – additional geotechnical survey. *Geotechnika*, **8**(2), 3–11 (in Czech).

Duffy, A. P. & O'Brien, P. F. 1996. A basis for evaluating the durability of new building stone. *In*: Smith, B. J. & Warke, P. A. (eds) *Processes of Urban Stone Decay*. Donhead Publishing Ltd., London, 253–260.

EN 12370, 2000. Natural stone test methods – Determination of resistance to salt crystallization. CEN, Brussels.

EN 12371, 2002. Natural stone test methods – Determination of frost resistance. CEN, Brussels.

Frangipane, A. 2004. The use of natural and artificial stone in the portlas of the town Udine (Italy). *In*: Přikryl, R. & Siegl, P. (eds) *Architectural and Sculptural Stone in Cultural Landscape*. Karolinum, Prague, 149–156.

Frangipane, A. 2010. Working for an electronic database of historical stone resources in Friuli Venezia Giulia (Italy). *In*: Přikryl, R. & Török, Á. (eds) *Natural Stone Resources for Historical Monuments*. Geological Society, London, Special Publications, **333**, 197–209.

Gomez-Heras, M., Alvarez de Buego, M., Fort, R., Hajpál, M., Török, Á. & Varas, M. J. 2006. Evolution of porosity in Hungarian building stones after simulated burning. *In*: Fort, R., Alvarez de Buergo, M., Gomez-Heras, M. & Vazquez-Calvo, C. (eds) *Heritage, Weathering and Conservation*. Vol. 1. Taylor & Francis Group, London, 513–519.

Goudie, A. 1999. Experimental salt weathering of limestones in relation to rock properties. *Earth Surface Processes and Landforms*, **24**, 715–724.

HÄFNER, F. 2007. On the perspective of the availability of ornamental rocks for the conservation of monuments. *Zeitschrift der Deutschen Gesellschaft für Geowissenschaften*, **158**(4), 695–700.

HIRSCHWALD, J. 1911. *Handbuch der Bautechnischen Gesteinsprüfung*. Teil I., Verlag von Gebrüder Borntraeger, Berlin.

KAMPFOVÁ, H. & PŘIKRYL, R. 2010. Electronic database of historical natural stones of the Czech Republic: structuring field and laboratory data. *In*: PŘIKRYL, R. & TÖRÖK, Á. (eds) *Natural Stone Resources for Historical Monuments*. Geological Society, London, Special Publications, **333**, 211–217.

KORPÁS, L. 2003. Basic pattern of Quaternary travertine: a review with special regard to the Hungarian deposits. *Acta Geologica Hungarica*, **46**, 131–148.

NIJLAND, T. G., VAN HEES, R. P. J. & BOLONDI, L. 2010. Evaluation of three Italian tuffs (Neapolitan Yellow Tuff, Tufo Romano, Tufo Etrusco) as compatible replacement stone for Römer tuff in Dutch built cultural heritage. *In*: PŘIKRYL, R. & TÖRÖK, Á. (eds) *Natural Stone Resources for Historical Monuments*. Geological Society, London, Special Publications, **333**, 119–128.

OGUCHI, C. T. & YUASA, H. 2010. Simultaneous wetting/drying, freeze/thaw and salt crystallization experiments of three types of Oya tuff. *In*: PŘIKRYL, R. & TÖRÖK, Á. (eds) *Natural Stone Resources for Historical Monuments*. Geological Society, London, Special Publications, **333**, 59–72.

PENTECOST, A. 2005. *Travertine*. Springer-Verlag, Berlin.

PŘIKRYL, R. 2004*a*. The natural stone industry in the Czech Republic. *Roc Maquina*, **53**, 32–35.

PŘIKRYL, R. 2004*b*. Which natural stone should be selected for the repair of Charles Bridge? *In*: *Charles Bridge Repair*, ČSSI, Šel, Praha, pp. 50–54 (in Czech).

PŘIKRYL, R. 2006*a*. 'New natural stone' for the reconstruction of Charles Bridge in Prague. *In*: FORT, R., ALVAREZ DE BUERGO, M., GOMEZ-HERAS, M. & VAZQUEZ-CALVO, C. (eds) *Heritage, Weathering and Conservation*. Vol. 1. Taylor & Francis Group, London, 23–29.

PŘIKRYL, R. 2006*b*. Assessment of rock geomechanical quality by quantitative rock fabric coefficients: Limitations and possible source of misinterpretations. *Engineering Geology*, **87**(3–4), 149–162.

PŘIKRYL, R. 2007. Understanding the earth scientist's role in the pre-restoration research of monuments: an overview. *In*: PŘIKRYL, R. & SMITH, B. J. (eds) *Building Stone Decay: From Diagnosis to Conservation*. Geological Society, London, Special Publications, **271**, 9–21.

PŘIKRYL, R. 2009. Geomorphosites and quarries. *In*: REYNARD, E. & CORATZA, P. (eds) *Geomorphosites*. Pfeil Verlag, München, 225–233.

PŘIKRYL, R., SVOBODOVÁ, J. & SIEGL, P. 2001. Search for historical resources of dimension stone in the Czech Republic. *In*: SANDRONE, R. (ed.) *Proceedings of the International Workshop 'Dimension Stones of the European Mountains'*, June 10–12, 2001, Luserna san Giovanni – Torre Pellice (Italy), pp. 307–309.

PŘIKRYL, R., SVOBODOVÁ, J. & SIEGL, P. 2003. Balanced use of building stone in the Czech Republic – an achievable goal? *Applied Earth Science (Transactions of the Institute of Mining and Metallurgy, Section B)*, **112**(2), B140–B142.

PŘIKRYL, R., SVOBODOVÁ, J. & SIEGL, P. 2004*a*. Historical dimension stone resources in the Czech Republic. *Roc Maquina*, **53**, 28–31.

PŘIKRYL, R., SVOBODOVÁ, J., ŽÁK, K. & HRADIL, D. 2004*b*. Anthropogenic origin of salt efflorescences on sandstone sculptures (Charles bridge, Prague, Czech Republic) – mineralogical and stable isotope geochemistry evidence. *European Journal of Mineralogy*, **16**(4), 609–617.

SCHAFFER, R. J. 1932. *The Weathering of Natural Building Stones*. Donhead facsimilie, 2004.

SELONEN, O., LUODES, H. & EHLERS, C. 2000. Exploration for dimensional stone – implications and examples from the Precambrian of southern Finland. *Engineering Geology*, **56**, 275–291.

SHADMON, A. 1989. *Stone: An Introduction*. Intermediate Technology Publications, London.

SIDRABA, I., NORMANDIN, K. C., CULTRONE, G. & SCHEFFLER, M. J. 2004. Climatological and regional weathering of Roman travertine. *In*: PŘIKRYL, R. & SIEGEL, P. (eds) *Architectural and Sculptural Stone in Cultural Landscape*. The Karolinum Press, Prague, 211–228.

SIEDEL, H. 2010. Alveolar weathering of Cretaceous building sandstones on monuments in Saxony, Germany. *In*: PŘIKRYL, R. & TÖRÖK, Á. (eds) *Natural Stone Resources for Historical Monuments*. Geological Society, London, Special Publications, **333**, 11–23.

SMITH, B. J., WHALLEY, B. & MAGEE, R. 1992. Assessment of building stone decay: a geomorphological approach. *In*: WEBSTER, R. G. M. (ed.) *Stone Cleaning and the Nature, Soiling and Decay Mechanisms of Stone*. Donhead, London, 249–257.

SMITH, B. J., TURKINGTON, A. V. & CURRAN, J. M. 2001. Calcium loading of quartz sandstones during construction: implications for future decay. *Earth Surface Processes and Landforms*, **26**, 877–883.

SMITH, B. J. & PŘIKRYL, R. 2007. Diagnosing decay: the value of medical analogy in understanding the weathering of building stones. *In*: PŘIKRYL, R. & SMITH, B. J. (eds) *Building Stone Decay: From Diagnosis to Conservation*. Geological Society, London, Special Publications, **271**, 1–8.

SNETHLAGE, R. 2005. *Leitfaden Steinkonservierung*. 2nd edn. Fraunhofer IRB Verlag, Stuttgart.

TAYLOR-FIRTH, A. & LAYCOCK, E. A. 1999. Weathering simulation for durability assessment of the in-service performance of construction materials. *Quarterly Journal of Engineering Geology*, **32**, 291–302.

THOMACHOT, C. & JEANNETTE, D. 2002. Evolution of the petrophysical properties of two types of Alsatian sandstone subjected to simulated freeze-thaw conditions. *In*: SIEGESMUND, S., WEISS, T. & VOLLBRECHT, A. (eds) *Natural Stone, Weathering Phenomena, Conservation Strategies and Case Studies*. Geological Society, London, Special Publications, **205**, 19–32.

TÖRÖK, Á. 2002. Oolitic limestone in polluted atmospheric environment in Budapest: weathering phenomena and alterations in physical properties. *In*: SIEGESMUND, S., WEISS, T. & VOLLBRECHT, A. (eds) *Natural Stone, Weathering Phenomena, Conservation Strategies and Case Studies.* Geological Society, London, Special Publications, **205**, 363–379.

TÖRÖK, Á. 2003*a*. Surface strength and mineralogy of weathering crusts on limestone buildings in Budapest. *Building and Environment*, **38**, 1185–1192.

TÖRÖK, Á. 2003*b*. Facies analysis and genetic interpretation of travertine. Buda Vár-hegy, Hungary. *Acta Geologica Hungarica*, **46**, 177–193.

TÖRÖK, Á. 2004. Comparison of the processes of decay of two limestones in a polluted urban environment. *In*: MITCHELL, D. J. & SEARLE, D. E. (eds) *Stone Deterioration in Polluted Urban Environments.* Science Publishers Inc., Enfield, USA, 73–92.

TÖRÖK, Á. 2006. Influence of fabric on the physical properties of limestones. *In*: KOURKOULIS, S. K. (ed.) *Fracture and Failure of Natural Building Stones.* Springer, Dordrecht, 487–495.

TÖRÖK, Á. 2007. Hungarian dimensional stones: an overview. *Zeitschrift der Deutschen Gesellschaft für Geowissenschaften*, **158**, 361–374.

TÖRÖK, Á. 2008. Black crust on travertine: factors controlling development and stability. *Environmental Geology*, **56**, 583–597.

TURKINGTON, A. V. & PARADISE, T. R. 2005. Sandstone weathering: a century of research and innovation. *Geomorphology*, **67**, 229–253.

WARKE, P. A. 1996. Inheritance effects in building stone decay. *In*: SMITH, B. J. & WARKE, P. A. (eds) *Processes of Urban Stone Decay*. Donhead, London, 32–43.

WARKE, P. A., CURRAN, J. M., TURKINGTON, A. V. & SMITH, B. J. 2003. Condition assessment for building stone conservation: a staging system approach. *Building and Environment*, **38**, 1113–1123.

WINKLER, E. M. 1997. *Stone in Architecture*. 3rd edn. Springer-Verlag, Berlin.

YU, S. & OGUCHI, C. T. 2010. Is sodium sulphate invariably effective in destroying any type of rock? *In*: PŘIKRYL, R. & TÖRÖK, Á. (eds) *Natural Stone Resources for Historical Monuments.* Geological Society, London, Special Publications, **333**, 43–58.

Alveolar weathering of Cretaceous building sandstones on monuments in Saxony, Germany

HEINER SIEDEL

Dresden University of Technology, Institute of Geotechnical Engineering, Chair of Applied Geology, D-01062 Dresden, Germany (e-mail: Heiner.Siedel@tu-dresden.de)

Abstract: The pattern of alveolar weathering on Cretaceous building sandstones from Saxony has been studied and related to the sandstone rock fabric and the spatial distribution of soluble salts at two monuments. The results demonstrate that uneven weathering of the stone surface is initiated by differences in porosity and mineral composition at macroscopic or microscopic scale within the heterogeneous material. Magnesium sulphate, the dominant salt, accumulates in areas with a higher content of clay minerals or ferric oxides/hydroxides, characterized by a higher amount of smaller pores. At an incipient stage, small pits in the surface are formed by salt weathering at these points. The stone dries more slowly from the base of these pits, thus accumulating more salt and leading to further material loss. Once deepened, the hole can shelter efflorescing salts from rain. The preferential weathering at the surface of the holes by salt crystallization and hydration pressure becomes a self-perpetuating process at this second stage, dependent only on changing moisture and climate. Building materials (mortars) and sulphur derived from air pollution are the main sources of the salts involved in the process of alveolar weathering on the investigated buildings. In contrast, salts that cause alveolar weathering of the Cretaceous sandstone bedrock in natural outcrops in Saxony mainly originate from long-term chemical weathering of the rock itself.

Alveolar or honeycomb weathering is a weathering phenomenon found in diverse environments throughout the world. It occurs on sandstones and other rock types in natural outcrops (Mustoe 1982), where these conspicuous, peculiar and sometimes bizarre weathering forms (Fig. 1) had been noted by scientists in the 19th century. Nevertheless, the origin of honeycomb weathering is still under discussion (Robinson & Williams 1994; Turkington & Paradise 2005). Alveolar weathering develops in a range of environments and as a result of a number of different weathering processes involving soluble salts in most cases. In coastal environments, the role of salty seawater and salt spray in the weathering process is emphasized (Mustoe 1982; Mottershead 1994; Pye & Mottershead 1995) whereas in humid, inland environments the irregular flow of solutions through porous rocks, case hardening and the presence of salts (Winkler 1994) seem to be the most important factors for honeycomb weathering. The wind might play a role in the initial formation of alveoli in salt-loaded rocks (Rodriguez-Navarro *et al.* 1999; Quayle 1992). Gravity and oriented pressure influence their shape (Mikuláš 2001). A crucial point in the discussion is the question of whether honeycomb patterns may develop randomly from macroscopically homogeneous rock fabrics (as demonstrated by a weathering experiment of Rodriguez-Navarro *et al.* 1999) or if incipient honeycomb formation is necessarily related to heterogeneities in rock fabric (as could be suggested from strata-bound development of alveoli, Fig. 1 Jeanette 1980; Alessandrini *et al.* 1992). A model calculation by Huinink *et al.* (2004) has demonstrated the evolution of an alveole from a small pit in the surface of a homogeneous rock solely by salt crystallization during wetting/drying cycles. This corresponds to the suggestion that the deepening of alveoli is in general (apart from any specific weathering mechanism) a self-reinforcing process (Turkington & Phillips 2004).

In Saxony, the honeycomb weathering of Cretaceous sandstone in natural outcrops in a humid inland environment (Fig. 1) was described as early as the middle of the 19th century (Gutbier 1858) and has been explained by combined salt attack (alum, gypsum) and case hardening (Beyer 1911; Lentschig-Sommer 1960).

The term alveolar or honeycomb weathering is also used in the description of weathering forms of building stones. Fitzner *et al.* (1995, p. 54) explained 'alveolar weathering' on building stones generally as 'relief in the form of closely spaced cavities (alveolae)'. Several authors used the term for a more or less regular system of back-weathered holes and pits in building stone surfaces: Quayle (1992, p. 110) characterized this weathering as the

From: PŘIKRYL, R. & TÖRÖK, Á. (eds) *Natural Stone Resources for Historical Monuments.*
Geological Society, London, Special Publications, **333**, 11–23.
DOI: 10.1144/SP333.2 0305-8719/10/$15.00

Fig. 1. Alveolar ('honeycomb') weathering of the Cretaceous sandstone bedrock at Bärenstein, Elbe Valley (Saxony). Right figure margin *c.* 2.5 m.

formation of 'small but often deep cavities (up to about 30 mm across and 20 mm in depth, but sometimes much larger) occurring either individually or in honeycomb-like clusters'. Livingstone (1994) presented a case study and called single, deep holes with a large distance between each other 'alveolar holes'. Alessandrini *et al.* (1992, p. 15) defined alveolar weathering (following the Italian standard 'Raccomandazione Normal 1/88') as 'degradation ... by formation of various shape and dimension cavities. The alveoli are often interconnected and without a uniform distribution'. These few examples show that there is some terminological confusion in the use of 'alveolar' or 'honeycomb weathering' for building stones. Because of morphological transitions, a distinction between alveolar weathering and weathering out of stone components or weathering out dependent on stone structure (in the sense of Fitzner *et al.* 1995) is problematic in some cases. The latter forms of weathering could be a first step in the development of honeycomb-like pattern, as shown for tuff by Siedel (1998) and for sandstone by Jeanette (1980).

With regard to Cretaceous building sandstones in Saxony, 'alveolar weathering' is generally used in this paper for a regular or irregular network of less- or non-weathered 'walls' (area where the original surface is preserved) between deep-weathered alveoli (small hollows with maximum dimension of the order centimetres). The aim of this paper is a detailed investigation of alveolar structures on Cretaceous sandstones from Saxony that have been used for building purposes with respect to:

- influence of the type and the spatial distribution of salts in alveoli and neighbouring stable areas or walls on the weathering process; and
- the control of incipient alveolar weathering by rock fabric heterogeneities of the sandstones at the microscopic or macroscopic scale.

Geological setting and sandstone material

Cretaceous sandstones used as historic building materials in Saxony mainly originate from different stratigraphic horizons (Cenomanian to Coniacian age) of the Cretaceous sediment basin of the Elbe Zone (around the city of Dresden), which consists of layers of clays, marlstones and sandstones up to several hundred metres thick (Fig. 2, area 1). Sandstones have been quarried there at least since the early 13th century.

The stones used for building purposes are mature, monotonous quartz sandstones (Götze & Siedel 2007), which are dominated by high quartz content (up to 99%). The feldspar content is rather low (5 to 10 wt % at maximum) and sometimes totally lacking. Heavy minerals are in general scarce (mostly <0.1 wt %), and the heavy fraction generally includes only the stable minerals zircon, tourmaline and rutile. Among the clay minerals, only kaolinite and illite occur in most samples, and mixed layer minerals (smectite) are rare. For building purposes the stone masons traditionally

Fig. 2. Schematic sketch map showing the occurrence of Cretaceous sandstones in Saxony: (1) Elbe Zone, (2) area south of Zittau (Zittauer Gebirge).

use the terms 'Posta type' and 'Cotta type' sandstones, which allow a rough practical differentiation of material with different technical properties independent of stratigraphic assignment. The medium to coarse-grained 'Posta type' sandstone is characterized by siliceous intergrain cement, high compressive strength and a high capillary water uptake, whereas the cementing material in the fine- to medium-grained 'Cotta type' sandstone partly consists of clay minerals. It therefore shows significantly lower compressive strength and capillary water uptake (Grunert 1986).

Another smaller region of historic sandstone quarrying is only of local importance. It is situated in the south east of Saxony (south of the city of Zittau) and represents the northern margin of the Bohemian Cretaceous Basin (Turonian to Coniacian age, Fig. 2, area 2). The building stones from that region are also mature quartz sandstones with high quartz contents (Michalski *et al.* 2002). Depending on the stratigraphic horizon, the grain size ranges from fine-grained to conglomeratic. Low contents of feldspar (sometimes kaolinized) are limited to some varieties and accessory minerals such as tourmaline, mica, ilmenite, rutile and zircon and, in some cases, scarce monazite and xenotime. The intergranular cement is mainly siliceous. Some varieties show pore-filling cement consisting of clay minerals and ferric oxides/hydroxides.

Case studies

Alveolar weathering is not the dominant weathering form of building sandstone in Saxony, but can be detected on some façades or walls of different age. Observations with the naked eye suggest that in these cases the weathering is dependent on salt load, as salt efflorescence in the alveoli is very common. Furthermore, the rock fabric of the affected stone blocks is mostly heterogeneous. Bedding associated with changes in mineralogy and grain size or cementation of parts of the rock by ferric oxides and hydroxides as well as trace fossils (foralites) are observed. That these features may facilitate initiation of alveolar weathering on particular stone blocks is suspected from experience, but the detailed mechanism of this type of weathering, and the exact influence of the sandstone fabric on it, is still poorly understood.

Two examples of alveolar weathering of building sandstones were chosen for more detailed investigations: the village church in Leuba and the church of St Thomas in Leipzig. The church of the village of Leuba, Upper Lusatia (Laue *et al.* 2004) was built between 1854 and 1857 from local Cretaceous sandstone of the Zittau region (upper Turonian age, quarries Waltersdorf and Hochwald, 20 km south of Leuba; cf. Fig. 2, area 2). The grain size of the building stone ranges from fine-grained (0.01 mm) to coarse-grained (1 mm). Depending on bedding structures, fine-grained and coarse-grained layers may occur in the same block within a distance of a few centimetres. In some blocks, the rock is affected by late diagenetic infiltration of ferric oxides and hydroxides. The sandstone has a yellowish to brownish colour, mostly turning to grey at the soiled surface. Alveoli sometimes develop along the bedding planes (with transitions to weathering out of bedding structures) but may also occur as an irregular network independent from the direction of bedding. In most cases, single stone blocks or only parts of them are affected, surrounded by intact blocks. Generally, the weathering intensity is independent of aspect and height above the ground. Salt efflorescence can be frequently found in the alveoli, some of which are more than 50 mm deep. For the purposes of this research, a typical coarser grained block (Leuba 1, south side, height above the ground 1.2 m) with a hole strictly following the bedding and a fine-to-medium-grained block (Leuba 2, tower, west side, height above ground 10 m, Fig. 3) were selected. Samples of salt efflorescence were taken from several alveoli at different parts of the building.

The northern entrance hall of the church St Thomas in Leipzig was added to the Gothic building in the late 19th century. It is built of Cotta type sandstone from the Elbe zone. The structure of the fine-grained rock shows foralites, that is, trace fossils originated by endobenthic animals living in a system of burrows (the burrows probably belong to the ichnogenera *Thalassinoides* and *Ophiomorpha*; Mikuláš 2008, personal communication). The most weathered stone blocks with alveoli are situated near the eaves and up to about 2 m from the ground, suggesting the influence of moisture and salts which normally concentrate in these

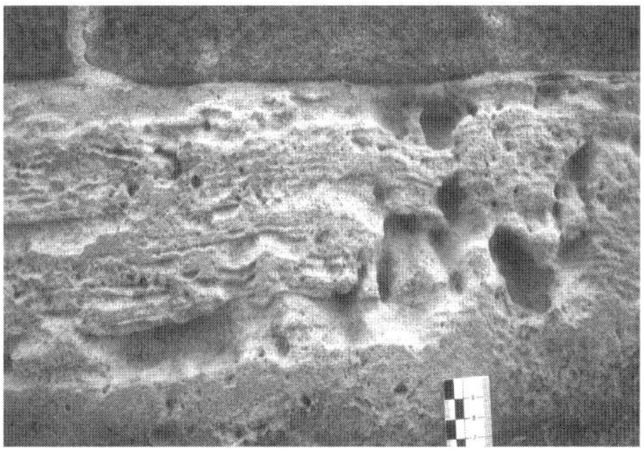

Fig. 3. Alveolar weathering on building sandstone of the Leuba church, block 2. Weathering is restricted to the upper part with fine lamination; the area below is not affected.

parts of a building. The surface of the yellowish to grey sandstone is mostly blackened in the stable parts of the blocks (walls between alveoli and non-weathered surfaces). Alveoli (up to 20 mm deep) and salt efflorescence could be particularly observed around the foralites in the 'matrix' of the sandstone, whereas the foralites themselves are not affected. If the foralites are cut perpendicular to their main axis by the surface of the stone block, they have often been protected from back weathering and form insular structures within the alveoli developing around them. If they are cut along their main axis, they weather out as the alveoli around them become deepened.

Samples were taken from a stone block at the east side of the entrance hall at a height of 1.10 m above the ground (Fig. 4).

Methods

Salt efflorescence in the alveolar holes was scratched off and the salt minerals were determined by X-ray diffraction (XRD, Siemens D5000). Samples for chemical analysis of the soluble salts were taken from the walls between alveoli or the stable original surfaces nearby, as well as from the alveoli themselves in profiles from surface to depth by collecting the stone powder from different depth sections of drilling holes (10 mm in diameter). The powder samples were dried (105 °C). The moisture content of the powder was determined only for the samples from Leipzig by weighing them immediately after sampling and again after drying. All samples were disaggregated in distilled water and filtered. Na^+ was determined by an ion sensitive electrode; other ions in the solution were analysed by a spectrophotometer (Hach) using standardized reagents. The results were referred to dry stone powder (wt. %).

Fig. 4. Alveolar weathering on Cotta type building sandstone of the church of St Thomas, Leipzig. Weathering starts in the area around trace fossils (foralites).

Petrographical investigations of the Leuba samples were carried out on drill cores (diameter 20 mm). The aim was to record differences in the fabric of the fresh stone that could be crucial for the initial development of alveolar weathering (alveoli versus walls). Primary (horizontal) bedding in the analysed stone blocks made it possible to obtain a sample comparable to the lost surface (without salt load, but in the same petrographically identical layer, some centimetres behind the surface) even in the case of back-weathered holes. At the coarser grained block Leuba 1, drill cores were taken just beside the drilling holes for salt analysis in a wall and from the bottom of an alveolar hole. At the block Leuba 2 (Fig. 3), one drill core was taken from the stable lower part of the block, where no alveolar weathering has developed at all. Another was taken from the upper, laminated part with incipient development of a relief parallel to bedding planes. In contrast to Leuba 1, drilling holes for salt analyses in Leuba 2 were located some centimetres away from the drill core in a mature, deeper alveolar hole and in the wall nearby. Thin sections for optical microscopy were prepared from all drill cores. A part of every drill core sample was used for measurements of mercury intrusion porosimetry, performed with the porosimeters 2000 WS – Carlo Erba for smaller pores (radius < 7 μm) and Pascal 140 – Fisons Instruments for pores with radii > 7 μm. Additionally, the specific surface of the samples was determined with a Sorptomatic 1990 system using N_2 gas. For conservation reasons, the author was not allowed to take drill cores at the St Thomas church in Leipzig. A broken piece of 'matrix' sandstone from the edge of an alveole and a non-weathered foralite from the middle of an alveole were therefore taken from the stone block for measurements of mercury intrusion porosimetry. To obtain a microscopic view of fabric, a thin section was made from a comparable, freshly quarried Cotta type sandstone containing foralites.

Results

Salt distribution and moisture content

The salt efflorescence found in alveoli from Leuba and Leipzig mainly consists of magnesium sulphates and occasionally small amounts of gypsum (Table 1). Depth profiles of the salt ion distribution show that the dominant salts are sulphates and nitrates (Figs 5a, 6a & 7). Chlorides in significant amounts occur only at the church of St Thomas in Leipzig (Fig. 7b). In Leuba, chloride content was very low ($<$0.03 wt. %) and could be neglected. The pattern of salt distribution is similar in both cases. Sulphate contents show a steep decline with

Table 1. *Minerals in salt efflorescence from alveoli, churches in Leuba and Leipzig (salts in brackets: small amount)*

St Thomas Church, Leipzig		Church of Leuba	
Sample from	Salts	Sample from	Salts
East side of the entrance hall, height 1.70 m	Starkeyite $MgSO_4 \cdot 4H_2O$ (gypsum $CaSO_4 \cdot 2H_2O$ picromerite $K_2Mg(SO_4)_2 \cdot 6H_2O$)	West side of the tower, height 10 m	Hexahydrite $MgSO_4 \cdot 6H_2O$
		West side of the tower, height 10 m (under moulding)	Hexahydrite $MgSO_4 \cdot 6H_2O$
East side of the entrance hall, inner side of the wall (on joint mortar, not from alveole), height 2.00 to 2.60 m	Hexahydrite, $MgSO_4 \cdot 6H_2O$ (gypsum $CaSO_4 \cdot 2H_2O$)	South side, buttress, west-exposed surface, height 1.70 m	Hexahydrite $MgSO_4 \cdot 6\ H_2O$

the depth in non-weathered areas and walls and also in alveoli. The spatial distribution of the cations calcium and magnesium is different (Figs 5b, 6b, 8a). Calcium (bound in gypsum, $CaSO_4 \cdot 2H_2O$) is concentrated near the non-weathered surface in walls, whereas its content near the surface in the alveoli is much lower (except for the example Leuba 1 (Fig. 5b) where the total content of gypsum is very low in general). In comparison, magnesium (bound in magnesium sulphate, $MgSO_4 \cdot nH_2O$) is always remarkably concentrated near the surface in the alveole and shows significantly lower content near the non-weathered surface of walls. The pattern of nitrate and chloride

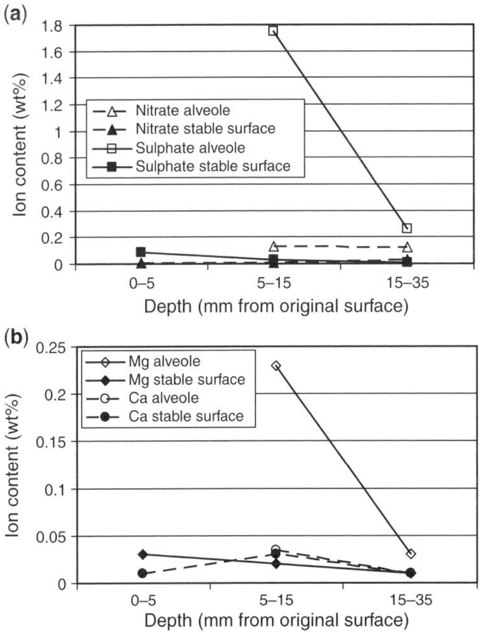

Fig. 5. Salt distribution from surface to depth in non-weathered area and behind alveole: (**a**) anions and (**b**) cations. Medium to coarse-grained sandstone from the Zittau region, church of Leuba, block 1. Depth of the alveole: 5 mm.

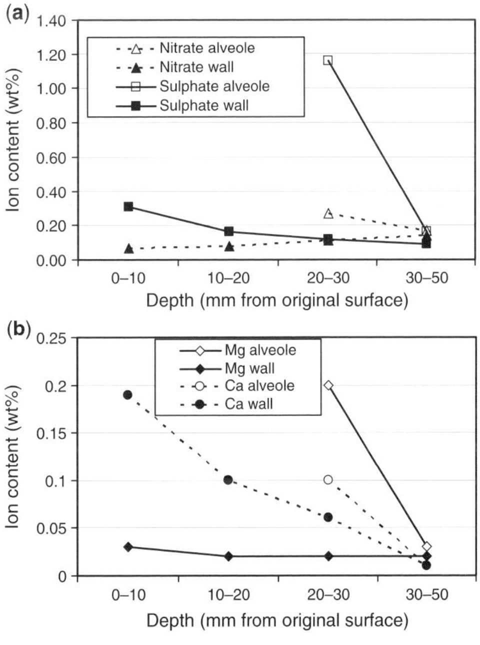

Fig. 6. Salt distribution from surface to depth in non-weathered wall and behind alveole: (**a**) anions and (**b**) cations. Fine to medium grained sandstone from the Zittau region, church of Leuba, block 2. Depth of the alveole: 20 mm.

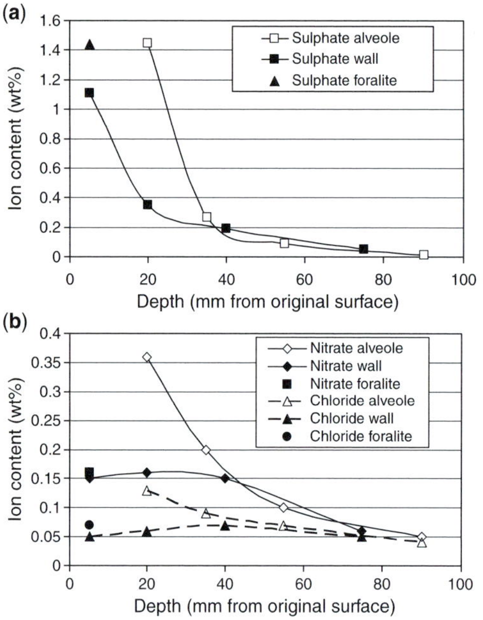

Fig. 7. Salt distribution (anions) from surface to depth in non-weathered wall, in a foralite (0 to 1 cm) and behind alveole, (**a**) sulphate and (**b**) chloride and nitrate. Cotta type sandstone, church of St Thomas in Leipzig. Depth of the alveole: 15 mm.

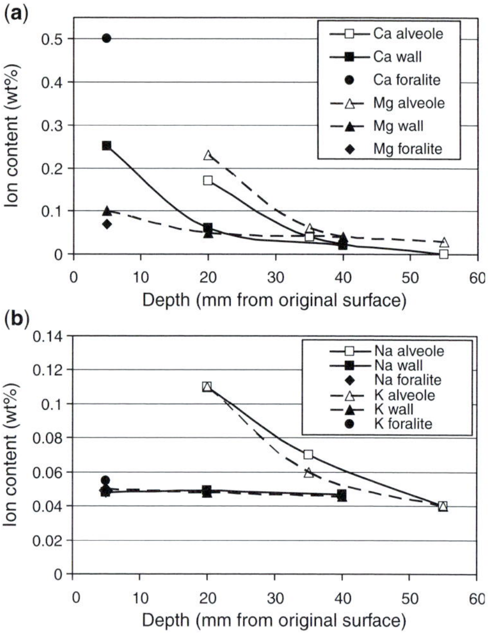

Fig. 8. Salt distribution (cations) from surface to depth in non-weathered wall, in a foralite (0 to 1 cm) and behind alveole: (**a**) alkaline earths and (**b**) alkalis. Cotta type sandstone, church of St Thomas in Leipzig. Depth of the alveole: 15 mm.

distribution shows the typical approximately even distribution of these highly soluble salts over the whole depth profile in the walls and in the stable areas (Figs 5a, 6a & 7b), which has been frequently found on other stonework objects (e.g. Siedel 1998). Only in the alveoli of the church of St Thomas, nitrates and chlorides are remarkably concentrated near the surface. Sodium and potassium distribution (Fig. 8b) follow the pattern of chloride and nitrate, thus reflecting the affinity of these ions to compounds like halite, niter? or nitrokalite. Ion content in the deeper parts of the profile reaches a comparably low level in both walls and alveoli. The ion content in the foralite are generally more consistent with the walls than with alveoli.

The moisture content was only determined for samples of the church of St Thomas in Leipzig (Fig. 9). At the time of sampling (February, therefore cold, dry climate) it was about 2.5 times higher behind the alveole than near the non-weathered surface.

Differences in the rock fabric

Differences in the primary rock fabric of the sandstones were studied in detail on the two selected blocks from the church of Leuba. The results of microscopic observations of thin sections are shown in Table 2. In block Leuba 1, differences between non-weathered surface and alveole could be detected in grain size distribution as well as in the content of ferric oxides/hydroxides. The latter are lacking in the poorly graded fabric of the non-weathered area but are present in the ill-sorted fabric behind the alveole. The diagenetic impregnation with ferric compounds (Fig. 10) has partly formed a secondary micropore structure resulting in a lower total porosity. The data obtained from mercury porosimetry (Fig. 11) and from

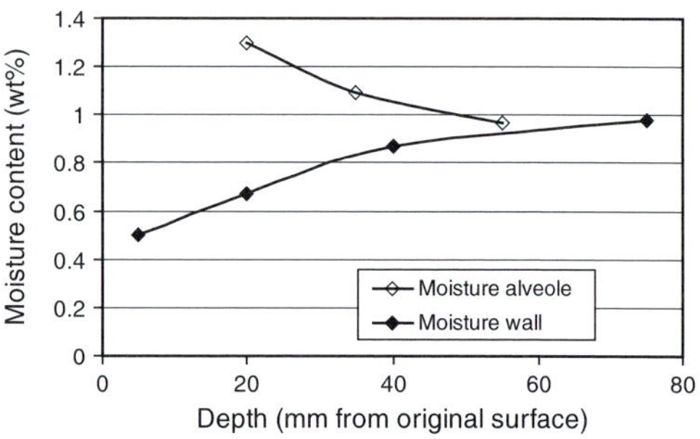

Fig. 9. Moisture content from surface to depth in non-weathered wall and behind alveole in February (cold, dry climate). Cotta type sandstone, church of St Thomas in Leipzig. Depth of the alveole: 15 mm.

Table 2. *Observations on thin sections, church of Leuba (sandstone from the Zittau region)*

Fresh, non-weathered material	Areas affected by alveolar weathering	Stable areas (walls, non-weathered areas)
Leuba block 1	Grain size 0.01 to 0.4 mm; well graded; grains angular to (rarely) subrounded; opaque ferric oxides and hydroxides fill parts of the pore space forming a micropore structure	Grain size 0.05 to 0.2 mm, (bigger grains up to 0.4 mm are scarce); poorly graded; grains angular to (rarely) subrounded; opaque ferric oxides and hydroxides are scarce
Leuba block 2	Grain size 0.1 to 0.2 mm (<10% 0.6 to 1 mm), grains angular to subrounded; opaque ferric oxides and hydroxides fill parts of the pore space forming a micropore structure	Grain size 0.1 to 0.2 mm (<10% 0.6–1 mm), grains angular to subrounded; opaque ferric oxides and hydroxides are scarce

determination of the specific surface (Table 3) are consistent with the microscopic observations. Within the same block, the total porosity is lower and the portion of smaller pores (Fig. 11b) as well as the specific surface is higher in the areas where alveoli are formed, compared to stable areas forming walls. The thin sections from block Leuba 2 show that grain size and grading in this block are very similar in the non-weathered lower part and in the fresh sandstone behind the area with incipient weathering in the upper part (Table 2, cf. Fig. 3). The only difference is the occurrence of ferric compounds in the pore space of some sandstone layers in the area with incipient weathering. Due to the fine lamination of the upper part of the block, the sample from this area included several thin iron-impregnated and non-impregnated sheets within the 20 mm diameter of the drill core, which were analysed together. Hence, the difference between the pore size distributions (Fig. 12) and the specific surfaces (Table 3) compared to the non-weathered part of the stone is not significant. However, a slight tendency to a higher amount of smaller pores and a higher specific surface for the area with incipient weathering could also be detected in block 2 (Table 3, Fig. 12b). This trend might have been more evident if samples had been taken from iron-impregnated sheets only, which was not possible because they were too small. At the macroscopic scale (Fig. 3, upper part), incipient back-weathering was detected preferentially in the (brown-coloured) iron-impregnated layers parallel to bedding planes.

Fig. 10. Thin section of the area behind an alveole of block 1 from Leuba church: the pore space (grey) between quartz grains (white) is partially impregnated with ferric hydroxides (black). Nicols II, right figure margin 2.8 mm.

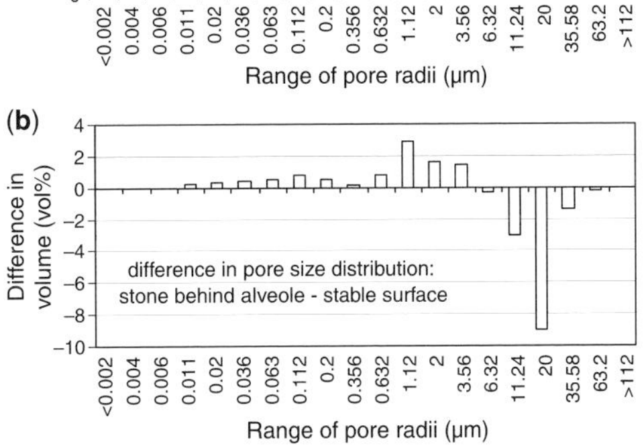

Fig. 11. Pore threshold radii according to mercury intrusion porosimetry in drill cores: (**a**) fresh sandstone from behind alveole and from stable area, block Leuba 1 and (**b**) difference between both samples.

Table 3. *Total pore volume (from Hg porosimetry) and specific surface for fresh sandstone samples from the Leuba church (taken from behind alveoli and stable areas)*

Sample	Total pore volume (vol. %)	Specific surface (m^2 g^{-1})
Leuba block 1, alveole	20.1	1.50
Leuba block 1, stable area	24.2	0.53
Leuba block 2, weathered*	23.0	0.64
Leuba block 2, stable area	25.0	0.44

*A deep alveole has not yet developed.

In case of the church of St Thomas in Leipzig, the texture of the Cotta type sandstone is characterized by trace fossils (foralites) with a cylindrical shape. Samples were taken both from the non-disturbed matrix (from a wall) and from a foralite (non-weathered, in the middle of an alveole) to compare their pore structure. The total sample porosity is not significantly different (21.7 vol.% for the matrix and 21.4 vol.% for the foralite). Although the pore size distributions have a similar pattern (Fig. 13a), the pore structure of the matrix has somewhat smaller pore radii compared to the foralite (Fig. 13b). A thin section from fresh quarried Cotta type sandstone (not from the church) shows the typical microstructure of a foralite embedded in the non-disturbed matrix (Fig. 14). It is surrounded by a thin layer of clay-rich material and has a texture with less clay mineral than the

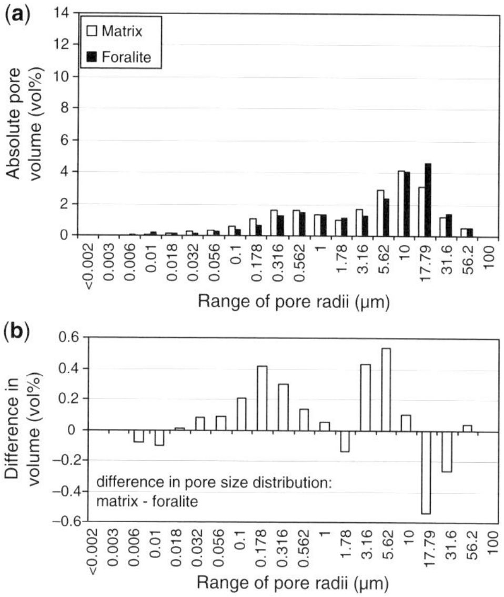

Fig. 13. (**a**) Pore threshold radii according to mercury intrusion porosimetry of matrix and foralite in Cotta type bioturbiditic sandstone and (**b**) difference between both samples.

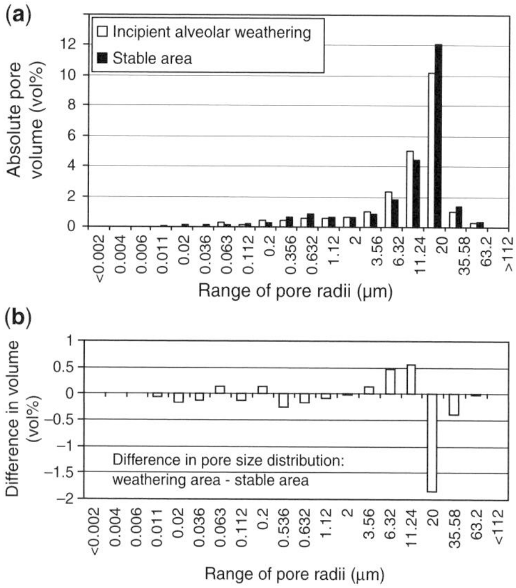

Fig. 12. Pore threshold radii according to mercury intrusion porosimetry in drill cores: (**a**) fresh sandstone from behind stable area and thin bedded area with incipient alveolar weathering, block Leuba 2 and (**b**) difference between both samples.

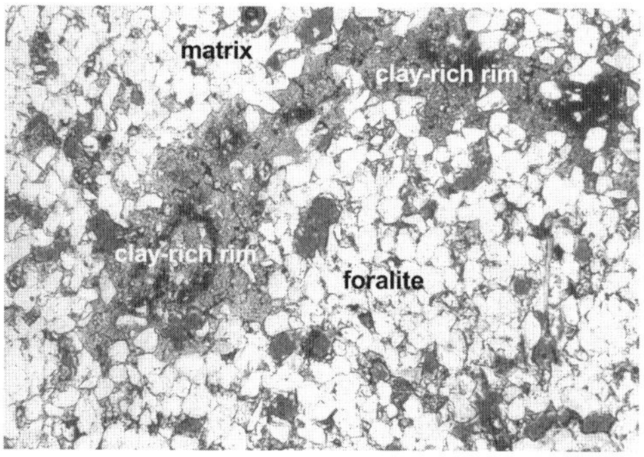

Fig. 14. Thin section of Cotta type sandstone with bioturbate texture: a rim of clay minerals (with ferric hydroxides, dark grey) marks the border of the foralite (right, lower half of the figure) with the matrix (left, upper half of the figure). Nicols II, right figure margin 2.8 mm.

matrix, whereas the grain size and the visible porosity of both matrix and foralite are the same. According to macroscopic observations, weathering starts in the clay-rich transition zone between foralite and matrix with small cavities surrounding the foralites, later developing into the matrix by the formation of larger alveoli.

Discussion

Two main aspects should be considered regarding alveolar weathering of Cretaceous building sandstones. These are first the influence of external factors of climate, moisture and salt availability and secondly the intrinsic factors of the rock texture and homogeneity/heterogeneity and the minerals of the stone block.

The dominant salts found in both case studies and in further examples of alveolar weathering on building stones studied in Saxony (Siedel 1998; Heckmann *et al.* 1994) were magnesium sulphate and gypsum. Other compounds such as nitrates and chlorides may additionally occur in varying, and sometimes very low, amounts. The sources of sulphate salts cannot be found in the natural stone itself (Siedel & Klemm 2001) but must be attributed to air pollution and environmental influences (Klemm & Siedel 2002). Both churches are situated in regions of high pollution due to combustion of lignite in the past, which resulted in high SO_2 levels. In Leuba, a large lignite power plant (Hagenwerder) was located only two kilometres from the church between 1958 and 1997. The city of Leipzig is situated within a large centre of industry with lignite power plants and chemical industry all around. The appearance of magnesium sulphate and gypsum as dominant salts is not surprising due to the widespread use of dolomitic lime for the joint mortars, which is a significant source of Mg^{2+} and Ca^{2+} ions. Dolomitic limestones of different geological age occur in some regions of Saxony and had been used for lime-burning from ancient times until the 20th century. The relation between such mortars and salt damage caused by magnesium sulphate has been observed on many historical buildings in Saxony (Siedel 2003; Siedel & Laue 2003). Magnesium sulphate is a highly soluble salt which can be easily dissolved and reprecipitated by wetting/drying cycles. Furthermore, it easily hydrates and dehydrates. Depending on the climatic situation while sampling, the hydrated forms found in the efflorescence were starkeyite ($MgSO_4 \cdot 4H_2O$) and hexahydrite ($MgSO_4 \cdot 6H_2O$). Epsomite ($MgSO_4 \cdot 7H_2O$) has also been found on building surfaces (Siedel 2003). Magnesium sulphates are among the most dangerous salts due to their changes in volume with the phase transition; hydration and dehydration occur under the rapidly changing microclimatic conditions that are quite normal for building surfaces (Klemm & Siedel 1996; Juling *et al.* 2004). According to Steiger (2000), the molar volume increases by nearly 10% with the transition from hexahydrite to epsomite and by about 28% with the starkeyite–hexahydrite transformation. These volume changes, induced by changes in relative humidity and/or temperature, also proceed in positions totally sheltered from direct rain attack. In contrast, gypsum is less soluble; it does not hydrate or dehydrate under normal conditions and is less mobile in the pore spaces of building materials.

The results of the investigations into sandstone fabrics have clearly demonstrated the significant differences in pore space between neighbouring areas at the (macroscopic) scale of centimetres for the block Leuba 1. Weathering is limited to areas with lower porosity and smaller pores. In these areas, drying of the wet stone takes more time and salts from the remnant solution can accumulate. For the other investigated blocks, heterogeneities were found only at the microscopic scale (some 100 μm thick, iron-impregnated sheets with lower porosity and smaller pore radii in case of the block Leuba 2 and some 100 μm thick, clay-rich rims of foralites in the case of Cotta type sandstone in Leipzig).

Huinink *et al.* (2004) have demonstrated by 2D model calculations based on the drying behaviour of a homogeneous rock that large, single alveoli (tafoni) can develop from a pit in the surface which is much smaller in dimension than the mature hole. The dominant process suggested by these authors is a repeated wetting/drying cycle of the stone surface with water and salts in the pores. In a slow drying process, the drying front (i.e. the front, where salts crystallize) loses contact with the surface in large parts during the long drying period. Only in the area around the small pit (beneath the surface) is the drying slower, and the wet stone is still in direct contact with the air. Consequently, accumulation of salt ions and precipitation of salts will mainly take place here. Supersaturation of the pore liquid results in crystallization pressure with subsequent damage of the stone fabric (Scherer 1999). The loss of material leads to deeper holes with a larger surface, where the process will continue further. The simulation of Huinink *et al.* (2004) is very general and does not accurately model the behaviour of a particular rock type or a particular salt or salt mixture. It gives an idea of how the growth of single alveoli might be controlled only by the drying behaviour of a salt-bearing rock, however. It is therefore discussed below in relation to the specific features of stone material and salt distribution found in this study.

At a first stage, the 'initial pits' needed in the model of Huinink *et al.* (2004) to develop single holes are formed by heterogeneities at the microscopic scale as observed in the stone blocks Leuba 2 and Leipzig. The clay-rich rims of the foralites in Cotta type sandstone have a higher weathering susceptibility (smaller pores, lower mechanical strength) compared to the framework of quartz grains in the surrounding matrix or in the foralite itself. They will therefore be the first parts of the surface that are affected by salt weathering. This is consistent with the macroscopic observations at the church of St Thomas in Leipzig (Fig. 4) and with observations and laboratory experiments of Alessandrini *et al.* (1992) on a limestone with foralites from Sicily. In the latter case, the development of alveoli also started in the small, inhomogeneous transition zone between foralite and matrix. In the case of block Leuba 2, small iron-impregnated sheets with a higher amount of small pores were first affected by salt weathering. Areas with smaller pores dry more slowly than areas with coarser pores and thus preferentially accumulate salt from the remaining, concentrated solution. According to Doehne (2002) and Scherer (1999), smaller pores also favour higher stresses of the crystallizing salts on the stone structure. The damaging forces of salts therefore concentrate immediately on the spots that are rich in iron or clay, which work as a binding agent in the structure. Preferential granular disintegration in these areas is the starting point of an uneven back weathering of the surface. As a result, a 'micro-relief' with stable and partly back-weathered areas is formed, as can be observed in the upper part of block Leuba 2 (Fig. 3).

At a second stage, small pits at the micrometre or millimetre scale deepen to alveoli at the centimetre scale during numerous wetting/drying cycles as described by Huinink *et al.* (2004). The reason for this is that most salt accumulates in the more sheltered areas of the rock surface with the lowest evaporation rates (at the bottom and on the walls of the pit or later the alveole). One measurement of the momentary spatial distribution of moisture on the church of St Thomas at the day of sampling (in a cold, dry climate situation in February) confirms the drying model of Huinink *et al.* (2004): the outermost zone of the wall is very dry whereas the surface zone of the alveole is rather wet at the same time (cf. Fig. 9).

Huinink *et al.* (2004) calculated their model using only one 'ideal' salt, which crystallizes at the boundaries of the wet regions where water evaporates. In this case study (and in most other cases on buildings) more than one salt is present in the weathering process. Every salt has a different solubility, that is, in a certain solution with different ions different salts will precipitate one after another during evaporation. This selective precipitation may lead to a spatial separation of different salts according to their solubility. Gypsum and magnesium sulphate are the dominant salts found in Leuba and Leipzig. Magnesium sulphate is a more soluble salt in a solution together with gypsum (Reeves *et al.* 2000). When the drying front retreats from the surface of the wet sandstone in a slow drying process, the less soluble gypsum is the first salt that is precipitated in the outermost zone. Most of the magnesium sulphate remains in solution, finally precipitating at the bottom of the alveoli. There it is in sheltered position and cannot be washed away during the next rain event. In the case of rapid drying after a rain event, both magnesium sulphate and gypsum are precipitated together near the surface. Due to differences in solubility, gypsum is not (or not totally) dissolved when the surface is moistened by the next rain event, while dissolved magnesium sulphate is washed away by rainwater or partly penetrates again further into the stone. This selective precipitation/dissolution behaviour leads to the salt distribution found in the alveolar holes and walls or stable areas in Leuba and Leipzig. The less soluble gypsum fills up the pores near the stone surface, sometimes resulting in encrustation (blackening of the surface in Leipzig). The capillary water uptake of the affected zones is reduced. This process leads to a temporary stabilization of the surface rather than to active weathering.

The hydration/dehydration behaviour of magnesium sulphate (see above) is an additional factor that leads to damage and material loss in the developing alveolar holes. In addition to crystallization pressure, volume changes of this salt cause expansion and shrinking of the stone in the affected zone (Juling *et al.* 2004) and accelerate the damage of the surface grain layers in the alveole. Since these processes are only dependent on the changing relative humidity of the air in the alveole and not on contact with liquid water, they can work even in dry periods without rain events. Weathering is limited to the outermost grain layers at the bottom and on the walls of the alveole with extremely high salt concentrations, gradually moving deeper each time after loss of the surface material. Detached sand grains mixed with salts can frequently be detected at the bottom of alveoli. Larger holes can finally coalesce (cf. Figs 3 & 4).

Obviously, the weathering dynamics of the second stage (deepening of the holes) becomes more and more independent of the stone fabric. This is indicated by the development of holes of greater dimensions (Leuba 2 and Leipzig) and by similar patterns of salt distribution in mature alveoli and walls in all cases investigated in this study.

Summary and conclusions

Analyses of salt content and salt distribution in Cretaceous building sandstones from two monuments in Saxony indicate that salt weathering is the main reason for development of alveolar structures in these sandstones. Magnesium sulphate was found to be the most active salt compound in the weathering process. At an incipient stage, the formation of small pits (of a dimension of the order millimetres) on the sandstone surface is controlled by rock fabric heterogeneities at the microscopic or macroscopic scale. Mobile, soluble salts such as magnesium sulphate accumulate in clay-rich or iron-impregnated zones with a higher amount of smaller pores, where the weathering process starts with the loss of some material from the surface. At a second stage, numerous wetting and drying cycles under slow solution transport and evaporation conditions result in further accumulation of highly soluble salts, mainly in the wet area behind the small pit (Huinink *et al.* 2004) leading to crystallization pressure and further material loss there. The less soluble gypsum is mainly deposited near the non-weathered surface. Once deepened, the hole will shelter accumulated salts on its walls and at its bottom from rainwater. Effloresing magnesium sulphate can hydrate and dehydrate under changing climatic conditions, resulting in the detachment of grains from the surface. The combined effect of crystallization pressure and hydration pressure of magnesium sulphate forms holes with dimensions of the order centimetres in diameter and depth that can also coalesce with neighbouring holes. Consequently, more salt can accumulate on the increasing, wet surface of the hole. The weathering process becomes self-perpetuating and self-reinforcing at the second stage, independent of whether it has developed from heterogeneities in a microscopic or macroscopic scale. This was described more generally as a dynamical unstable system by Turkington & Phillips (2004).

Considering the building ages and maximum depth of the holes, the average rate of erosion in the alveoli over the last 100–150 years has exceeded 2 mm a^{-1} in some places. It is not assumed that the holes have developed at a constant rate from initialization. There might be an initial accumulation of salts in the holes to a 'critical concentration', followed by an acceleration of the weathering process during a second stage.

Sulphate salts play a dominant role in the development of alveolar weathering on buildings and also on rocks in natural outcrops. On building surfaces, however, they originate under the influence of intensive environmental pollution (Klemm & Siedel 2002) within some decades. In contrast, the 'honeycomb weathering' on sandstone rocks in the field is a long-term geological process which was already recognized in the middle of the 19th century (Gutbier 1858) before the period of rapid industrialization.

Potassium-ammonium alum is a widespread and common salt in efflorescences on honeycomb-weathered sandstone bedrock in the Cretaceous of Saxony and the Czech Republic, but was never found on building sandstone affected by alveolar weathering. Williams & Robinson (2000) have demonstrated that alum is highly destructive, especially if gypsum is present. The occurrence of this salt mineral indicates the mobilization of aluminium from the low contents of clay minerals in the rock itself (Soukupová *et al.* 2002), a process that requires the infiltration of a large rock volume by acidic water over a long period of time. Beyer (1911) discussed the origin of sulphate salts from sulphuric acid as a result of chemical weathering of the rock (oxidation of pyrite, deterioration of clay minerals in the binding agent) and of the influence of vegetation and humus, that is, of natural factors. Recent investigations (Siedel & Klemm 2001) have shown that the ongoing natural process of chemical weathering might have been intensified by the deposition of anthropogenic sulphur in the last 150 years. Nevertheless, the weathering process in the field could also occur in the absence of any anthropogenic influence. In contrast, the widespread occurrence of magnesium sulphate as the active salt in alveolar weathering demonstrates the different situation on buildings, where other building materials such as mortars are a source of agents involved in the weathering process.

Case hardening by amorphous silica has never been detected on Cretaceous sandstones in historic buildings in Saxony, but was suggested by several authors to cause the formation of honeycombs in interaction with salt attack in natural outcrops of Cretaceous sandstones (Lentschig-Sommer 1960; Vařilová 2002). Additional influences such as gravity and primary rock structures at a bigger scale could also force the weathering process in the field, thus producing a greater variability of shapes (Mikuláš 2001). Hence, the processes forming alveolar structures seem to be different in nature and on buildings at least at an incipient stage. However, if salts are involved, the further growth of alveoli on the bedrock at the second stage may be controlled by self-reinforcing salt weathering processes, similar to those described for building sandstones above.

The results obtained so far provide the first quantitative information about how soluble salts are distributed in alveolar structures on building sandstones. Furthermore, they prove the influence of local petrographic variations on incipient alveolar weathering in all cases investigated. The investigation method applied to building sandstones could also be useful for investigations of alveolar structures on bedrock in the field.

Thanks are due to the German Federal Foundation for the Environment for funding parts of the investigation on the church of Leuba (Az 18727), to Simone Hempel for measuring pore radii of sandstones and to Steffen Laue for fruitful discussions. The author also thanks Derek Mottershead and Radek Mikuláš for their critical comments, which helped to improve the manuscript.

References

ALESSANDRINI, G., BOCCI, A., BUGINI, R., EMMI, D., PERUZZI, R. & REALINI, M. 1992. Stone materials of Noto (Siracusa) and their decay. *In*: DELGADO RODRIGUES, J., HENRIQUES, F. & JEREMIAS, F. T. (eds) *Proceedings of the 7th International Congress on Deterioration and Conservation of Stone*. Laboratório Nacional de Engenharia Civil, Lisbon, **1**, 11–20.

BEYER, O. 1911. Alaun und Gips als Mineralneubildungen und als Ursachen der chemischen Verwitterung in den Quadersandsteinen des sächsischen Kreidegebietes. *Zeitschrift der Deutschen Geologischen Gesellschaft*, **63**, 429–467.

DOEHNE, E. 2002. Salt weathering: a selective review. *In*: SIEGESMUND, S., WEISS, T. & VOLLBRECHT, A. (eds) *Natural Stone. Weathering Phenomena, Conservation Strategies and Case Studies*. Geological Society, London, Special Publications, **205**, 51–64.

FITZNER, B., HEINRICHS, K. & KOWNATZKI, R. 1995. Weathering forms – classification and mapping. *In*: SNETHLAGE, R. (ed.) *Denkmalpflege und Naturwissenschaft. Natursteinkonservierung I*. Ernst & Sohn, Berlin, 41–88.

GÖTZE, J. & SIEDEL, H. 2007. A complex investigation of building sandstones from Saxony (Germany). *Materials Characterization*, **58**, 1082–1094.

GRUNERT, S. 1986. Der Sandstein der Sächsischen Schweiz. *Abhandlungen des Staatlichen Museums für Mineralogie und Geologie Dresden*, **34**, 1–155.

GUTBIER, A. V. 1858. *Geognostische Skizzen aus der Sächsischen Schweiz*. Leipzig.

HECKMANN, F., MASCH, L., WENDLER, E. & SNETHLAGE, R. 1994. Untersuchung der Alveolarverwitterung an Sandsteinen der Burgruine Tanstein (Dahn/Pfalz) und der Marienkirche in Marienberg (Sachsen). *Berichte der Deutschen Mineralogischen Gesellschaft, Beiheft European Journal of Mineralogy*, **6**(1), 98.

HUININK, H. P., PEL, L. & KOPINGA, K. 2004. Simulating the growth of tafoni. *Earth Surface Processes and Landforms*, **29**, 1225–1233.

JEANNETTE, D. 1980. Les grès du château du Landsberg: exemple d'évolution des 'grès Vosgiens' en milieu rural. *Science Géologique Bulletin*, **33**(2), 111–118.

JULING, H., KIRCHNER, D., BRÜGGERHOFF, S., LINNOW, K., STEIGER, M., EL JARAD, A. & GÜLKER, G. 2004. Salt damage of porous materials: A combined theoretical and experimental approach. *In*: KWIATKOWSKI, D. & LÖFVENDAHL, R. (eds) *Proceedings of the 10th International Congress on Deterioration and Conservation of Stone*. ICOMOS Sweden, Stockholm, **1**, 187–194.

KLEMM, W. & SIEDEL, H. 2002. Evaluation of the origin of sulphate compounds in building stone by sulphur isotope ratio. *In*: SIEGESMUND, S., WEISS, T. & VOLLBRECHT, A. (eds) *Natural Stone. Weathering Phenomena, Conservation Strategies and Case Studies*. Geological Society, London, Special Publications, **205**, 419–430.

KLEMM, W. & SIEDEL, H. 1996. Sources of sulphate salt efflorescences at historical monuments – a geochemical study from Freiberg, Saxony. *In*: RIEDERER, J. (ed.) *Proceedings of the 8th International Congress on Deterioration and Conservation of Stone*. Möller Druck and Verlag, Berlin, **1**, 489–495.

LAUE, S., SIEDEL, H. & PFEFFERKORN, S. 2004. Alveolar (honeycomb) weathering of Cretaceous building sandstone on the church of Leuba (Upper Lusatia, Germany): causes, processes, damages. *In*: KWIATKOWSKI, D. & LÖFVENDAHL, R. (eds) *Proceedings of the 10th International Congress on Deterioration and Conservation of Stone*. ICOMOS Sweden, Stockholm, **1**, 211–218.

LENTSCHIG-SOMMER, S. 1960. Petrographische Untersuchung der Wabenverwitterung des Elbsandsteins. *Jahrbuch des Staatlichen Museums für Mineralogie und Geologie Dresden*, **1960**, 111–126.

LIVINGSTONE, R. A. 1994. Influence of evaporite minerals on gypsum crusts and alveolar weathering. *In*: FASSINA, V., OTT, H. & ZEZZA, F. (eds) *Proceedings of the Third International Symposium on the Conservation of Monuments in the Mediterranean Basin*. Venezia, 22–25 June, 1994, Soprintendenza ai Beni Artistici e Storici di Venezia, Venezia, 101–107.

MICHALSKI, S., GÖTZE, J., SIEDEL, H., MAGNUS, M. & HEIMANN, R. B. 2002. Investigations into provenance and properties of ancient building sandstones of the Zittau/Görlitz region (Upper Lusatia, Eastern Saxony, Germany). *In*: SIEGESMUND, S., WEISS, T. & VOLLBRECHT, A. (eds) *Natural Stone. Weathering Phenomena, Conservation Strategies and Case Studies*. Geological Society, London, Special Publications, **205**, 283–297.

MIKULÁŠ, R. 2001. Gravity and orientated pressure as factors controlling "honeycomb weathering" of the Cretaceous castellated sandstones (northern Bohemia, Czech Republic). *Bulletin of the Czech Geological Survey*, **76**(4), 217–226.

MOTTERSHEAD, D. N. 1994. Spatial variations in intensity of alveolar weathering of a dated sandstone structure in a coastal environment, Weston-super-Mare, UK. *In*: ROBINSON, D. A. & WILLIAMS, R. B. G. (eds) *Rock Weathering and Landform Evolution*. John Wiley & Sons, New York, 151–175.

MUSTOE, G. E. 1982. The origin of honeycomb weathering. *Geological Society of America Bulletin*, **93**, 108–115.

PYE, K. & MOTTERSHEAD, D. N. 1995. Honeycomb weathering of Carboniferous sandstone in a sea wall at Weston-super-Mare, UK. *Quarterly Journal of Engineering Geology*, **28**, 333–347.

QUAYLE, N. T. J. 1992. Alveolar decay in stone – its possible origin. *In*: DELGADO RODRIGUES, J., HENRIQUES, F. & JEREMIAS, F. T. (eds) *Proceedings of the 7th International Congress on Deterioration and Conservation of Stone*. Laboratório Nacional de Engenharia Civil, Lisbon, **1**, 109–118.

REEVES, N. J., CLEGG, S. L. & BRIMBLECOMBE, P. 2000. Data evaluation and molality-based parameterisation. *In*: PRICE, C. (ed.) *An Expert Chemical Model for Determining the Environmental Conditions Needed to Prevent Salt Damage in Porous Materials*. European Commission, Project ENV4-CT95-0135 (1996–2000), Research Report, **11**, 65–116.

ROBINSON, D. A. & WILLIAMS, R. B. G. 1994. Sandstone weathering and landforms in Britain and Europe. *In*: ROBINSON, D. A. & WILLIAMS, R. B. G. (eds) *Rock Weathering and Landform Evolution*. John Wiley & Sons, New York, 371–391.

RODRIGUEZ-NAVARRO, C., DOEHNE, E. & SEBASTIAN, E. 1999. Origins of honeycomb weathering: The role of salt and wind. *Geological Society of America Bulletin*, **111**(8), 1250–1255.

SCHERER, G. W. 1999. Crystallization in pores. *Cement and Concrete Research*, **29**, 1347–1358.

SIEDEL, H. 2003. Dolomitkalkmörtel und Salzbildung an historischer Bausubstanz. *In*: LEITNER, H., LAUE, S. & SIEDEL, H. (eds) *Mauersalze und Architekturoberflächen*. University of Fine Arts, Dresden, 57–64.

SIEDEL, H. 1998. Zur Verwitterung des Rochlitzer Porphyrtuffs an der Kunigundenkirche in Rochlitz. *Jahresberichte Steinzerfall-Steinkonservierung*, **6**, 335–344.

SIEDEL, H. & LAUE, S. 2003. Herkunft, Kristallisation und Hydratstufenwechsel von Magnesiumsulfat an Bauwerken. *In*: *Umweltbedingte Gebäudeschäden an Denkmälern durch die Verwendung von Dolomitkalkmörteln*. Report no. 16, Institut für Steinkonservierung Mainz, 31–38.

SIEDEL, H. & KLEMM, W. 2001. Sulphate salt efflorescence at the surface of sandstone monuments and at the sandstone bedrock in outcrops – natural or anthropogenic reasons? *Geologica Saxonica*, **46/47**, 203–208.

SOUKOPOVÁ, J., HRADIL, D. & PŘIKRYL, R. 2002. Chemical weathering of clay-rich sandstone matrix – control and case studies. *In*: PŘIKRYL, R. & VILES, H. A. (eds) *Understanding and Managing Stone Decay*. The Karolinum Press, Prague, 263–271.

STEIGER, M. 2000. Total volumes of crystalline solids and salt solutions. *In*: PRICE, C. (ed) *An Expert Chemical Model for Determining the Environmental Conditions Needed to Prevent Salt Damage in Porous Materials*. European Commission, Project ENV4-CT95-0135 (1996–2000), Research Report, **11**, 53–63.

TURKINGTON, A. V. & PARADISE, T. R. 2005. Sandstone weathering: a century of research and innovation. *Geomorphology*, **67**, 229–253.

TURKINGTON, A. V. & PHILLIPS, J. D. 2004. Cavernous weathering, dynamical instability and self-organization. *Earth Surface Processes and Landforms*, **29**, 665–675.

VAŘILOVÁ, Z. 2002. A review of selected sandstone weathering forms in Bohemian Switzerland National park, Czech Republic. *In*: PŘIKRYL, R. & VILES, H. A. (eds) *Understanding and Managing Stone Decay*. The Karolinum Press, Charles University Prague, 233–242.

WINKLER, E. M. 1994. *Stone in Architecture. Properties, Durability*. Springer, Berlin.

WILLIAMS, R. G. B. & ROBINSON, D. A. 2000. Weathering of sandstone by alunogen and alum salts. *Quarterly Journal of Engineering Geology*, **31**, 369–373.

Black-crust growth and interaction with underlying limestone microfacies

GILLES FRONTEAU[1]*, CÉLINE SCHNEIDER-THOMACHOT[1], EDITH CHOPIN[1], VINCENT BARBIN[1], DOMINIQUE MOUZE[2] & ANDRÉ PASCAL[1]

[1]*Groupe d'Etude des Géomatériaux et des Environnements Naturels et Anthropiques (GEGENA), EA 3975, University of Reims Champagne-Ardenne, CREA, 2 esplanade Roland Garros, 51100 Reims, France*

[2]*Dynamique des Transferts aux Interfaces, EA 3803, Université de Reims Champagne-Ardenne, UFR Sciences Exactes et Naturelles, BP 1039, 51 687 REIMS Cedex 2, France*

**Corresponding author (e-mail: gilles.fronteau@univ-reims.fr)*

Abstract: Black crust growth mechanisms on three French building stones are described using diagenetic models that reveal the close links between the crust–stone interfaces and the microfacies of the host limestone. Each limestone is representative of a specific sedimentary facies and displays mixed pore structure: crinoidal limestone (Euville limestone), oolitic limestone (Savonnières limestone) and bioclastic matrix-supported limestone (Courville limestone). The crinoidal limestone is mainly made of well-developed calcitic cement (spar syntaxial calcite) with low macrocroporosity (15–20 vol. %). The oolitic limestone is macroporous (30–40 vol. %), oolite nucleus being partially or completely dissolved. The third building stone studied is less porous (14 vol. %) but presents a significant microporosity.

Weathering of the Euville limestone proceeds primarily through preferential exploitation of cleavages and microcracks and secondly by progressive recrystallization in the areas separated by previous gypsum fill-in (micro-box work). In the Savonnières limestone (oolitic limestone), gypsum recrystallization could occur without microcracks: elements are sometimes nearly totally weathered, while the palisadic calcitic cement surrounding the oolites was still preserved. In the matrix-supported limestone (Courville limestone), weathering could deeply affect the matrix while elements are not weathered. When a layer of microcrystalline calcite is observed on the surface of the limestone, however, the black crust growth seems to be limited to the external part of the stone.

Porous characteristics of limestones directly depend on sedimentary and diagenetic phases developed. The pore network controls moisture movement and also determines the reactivity of the stone to gypsum recrystallization.

Exposure to atmospheric conditions including air pollution results in the complex and natural process of stone ageing (Winkler 1994; Colston *et al.* 2001; Andriani & Walsh 2007). In urban environments, among the various weathering effects on stone, black crusts are certainly the most visible and the most studied (Jeannette 1981; Camuffo *et al.* 1983; Ausset *et al.* 1991; Schiavon 1992; Fassina *et al.* 2002; Török & Rozgonyi 2004).

Black crusts are extremely common in polluted urban environments; these sulphate encrustations can develop on all types of materials including metals and glass (Winkler 1994; Sabbioni 1995; Lefèvre & Ausset 2001). They grow on surfaces sheltered from rain and run-off, the accumulation of atmospheric particles and the development of salts crystals and micro-organisms (Rodriguez-Navarro & Sebastian 1996; Machill *et al.* 1997; Siegesmund *et al.* 2007).

Black crusts are mainly composed of newly formed gypsum crystals, but they also include other salts and atmospheric particles related to the prevailing environmental conditions: fly-ash (Hutchinson *et al.* 1992; Maravelaki-Kalaitzaki & Biscontin 1999; Potgieter-Vermaak *et al.* 2005), wind quartz, micro-organisms (bacteria, algae, mushrooms) and remains from the subjacent rock (Galleti *et al.* 1997; Ghedini *et al.* 2000).

The significant role of SO_2, air pollution (particulate or gas) and particles resulting from combustion (wood, coal, fuel or exhaust fumes) has been demonstrated (Rodiguez-Navarro & Sebastian 1996). The rate of black crust formation seems to be very variable according to the exposure, SO_2 concentration and unevenness of the crust thickness: from 20–600 μm for a same crust (Sabbioni 1995; Maravelaki-Kalaitzaki & Biscontin 1999; Bugini *et al.* 2000). Various stages in the formation of the

From: PŘIKRYL, R. & TÖRÖK, Á. (eds) *Natural Stone Resources for Historical Monuments.* Geological Society, London, Special Publications, **333**, 25–34.
DOI: 10.1144/SP333.3 0305-8719/10/$15.00

black crusts and crust morphologies are also distinguished: grey crusts, dendritic or compact black crusts.

Microscopic analysis identifies the complexity of black crusts which comprise a dark superficial upper layer and a lighter internal layer. The upper layer is full of opaque particles (fly ashes, soot) and micro-organisms; these elements are inserted in a matrix of gypsum crystals (Weber *et al.* 2007). The internal layer is more transparent and contains very few dark atmospheric elements. It tends to be composed of a dense gypsum crystallization and sometimes of other salts (Camuffo *et al.* 1983). The transition between the encrustation and the host stone can be progressive, with the underlying limestone being finely fractured, crumbly and exhibiting elevated salt concentrations. Interface morphologies can be linked to characteristics of the underlying microfacies: calcitic cements or micritic matrix, for example (Schiavon 1992; Török & Rozgonyi 2004).

When a black crust grows on a carbonate stone, weathering includes an interface with the subjacent stone where recrystallization process occur (Bromblet & Vergès-Belmin 1996; Thomachot & Jeannette 2000; Fassina *et al.* 2002). In this paper, we choose to focus on the internal part of the black crust and on this interface with the subjacent stone.

Materials and methods

Using approach and observation methods generally used for carbonate diagenesis studies (Moore 1989; Tucker & Wright 1990), the structural characteristics of black crusts at different stages of growth were identified on different limestones microfacies, each with very distinct and well-identified microstructural fabrics: crinoidal limestone, porous oolitic limestone and bioclastic limestone with micritic matrix.

The three building stones selected for this study are well-known limestones from the east of the Paris Basin. They were largely quarried and used for building, not only in France but also in other countries (Belgium, Germany and USA). They can also be found in prestigious monuments (Noël 1970) such as those in Stanislas Square in Nancy (Euville limestone), various churches, basilicas or modern buildings as the Paris East railway station (Savonnières limestone) and the cathedral of Rheims (Courville limestone).

In order to determine the characteristics and the natural variability of these limestones and the potential impact of these for weathering (Benavente *et al.* 2007; Rothert *et al.* 2007; Török *et al.* 2007), sedimentological and diagenetic analyses were carried out on the three stone deposits (Fronteau 2000*a*). Various microfacies categories were defined for the main stones recognized in the quarries. For this study, microfacies characterization included textures and recognition of elements using micropalaeontological and sedimento-diagenetic classifications (Moore 1989; Tucker & Wright 1990), as well as analysis of the porous network according to the Choquette and Pray (1970) or Tucker (1988) classifications (Table 1).

The encrusted samples of weathered limestones were taken directly from walls of various buildings from the Champagne-Ardenne area (mainly in Rheims), if possible from equivalent architectural positions: in sheltered areas under a raised edge, without capillary risings. About twenty black crust samples were taken for each building stone. Samples were hardened with a fluid epoxy resin (Geofix) before and during the thin-section process, in order to avoid any disorder or disturbance of the microstructures. In order to complete petrographic characterizations, analyses were performed using an Olympus BX60 epifluorescence microscope equipped with polarizing optical supplies. This UV radiation observation mode, already used in some carbonate diagenetic studies (Dravis & Yukewicz 1985), was extremely useful since it highlighted the presence of organic matter (e.g. endobiotas). Observations with a Scanning Electron Microscope (SEM) (JEOL JSM 6460L) equipped with a quantitative Energy Dispersive Spectrometer (EDS) were obtained and X-Ray Diffraction (XRD) analyses (Brucker AXS D8 Advance) were carried out to confirm the mineralogical nature of the various characterized phases.

The general morphology of these black crusts corresponded, on the whole, to what was previously described in the literature. Sulphated crusts have been shown to be composed of several layers: an upper dark layer containing numerous atmospheric particles and an internal layer which is less opaque and composed almost exclusively of gypsum crystals.

The black crusts described in this paper show thickness sufficient to enable the clear distinction of the various layers. Analytical emphasis was placed on the internal layer of the encrustations, that is, the interface between the black crust and the underlying limestone. Crust microstructures and underlying limestone were characterized and the various phases of the weathering fabrics were ordered chronologically using the microstratigraphic principles of diagenetic sequencing (Tucker 1988). The aim was to show the relationships between black crust growth and subjacent limestone nature, and to link these observations to the characteristics of limestone microfacies (Fronteau *et al.* 1999).

Table 1. *Main characteristics of the studied building stones. Limestone classification according to Folk & Dunham classifications (Tucker & Wright 1990), porosity fabrics according to Tucker (1988)*

Stone name	Limestone classification	$CaCO_3$ (%)	Accessory minerals	Total porosity (%)	Porosity fabrics
Euville stone *'Euville marbrier'*	Crinoidal limestone Biosparite/ Grainstone	99.8	/	10–15%/ 15–30%*	Few cases of intergranular macroporosity. Few cases of corroded pelletoïds (microporosity).
Savonnières stone *'Savonnières ½ fine'*	Oolitic limestone Oosparite/ Grainstone	98	Dolomite	30–40%	Intergranular macroporosity. Corroded oolites nuclei (macroporosity). Intragranular microporosity.
Courville stone *'Liais de Courville'*	Bioclastic limestone Biomicrite/ Wackestone/ Packstone	91.2	Quartz, glauconite, iron oxides	18–22%	Intergranular microporosity. Few intragranular macroporosity.

* Lower limit for compact Euville stone, formerly quarried; upper limit for modern quarried Euville stones.

Results

Black crust on crinoidal limestone with low macroporosity (Euville limestone)

The Euville limestone was almost exclusively composed of crinoidal fragments with their syntaxial cement. Its average porosity was about 15–20 vol. % which mainly reflected its intergranular macroporosity. The black crust shown in Figure 1 was taken from the St André Church (Rheims). It sometimes exceeded 500 μm in thickness and displayed a very simple structure with an upper almost opaque layer and a transparent internal layer which contained elements from the underlying limestone. The surface layer had a relatively constant thickness of approximately 150 μm composed of opaque particles, rare crystals of quartz and micro-organisms (visible under fluorescence light). The transparent internal layer varied from 100 to 350 μm in thickness and was almost exclusively composed of gypsum crystals (Figs 1a, b). The underlying limestone shows evidence of gypsum crystallization within microcracks and recrystallization of the crinoidal ossicles (Figs 1c, d, Figs 2a, b).

Black crust on vacuolar oolitic limestone (Savonnières limestone)

The second type of limestone came from the Vacuolar Oolite, a sedimentary formation outcropping at the boundaries of the Champagne and Lorraine regions. The most famous building stone quarried from this Tithonian stage limestone is the Savonnières stone, often called 'the Savonnières' (Blows *et al.* 2003). The elements of this oolitic limestone were only bonded by an isopachous fibrous spar cement. Porosity in this limestone is high (up to 30 vol. %), mainly due to an open intergranular macroporosity and to partial or quasi-total dissolution of 75% of the oolitic nuclei leading to an intragranular macroporosity (Fronteau 2000*b*; Roels *et al.* 2003).

By observing in detail diverse black crusts in which the lower layer penetrates deeply within the Savonnières limestone (Fig. 3), we studied the recrystallization phenomenon of oolites into gypsum and tried to establish its different stages of growth. The black crust (schematically represented in Fig. 3) from the St Martin de Gigny church in St Dizier shows weathered and recrystallized limestone. Some oolites were still partially visible in this encrusting whereas others were already completely weathered (Figs 3a & 4). The analysis of elements achieved by SEM (EDS analysis) confirmed the presence in the crust of Si-, Al-, Fe-, and P-rich atmospheric aluminosilicate particles, as well as the presence of K- and Cl-rich salts (sylvite) (Fig. 3b). The recrystallized oolites were mainly composed of gypsum (Fig. 3c). On the other hand, the practically unweathered subjacent limestone had a composition close to pristine limestone, without significant sulphation (Fig. 3d).

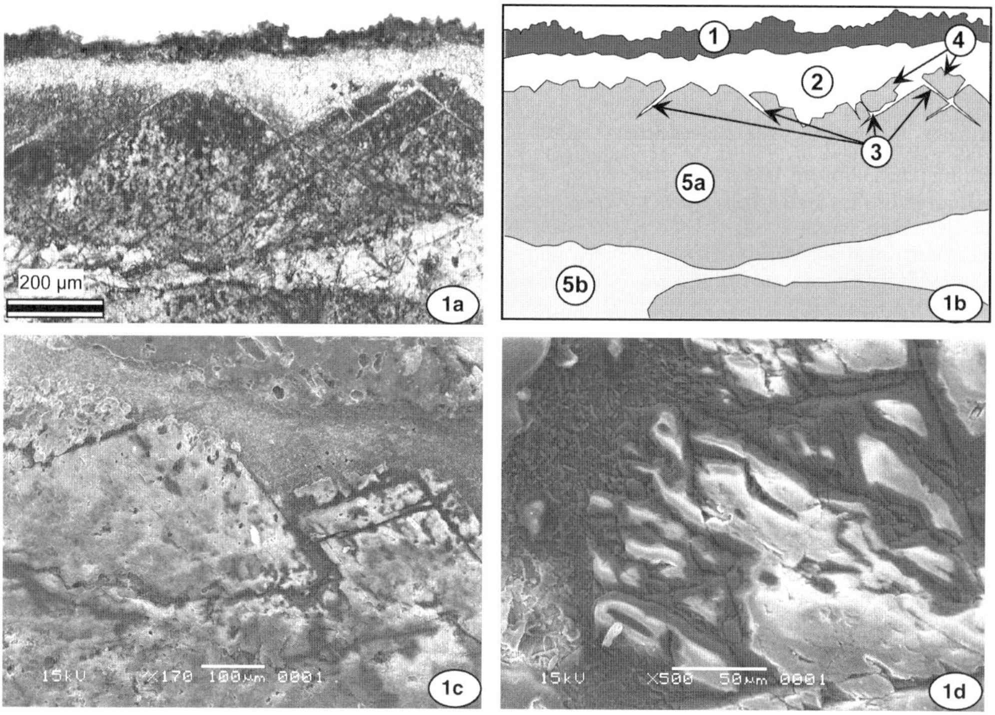

Fig. 1. Black crust on Euville Limestone (St André Church, Rheims, France). (**a**, **b**) Image under plain polarized light and interpretation (1, superficial crust with atmospheric particles; 2, internal crust mainly made of gypsum crystals; 3, advance of recrystallization front by cleavages of calcite crystals; 4, part of element almost isolated in the gypsum crust; 5a, crinoidal ossicle; 5b, syntaxial spar cement). (**c**, **d**) SEM micrograph view of weathered Euville limestone.

The composition of this limestone was relatively simple (only oolite and calcite fibrous cement) with some dolomite crystals. It allowed the main petrographical characteristics of the weathering process (Figs 4 & 5) to be reconstructed and various stages to be highlighted which were further compared to the calcite-gypsum pseudomorphosis mechanisms described for the crinoidal limestone (Euville Stone).

The recrystallization began at the periphery of the oolite's cortex, just under the spar cement (Fig. 5a). It then spread laterally mainly affecting

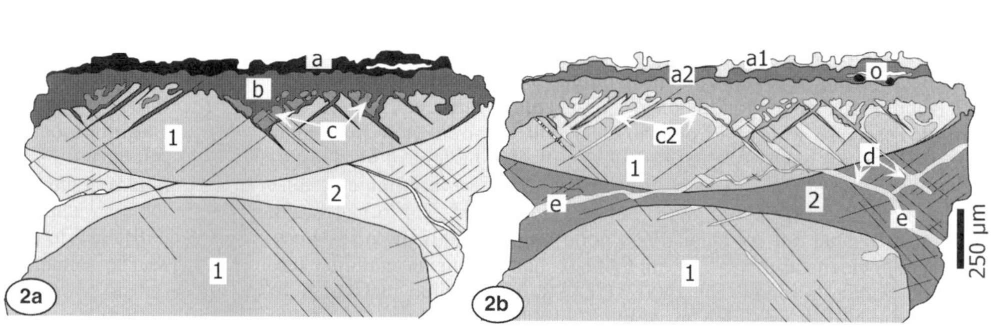

Fig. 2. (**a**) Micromorphology of a sulphated encrustation section developed on crinoidal limestone from observations under plain polarised light and cross-polarized light and (**b**) from observations under epifluorescence. 1, crinoidal ossicles (partly recrystallized into gypsum); 2, cracked syntaxial spar cement; a1, superficial crust with micro-organism; a2, lower part of the black layer with opaque particles; b, internal part of the encrustation, made of gypsum crystals; c, area partially recrystallized into gypsum; c2, luminescent areas in front of the recrystallization front; d, microcracks and cleavages filled with neogenic gypsum; o, endolithic micro-organisms.

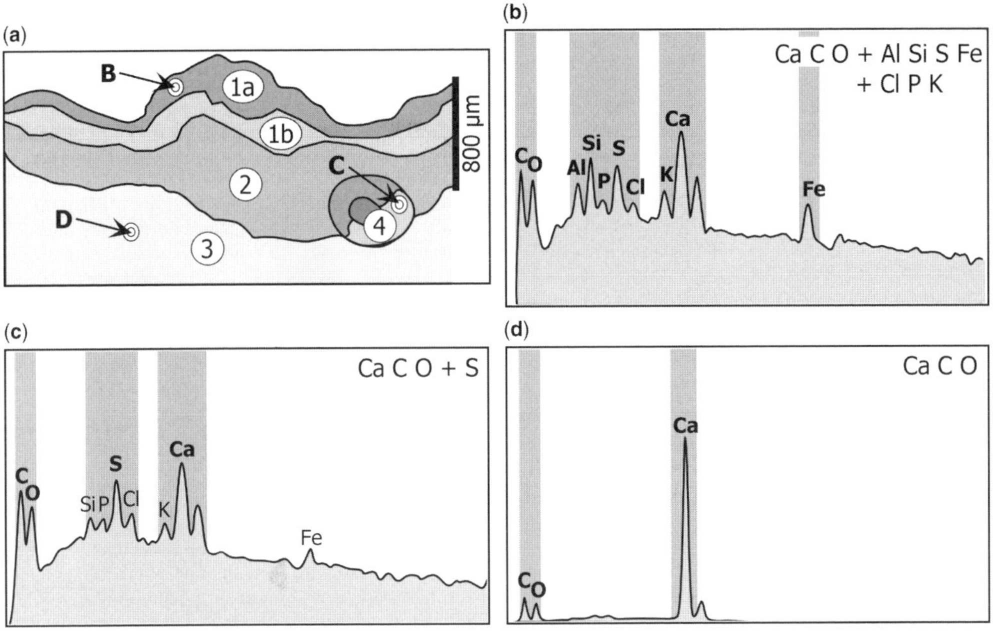

Fig. 3. Schematic representation and chemical composition of a black crust developed on oolitic limestone (Savonnières Stone, St Dizier, France). (**a**) Black crust made of two layers (1a, b); 2, very weathered and recrystallized limestone; 3, slightly cracked limestone; 4, oolite partially recrystallized into gypsum, (**b**–**d**) microprobe analysis.

the radial-fibrous part of the cortex (Fig. 5b). The concentric laminae appeared more resistant to weathering and could be observed as relics in areas largely transformed into gypsum (Fig. 5c). The presence of macroporosity and euhedral crystals in the broad gypsum area (Fig. 5e) shows that gypsum dissolution and crystallization cycles occur in the oolite (Fig. 5d), while the gypsum recrystallization front was still penetrating more deeply into the centre of the oolite. (Figs 5d–f). In addition, epifluorescence observations indicated that the dolomite rhombohedrons, present at the external side of the calcite fibrous cement (Fig. 5f), remained almost intact.

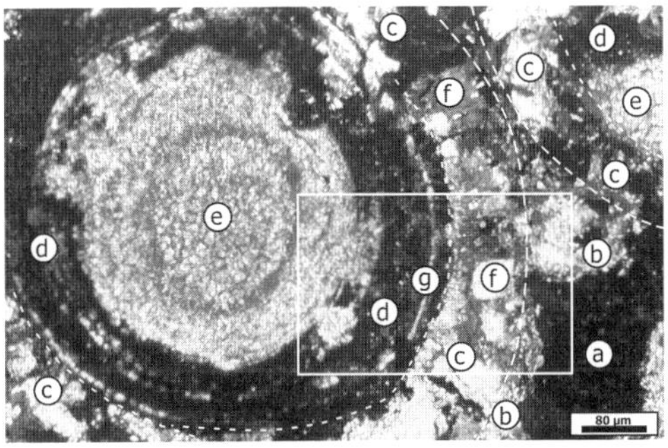

Fig. 4. Photomicrograph (cross-polarized light) showing an oolite partially recrystallized into gypsum (Savonnières Stone, St Dizier, France). a, intergranular pore; b, neogenic gypsum crystals on pore edges; c, non-recrystallized isopachous spar cement; d, neogenic gypsum areas; e, area still calcitic in the oolite; f, unaltered dolomite crystals; g, remains of oolitic lamina. White rectangle shows the area used as a diagenetic model (gypsum recrystallization process of an oolite in Savonnières limestone) in Figure 5.

Black crust on bioclastic limestone with micritic matrix (Courville limestone)

The last building stone studied, referred to as 'Liais' of Courville by quarrymen, was characteristic of the micritic matrix limestones used around the Rheims area (Blanc *et al.* 1985; Fronteau *et al.* 2002). This facies, mainly composed of foraminifera and other bioclasts, came from the middle of the Paris Basin where Lutetian Stage limestones, such as St Maximin or St Pierre-Aigle stones, are the main building stones present in heritage structures. The fine, micritic matrix of the Courville stones was partially recrystallized into microspar or even large spar crystals. In proportion to the micritic content, the porosity demonstrated a large range of values: up to 40 vol. % for the most friable bed or as low as 13–15 vol. % for the two beds used for

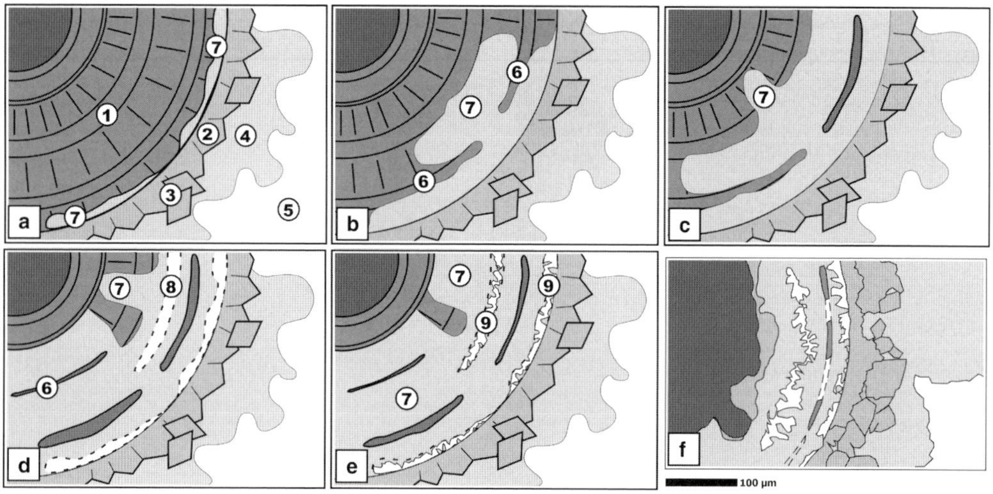

Fig. 5. Oolite gypsum recrystallization process in a black crust on Savonnières stone (St Dizier, France). (**a**–**f**) chronological changes: 1, oolitic cortex made of concentric laminae and radial fibrous areas; 2, isopachous spar cement; 3, dolomite crystals; 4, neogenic gypsum crystals on pore edges; 5, intergranular pore; 6, remains of concentric lamina; 7, penetration of the recrystallization front into the oolite cortex; 8, secondary porosity appearing into gypsum areas; 9, automorphic gypsum crystals.

construction. This porosity was mainly composed of intra-matrix micropores.

In general, these fine facies with micritic matrix were more sensitive to salt crystallizations than the facies with sparitic cementing and dominant macroporosity. In the example given in this study, the 'Courville stone' appears to be a fine limestone with micritic matrix, milliolid (foraminifera) and some quartz grains (Fig. 6). It was worth noticing that the surface of the limestone was relatively well-preserved and the stone seems to be relatively unweathered (Fig. 6, area 3). The black crust was composed of two thin layers, both rich in gypsum and opaque particles (Fig. 6, area 1 and 2). Under the crust, the external surface of the stone (Fig. 6, area 5) was covered by a dark film formed of microcrystalline calcite. This thin '*calcin*' layer may have protected the underlying limestone from weathering penetration.

However, epifluorescence microscopy showed two luminescent zones (Fig. 6, area 4) around a concavity located on the stone surface, possibly due to accidental marks of impact (such as a tool mark, for example). Towards the left of the image, luminescence indicated the possible presence of a faded zone. On the right of the depression,

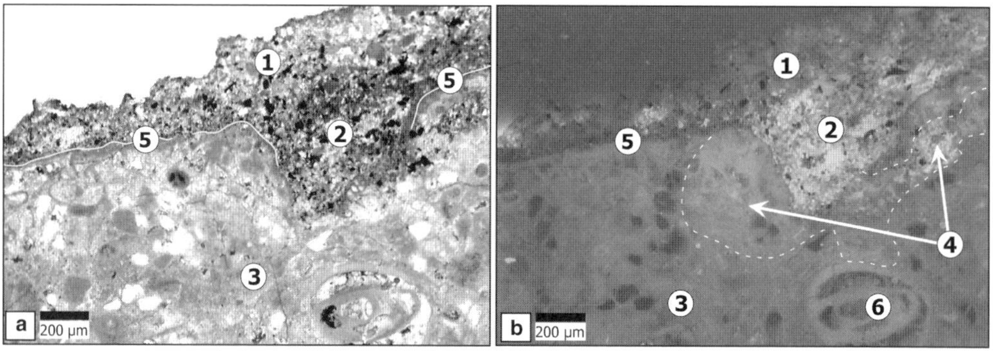

Fig. 6. Photomicrograph under (**a**) plain polarized light and (**b**) epifluorescence showing a black crust on micritic limestone (Courville Stone, Rheims, France). 1 and 2, outer and inner layers of the black crust; 3, pristine limestone; 4, luminescent areas highlighting the lateral penetration of weathering; 5, protective 'calcin' layer; 6, non-weathered foraminifera (milliolid).

however, examination identified the presence of gypsum crystallization. This crystallization feature meant that weathering had actually already developed under the microcrystalline coating, spreading laterally from the unprotected zone. It was also noted that none of the elements (foraminifera or bioclasts) were affected by a pseudomorphosis or recrystallization from calcite to gypsum (Fig. 6, area 6), and that only the fine calcitic matrix was weathered.

Discussion

Crinoidal limestone (Euville Stone)

In crinoidal limestone such as Euville stone which exhibits a low macroporosity, the transition zone between the stone and the black crust was composed of gypsum which crystallized inwards. This crystallization developed initially within cleavages or microcracks and penetrated within the calcite crystals (i.e. crinoidal fragments and associated spar syntaxial cement).

The formation and development of this black crust resulted from at least three mechanisms: (1) accumulation; (2) crystallization; and (3) recrystallization. During accumulation, atmospheric particles (wind-blown quartz, carbon microsoots and fly-ashes) settle and accumulate on surfaces sheltered from rainwater and rainwash. Micro-organisms could also contribute to fixing particles and to the outwards growth of the crust. The crystallization of gypsum takes place in microcracks. It frequently shows acicular morphologies which testified to its crystallization in an open space. During recrystallization, indentation of the lower limit of the internal gypsum layer showed weathering advancing towards the interior of the stone. Penetrating by way of intragranular cleavages and microcracks, the gypsum tended to replace the whole of the calcite by pseudomorphosis. These three mechanisms can occur in a same crust, certainly at the same time.

Due to crystallization and recrystallization, elements within the limestone (parts of crinoid ossicles and of syntaxial cement) were progressively integrated into the crust, while the weathering front penetrated into the limestone. The weathering process, observed and detailed here with the help of observation under an epifluorescence microscope, appeared identical to that described for statues made of Carrara marble by Vergès-Belmin (1994). The gypsum entered the interior of the ossicles and the sparitic cement by way of cleavages and microcracks (micro-box process). Subsequently, within the cells differentiated by weathering, a progressive recrystallization led to the emergence of gypsum-calcite pseudomorphosis (Fig. 1). The observed intense blue fluorescence corresponding to the weathering front allowed us to hypothesize that micro-organisms played a role in this process (area C2 on Fig. 2b).

For typical Euville stone microfacies, crinoid ossicles and cements formed large edge-to-edge calcite crystals, a texture which seems to restrict black crust development. Penetration of weathering agents into these non-porous crystals was slow and provided the stone with a good durability against calcite-gyspum recrystallization. However, microcracks and cleavages in syntaxial cement favoured weathering by accelerating the micro-box process (Delvigne 1998).

Oolitic limestone (Savonnières Stone)

In Savonnières limestone, which has a high macroporosity, the development of sulphated encrustation and the progression of gypsum recrystallization can be either favoured or limited by the sedimentary/diagenetic characteristics of this limestone.

The fibrous calcite cement does not completely close the intergranular porosity and does not protect elements within the limestone from recrystallization. For this limestone, the cementation does not isolate the oolites cortex from weathering fluids (water transfers from the inner or outer part of the crust), which can easily penetrate because of the residual intergranular macroporosity. Furthermore, this porosity can allow the growth of endolithic micro-organisms (Polh & Schneider 2002) and penetration of atmospheric particles into the pristine interior stone. The later are thought to act as catalysts in mineralogical recrystallizations (Ausset *et al.* 1999).

Formation of gypsum begins at the periphery of the oolite even when cements are largely intact. The cortex of the micritic oolites is more easily affected by weathering than the spar cement. In some extreme cases, oolites were entirely transformed into gypsum whereas the calcitic cement remained unaltered.

The dolomite crystals spread over the external surface of the calcitic cement of the Vacuolar Oolithe appeared to be resistant to gypsum recrystallization. Their presence should therefore reduce the damage caused by salt crystallization (Angeli *et al.* 2007) and the alterability of the limestone as a result of gypsum recrystallization.

In the case of the Savonnières limestone, growth of the lower layer of the black crust towards the pristine limestone does not follow micro-cracks or crystal cleavages as observed in Euville limestone. Instead, the weathering process was mainly guided by the intergranular macroporosity linked to the amount of palissadic cementation. The contrasting behaviour of oolitic elements and of sparitic or dolomitic should be noted; it can lead to a total recrystallization of oolites in gypsum when palissadic cements remain in calcite.

Fine limestone (Courville Stone)

In the case of micritic matrix-supported limestones such as Courville limestone, the development of the sulphated encrustation can show different morphologies according to the presence or absence of '*calcin*' and depending on the proportion of the various crystals within in the matrix (micrite, microspar and sometimes spar).

For this facies, the recrystallization front can either penetrate deeply into the rock or be restricted to the surface of the limestone. In the last case, when a surface layer of microcrystalline calcite protects the surface of the stone, weathering was restricted to areas where this thin layer was absent. The presence of surface abrasion marks (tool marks or impact) allowed weathering to penetrate the stone, progressing laterally under the coating of '*calcin*'.

The development of sulphated crusts and the progression of gypsum recrystallization on Courville stone, a fine matrix-supported bioclastic limestone, appeared to be favoured or limited by sedimentary or diagenetic fabrics specific to this limestone. The proportion of micritic matrix compared to partially recrystallized calcite matrix (microspar and spar) rendered the limestone very sensitive to salt-related weathering: for example, efflorescences, flaking and contour scaling or crust detachment. In Courville stone, the micritic matrix showed important microporosity (approximately two-thirds of total porosity). The elevated porous surface area in limestones with dominant micro- and nanoporosity are more sensitive to gypsum recrystallization than large sparite crystals found in marbles or crinoidal limestones.

On the other hand, rock fabric elements (bioclasts) in Courville stone were more resistant to weathering than the matrix. The opposite behaviour to that of Savonnieres stone, in which we observed that oolites were altered before the binding phase (a fibrous spar cement), was observed. This phenomenon, already observed on English oolitic limestones (Schiavon 1992), highlighted the important difference between the behaviour of sparitic cement limestones (e.g. Euville or Savonnières stones) and matrix-supported limestones (e.g. Courville Stone). For the former, the cement was more resistant to gypsum recrystallization than the elements whereas for the latter (micritic limestones), the matrix showed a higher alterability than the elements.

Conclusion

These three examples of black crust growth development illustrate the relationship between building stone microfacies and morphology of the lower layer of sulphated encrusting. Numerous characteristics of building stone nature could affect its weathering, especially since the porous system of sedimentary rocks directly depends on the sedimento-diagenetic facies. As shown, weathering behaviour towards calcite-gypsum pseudomorphosis of the Euville stone was similar to Carrara marble (Vergès-Belmin 1994; Bugini *et al.* 2000; Weber *et al.* 2007), obviously because these two stones were essentially made of large side-by-side calcite crystals. For the two other building stones, Savonnières and Courville, this weathering process may advance over healthy limestone without microcracks because of the large amount of macroporosity or, on the contrary, of fine micritic matrix. In these two microfacies, microcracks develop after gypsum crystallization or are linked to other stone deterioration patterns (exfoliation or granular disintegration, for example; Fronteau *et al.* 1999). The next step of our work is to define precise links between limestone microfacies and weathering behaviour, with sedimento-diagenetic analysis and petrophysical measurements. In the same way, more exhaustive studies must be carried out to identify the real nature of the thin layer of microcrystalline calcite which was observed only on samples of Courville limestone. If the limestone is more resistant to weathering with this protective layer, restoration and cleaning work is needed to preserve or imitate it.

The weathering area, where carbonate stone was partly or totally replaced by gypsum pseudomorphosis, varies greatly in thickness and morphology according to the type of the underlying elements (oolites, foraminifera, crinoid ossicle) or matrix and cements. Quality and roughness of the surface may also have an influence on the growth of the gypsum recrystallization in the limestone. All these parameters may also control the rate growth of the encrustation. Bugini *et al.* (2000) calculate an average growth rate of 2–5 μm/annum on white marble; in our study, growth rates vary from less than 1 μm/annum to 10 μm/annum. Quantitative values on black crust thicknesses and rates also seem inaccurate depending on the lower limit taken into account in the measurement, since gypsum pseudomorphosis inside the limestone is linked to the black crust growth but also contains the previous surface of the pristine stone.

The original limestone surface was therefore found inside the lower layer of the encrustation and not immediately under the atmospheric particle-rich layer (upper dark layer). This lower layer, partially formed by superficial growth of gypsum but also by limestone recrystallization, could create some problems during restoration or cleaning work (Bromblet & Vergès-Belmin 1996).

Before cleaning formerly blackened walls (Grossi & Brimblecombe 2007), it is important to characterize and understand black crust growth processes and their relationship with subjacent stones. If the gypsum crust is totally removed, part of the original stone could be destroyed with eventual damage to the painting or decoration.

The authors would like to acknowledge the two reviewers and especially P. Warke for their valuable comments.

References

ANDRIANI, G. F. & WALSH, N. 2007. The effects of wetting and drying, and marine salt crystallization on calcarenite rocks used as building material in historic monuments. *In*: PŘIKRYL, R. & SMITH, B. J. (eds) *Building Stone Decay: From Diagnosis to Conservation.* Geological Society, London, Special Publications, **271**, 179–188.

ANGELI, M., BIGAS, J.-P., BENAVENTE, D., MENÉNDEZ, B., HÉBERT, R. & DAVID, C. 2007. Salt crystallization in pores: quantification and estimation of damage, *Environmental Geology*, **52**, 205–213.

AUSSET, P., LEFÈVRE, R. A., PHILIPPON, J. & VENET, C. 1991. Large scale distribution of fly-ash particles inside weathering crust on calcium carbonate substrates. Some examples on French monuments. *In*: DECROUEZ, D., CHARMAY, J. & ZEZZA, F. (eds) *La Conservation Des Monuments Dans Le Bassin Mediterraneen: Actes du 2éme Symposium International* Genéva, 19–21 Novembre 1991, 121–139.

AUSSET, P., DEL MONTE, M. & LEFÈVRE, R. A. 1999. Embryonic sulphated black crust in Atmospheric Simulation Chamber and in the field: role of the carbonaceous fly ash. *Atmospheric Environment*, **33**, 1525–1534.

BENAVENTE, D., CUETO, N., MARTINEZ, J., GARCIA DEL CURA, M. A. & CANAVERAS, J. C. 2007. The influence of petrophysical properties on the salt weathering of porous building rocks. *Environmental Geology*, **52**, 215–224.

BLANC, A., BOUINEAU, A., NOËL, J., JATON, C. & ROCARD, J. 1985. Les pierres employées pour la construction et la restauration des monuments du nord-est de la France, leur comportement en oeuvre. 5ème *Congrès Int. sur l'Altération et la Conservation de la Pierre*, Lausanne, 985–995.

BLOWS, J. F., CAREY, P. J. & POOLE, A. B. 2003. Preliminary investigations into Caen Stone in the UK; its use, weathering and comparison with repair stone. *Building and Environment*, **38**, 1143–1149.

BROMBLET, P. & VERGÈS-BELMIN, V. 1996. L'élimination des sulfates sur la statuaire calcaire de plein air: une habitude discutable. *In*: *Le dessalement des Matériaux Poreux.* 7èmes *Journées d'Études de la SFIIC*, 55–66.

BUGINI, R., TABASSO, M. L. & REALINI, M. 2000. Rate of formation of black crusts on marble. A case study. *Journal of Cultural Heritage*, **1**, 111–116.

CAMUFFO, D., DEL MONTE, M. & SABBIONI, C. 1983. Origin and growth mechanisms of the sulfated crusts on urban limestone. *Water, Air and Soil Pollution*, **19**, 351–359.

CHOQUETTE, P. W. & PRAY, L. C. 1970. Geologic nomenclature and classification of porosity in sedimentary carbonates. *Bulletin of American Association of Petroleum Geologists*, **54**, 207–250.

COLSTON, B. J., WATT, D. S. & MUNRO, H. L. 2001. Environmentally-induced stone decay: the cumulative effects of crystallization–hydration cycles on a Lincolnshire oopelsparite limestone. *Journal of Cultural Heritage*, **2**, 297–307.

DELVIGNE, J. E. 1998. Atlas of micromorphology of mineral alteration and weathering. *The Canadian Mineralogist. Special publication*, **3**, Ottawa.

DRAVIS, J. J. & YUKEWICZ, A. 1985. Enhanced carbonate petrography using fluorescence microscopy. *Journal of Sedimentary Petrology*, **55**, 795–804.

FASSINA, V., FAVARO, M. & NACCARI, A. 2002. Principal decay patterns on Venetian monuments. *In*: SIEGESMUND, S., WEISS, T. & VOLLBRECHT, A. (eds) *Natural Stone, Weathering Phenomena, Conservation Strategies and Case Studies.* Geological Society, London, Special Publications, **205**, 381–391.

FRONTEAU, G. 2000*a*. *Comportements télogénétiques des principaux calcaires de Champagne-Ardennes en relation avec leur faciès de dépôt et leur séquençage diagénétique.* PhD Thesis. University of Reims Champagne-Ardenne, Reims.

FRONTEAU, G. 2000*b*. L'Oolithe vacuolaire: caractéristiques, mise en oeuvre et altération. *Bulletin d' Information du Bassin de Paris*, **37**(3), 34–39.

FRONTEAU, G., BARBIN, V. & PASCAL, A. 1999. Impact du faciès sédimento-diagénétique sur l'altération en oeuvre d'un géomatériau calcaire. *Comptes Rendus de l'Academie de Science de Paris*, **328**, 671–677.

FRONTEAU, G., PASCAL, A. & BARBIN, V. 2002. Caractérisation de la Pierre de Courville (Lutétien, Marne, France). Relations entre diagenèse et propriétés pétrophysiques. *Géologie de la France*, **4**, 51–59.

GALLETI, G. C., BOCCHINI, P., CAM, D., CHAIVARI, G. & MAZZEO, R. 1997. Chemical characterization of the black crust present on the stone central portal of St Denis abbey. *Fresenius Journal of Analytical Chemistry*, **357**, 1211–1214.

GHEDINI, N., GOBBI, G., SABBIONI, C. & ZAPPIA, G. 2000. Determination of elemental and organic carbon on damaged stone. *Atmospheric Environment*, **34**, 4383–4391.

GROSSI, C. M. & BRIMBLECOMBE, P. 2007. Effect of long-term changes in air pollution and climate on the decay and blackening of European stone buildings. *In*: PŘIKRYL, R. & SMITH, B. J. (eds) *Building Stone Decay: From Diagnosis to Conservation.* Geological Society, London, Special Publications, **271**, 117–130.

HUTCHINSON, A., JOHNSON, J. B., THOMPSON, G. E. & WOOD, G. C. 1992. The role of fly-ash particulate material and oxide catalysts in stone degradation. *Atmospheric Environment*, **26A**(15), 2795–2803.

JEANNETTE, D. 1981. Modifications superficielles de grès en oeuvre dans des monuments alsaciens. *Science Géologique Bulletin*, **34**(1), 37–46.

LEFÈVRE, R.-A. & AUSSET, P. 2001. Les effets de la pollution atmosphérique sur les matériaux du patrimoine bâti: la pierre et le verre. *Pollution Atmosphérique*, **72**, 571–588.

MACHILL, S., ALTHAUS, K., KRUMBEIN, W. E. & STEGER, W. E. 1997. Identification of organic compounds extracted from black weathered surfaces of Saxonean sandstones, correlation with atmospheric input and rock inhabiting microflora. *Organic Geochemistry*, **27**, 79–97.

MARAVELAKI-KALAITZAKI, P. & BISCONTIN, G. 1999. Origin, characteristics and morphology of weathering on Istria stone in Venice. *Atmospheric Environment*, **33**, 1699–1709.

MOORE, C. H. 1989. *Carbonate Diagenesis and Porosity*. Developments in Sedimentology, **46**, Elsevier, Amsterdam.

NOËL, P. 1970. Les carrières françaises de pierre de taille. *Société de Diffusion des Bâtiments et Travaux publics*. Paris.

POLH, W. & SCHNEIDER, J. 2002. Impact of endolithic biofilms on carbonate rock surfaces. *In*: SIEGESMUND, S., WEISS, T. & VOLLBRECHT, A. (eds) *Natural Stone, Weathering Phenomena, Conservation Strategies and Case Studies*. Geological Society, London, Special Publications, **205**, 177–194.

POTGIETER-VERMAAK, S. S., GODOI, R. H. M., VAN GRIEKEN, R., POTGIETER, J. H., OUJJA, M. & CASTILLEJO, M. 2005. Micro-structural characterization of black-crust and laser cleaning of building stones by micro-Raman and SEM techniques. *Spectrochimica Acta, Part A*, **61**, 2460–2467.

RODIGUEZ-NAVARRO, C. & SEBASTIAN, E. 1996. Role of particulate matter from vehicle exhaust on porous building stones (limestones) sulfation. *The Science of the Total Environment*, **187**, 76–91.

ROELS, S., CARMELIET, J. & HENS, H. 2003. Modelling unsaturated moisture transport in heterogeneous limestone (Part 1. A mesoscopic approach). *Transport in Porous Media*, **52**, 333–350.

ROTHERT, E., EGGERS, T., CASSAR, J., RUEDRICH, J., FITZNER, B. & SIEGESMUND, S. 2007. Stone properties and weathering induced by salt crystallization of Maltese Globigerina Limestone. *In*: PŘIKRYL, R. & SMITH, B. J. (eds) *Building Stone Decay: From Diagnosis to Conservation*. Geological Society, London, Special Publications, **271**, 189–198.

SABBIONI, C. 1995. Contribution of atmospheric deposition to the formation of damage layer. *The Science of the Total Environment*, **167**, 49–55.

SCHIAVON, N. 1992. BSEM study of decay mechanisms in urban limestones. *European Cultural Heritage*, **6**, 35–46.

SIEGESMUND, S., TÖRÖK, A., HÜPERS, A., MÜLLER, C. & KLEMM, K. 2007. Mineralogical, geochemical and microfabric evidences of gypsum crusts: a case study from Budapest. *Environmental Geology*, **52**, 385–397.

THOMACHOT, C. & JEANNETTE, D. 2000. Petrophysical properties modifications of Strasbourg's Cathedral sandstone by black crusts. *In*: FASSINA, V. (ed.) *Proceeding of the 9th International Congress on Deterioration and Conservation of Stone*. Venice, June 19–24, 2000, Elsevier, Amsterdam, 265–273.

TÖRÖK, Á. & ROZGONYI, N. 2004. Morphology and mineralogy of weathering crusts on highly porous oolitic limestones, a case study from Budapest. *Environmental Geology*, **46**, 333–349.

TÖRÖK, Á., SIEGESMUND, S., MÜLLER, C., HÜPERS, A., HOPPERT, M. & WEISS, T. 2007. Differences in texture, physical properties and microbiology of weathering crust and host rock: a case study of the porous limestone of Budapest (Hungary). *In*: PŘIKRYL, R. & SMITH, B. J. (eds) *Building Stone Decay: From Diagnosis to Conservation*. Geological Society, London, Special Publications, **271**, 261–276.

TUCKER, M. E. 1988. *Techniques in Sedimentology*. Blackwell, London.

TUCKER, M. E. & WRIGHT, V. P. 1990. *Carbonate Sedimentology*. Blackwell Scientific Publications, London.

VERGÈS-BELMIN, V. 1994. Pseudomorphism of gypsum after calcite, a new textural feature accounting for the marble sulphation mechanism. *Atmospheric Environment*, **28**, 295–304.

WEBER, J., BESELER, S. & STERFLINGER, K. 2007. Thin-section microscopy of decayed crystalline marble from the garden sculptures of Schoenbrunn Palace in Vienna. *Materials Characterization*, **58**, 1042–1051.

WINKLER, E. M. 1994. *Stone in Architecture. Properties, Durability*. 3rd edn. Springer-Verlag, Berlin.

Influence of temperature and salt concentration on the salt weathering of a sedimentary stone with sodium sulphate

MATTHIEU ANGELI[1,2]*, RONAN HÉBERT[1], BEATRIZ MENÉNDEZ[1], CHRISTIAN DAVID[1] & JEAN-PHILIPPE BIGAS[3]

[1]*Université de Cergy-Pontoise, Département de Sciences de la Terre et Environnement CNRS UMR 7072, 5 mail Gay-Lussac, Neuville-sur-Oise, 95031 Cergy-Pontoise cedex, France*

[2]*Norges Geotekniske Institutt, Sognsveien 72, Postboks 3930 Ullevål Stadion, 0806 Oslo, Norway*

[3]*CHRYSO, 7 rue de l'Europe, 45300 Sermaises, France*

**Corresponding author (e-mail: matthieu.angeli@ngi.no)*

Abstract: The aim of this study is to evaluate how the ambient temperature and salt concentration affect the salt decay of a sedimentary stone. Samples of a detritic limestone which have experienced cycles of accelerated ageing at 5, 25 (room temperature, RT) and 50 °C with brines which had different sodium sulphate concentration were analysed. The weight of the samples and of the pieces fallen off during the cycles was monitored. The results show that the damage is more important at 5 °C than at RT. The samples at 50 °C were intact at the end of the experiment. Second, the size of the pieces fallen from the samples is significantly smaller for low temperatures: at 5 °C, the decay produces a fine powder; at RT, the pieces fallen off are of millimetre to centimetre scale. The weathering patterns are therefore different at these two temperatures: fine crumbling at 5 °C; coarse crumbling and contour scaling at RT. The salt uptake seems quite similar for a given concentration whatever the temperature. The decay also seems to be of a different kind for each concentration at RT: crumbling at low concentration, contour scaling at high concentration. Crystallization seems to take place deeper inside the porous network of the stone when the concentration of salts in the brine is higher, that is to say when the brine viscosity is higher.

Salts, and particularly sodium sulphate, are known to be among the most destructive agents in porous stones, concrete or brick weathering. The study of its crystallization mechanism is therefore very important to fully understand its damaging effect on porous networks and, in the future, to find a way to prevent or limit it. Recent studies attribute the importance of sodium sulphate decay to the salt crystallization pressure of its decahydrate phase (mirabilite $Na_2SO_4 \cdot 10H_2O$), rather than to its anhydrous phase (thenardite Na_2SO_4) (Rodriguez-Navarro & Doehne 1999; Tsui *et al.* 2003; Angeli 2007). Thermodynamical studies (Benavente *et al.* 1999; Flatt 2002; Scherer 2004; Steiger 2005*a*, *b*; Coussy 2006) as well as experimental studies (Goudie 1986, 1993; Rodriguez-Navarro & Doehne 1999; Angeli 2007; Angeli *et al.* 2007) show that damage depends on the quantity of salt in the stone and the characteristics of the porous network, as well as on the environmental conditions (e.g. temperature and relative humidity).

This study aims to understand what role the temperature and the salt content in the sample play in the decay of a sedimentary stone, regarding the type and intensity of damage. For this, accelerated ageing tests are performed on only one type of stone, a detritic limestone, and with only one salt, sodium sulphate, but in different thermodynamic conditions (temperature and salt concentration of the brine).

Materials and methods

Materials

Twelve cubic (*c.* 7 × 7 × 7 cm) samples of a Lutetian limestone from the Parisian Basin commercially known as 'Roche fine' have been used in this study. This rock has been chosen for several reasons: its detritic fabric with unimodal porosity is much simpler to study than other bioconstructed or crystallized limestones; it has been demonstrated in previous tests (Angeli *et al.* 2006, 2007) that it is not very resistant to salt decay (hence the tests are quite fast); it is homogeneous and very regular from one sample to the other, which allows a very good reproducibility of the tests; it has been widely used for construction in Paris in the past (buildings and historical monuments); and it is one of the rocks imposed for restoration and

From: PŘIKRYL, R. & TÖRÖK, Á. (eds) *Natural Stone Resources for Historical Monuments.* Geological Society, London, Special Publications, **333**, 35–42.
DOI: 10.1144/SP333.4 0305-8719/10/$15.00

construction in the protected areas of the Ile de France region. It is a fine-grained detritic limestone made of calcite (90%) and quartz (10%) with a high porosity (37.2%) and a very low tensile strength (1.5 MPa). Its pore distribution is unimodal which makes the study of its modification easier. Complete hydromechanic properties are given in Table 1.

Methods

The twelve samples were subjected to accelerated ageing tests adapted from the European standard EN 12370. These tests were run under three different ambient temperatures: 5 °C, room temperature (RT *c.* 25 °C), and 50 °C. It is important to notice that two ambient conditions are below the temperature limit of mirabilite stability (32.4 °C), the latter being above which means that no mirabilite crystallization will occur for this experimental condition.

Three different concentrations of salt in solution have been used in this study: 5, 12 and 25% of anhydrous sodium sulphate (thenardite) in weight. The tests have only been performed under six thermodynamic conditions. Indeed, the three following conditions (12%/5 °C, 25%/5 °C and 25%/RT) are supersaturated with respect to mirabilite, that is to say under these conditions a liquid solution would be metastable, that is, not suitable for capillary imbibitions. The six thermodynamic conditions tested here are represented as stars on the phase diagram (Fig. 1): 5%/5 °C; 5%/RT; 5%/50 °C; 12%/RT; 12%/50 °C; 25%/50 °C.

As can be seen in the figure, these conditions correspond to various saturation conditions of the sodium sulphate brine. This saturation ranges from 0 for pure water to 1 for full saturation. The closer the solution is to the full saturation conditions, the faster it will reach supersaturation and hence weathering of the sample. This will happen after fewer imbibitions for a higher saturation (Flatt 2002; Coussy 2006). The six saturation conditions tested in this study are as follows

(1) T = 5 °C; 5% in weight – around 0.85 saturation;
(2) T = 25 °C; 5% in weight – around 0.25 saturation;

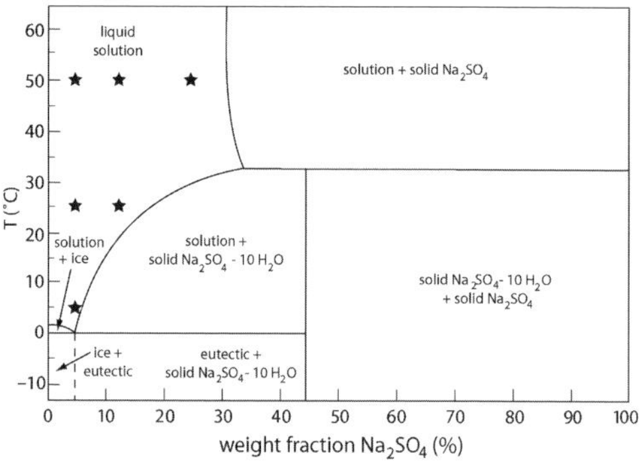

Fig. 1. Sodium sulphate phase diagram (Hougen *et al.* 1954). The black stars represent the six thermodynamic conditions studied.

(3) T = 25 °C; 12% in weight – around 0.60 saturation;
(4) T = 50 °C; 5% in weight – around 0.17 saturation;
(5) T = 50 °C; 12% in weight – around 0.40 saturation; and
(6) T = 50 °C; 25% in weight – around 0.83 saturation.

These tests are composed of cycles which have a duration of 24 hours. The cycles comprise three different stages: imbibition at ambient temperature; drying; and return to initial ambient temperature. Timing for each stage is modified from the EN 12370 standard regarding stone resistance to crystallization of salts in pores.

Let us describe the three steps in more detail

(1) Two hours imbibition with a brine (concentration of Na_2SO_4 is fixed by the six thermodynamic conditions) is realized for the three different ambient temperature conditions. Note that temperatures of imbibition solution are the same as ambient temperatures, that is, 5 °C, RT and 50 °C. Samples soak in a container with 1 cm height of solution. The level of solution is maintained during that stage. Imbibition solution was prepared with sodium sulphate decahydrate and demineralized water. The

Table 1. *Hydromechanic properties of the 'Roche Fine' (from Angeli* et al. *2007)*

	Porosity (vol. %)	Absorption (wt %)	Bulk density (g cm^{-3})	Evaporation coefficient (g m^{-2} s$^{-1/2}$)	Capillary coefficient (g m^{-2} s$^{-1/2}$)	Median pore radius (μm)	P-wave velocity (m s^{-1})	Tensile strength (MPa)
Roche fine (St Maximin)	37.2	75.9	1.7	67	1106	12	2898	1.5

cycles start at 16 h in the afternoon; the weighing at the beginning of this step will therefore be referred to as 16 h weighing.

(2) Drying: All the samples are placed in a drying oven at 105 °C for 16 hours. The weighing performed at the beginning of this step will be referred to as the 18 h weighing. This step ends at 10 h in the morning.

(3) Return to initial ambient temperature, that is, 6 hours of cooling from 105 °C to 5 °C, RT and 50 °C. The weighing performed at the beginning of this step is the 10 h weighing.

Sixteen full cycles were performed during which all the samples were weighed three times per cycle (before imbibition, after imbibition and after drying). The parts of the stones which fell during the experiments with size from *c.* 0.5 mm were collected and weighed every cycle after drying. They will be referred to as 'remains'. The only remains which were not collected consist of powder resulting from crumbling.

A cooled incubator is used for the 5 °C ambient environment. Ambient temperature of 50 °C is obtained using a drying oven, in which the drying stage takes place. Temperature is set up with a precision of ± 0.1 °C.

Results

General observations

The first thing to notice is that all the samples demonstrate a different level of decay from one thermodynamic condition to the other. Some pictures of each sample, characteristic of the decay, are presented in Figure 2. Future references to this figure will refer precisely to the experimental condition and the number of the cycle.

At 50 °C, no damage was evident on any of the six samples, only harmless efflorescence on the sides of the samples (Fig. 2; 50 °C, 15th cycle). At 5 °C, the damage begins immediately as the samples contain salt, that is, during the 2nd cycle. After this cycle, a few signs of very thin contour scaling appear (a few tenths of a millimetres). During the 3rd cycle, the thin contour scaling almost ends and is followed by crumbling. This crumbling is very regular throughout the cycles until the end of the test and the particles lost form a very fine powder.

At RT, the decay processes are a little different. During the first 5–6 cycles, the samples are slowly damaged. Efflorescences are observed as well as a decay of the lower parts of the samples, which

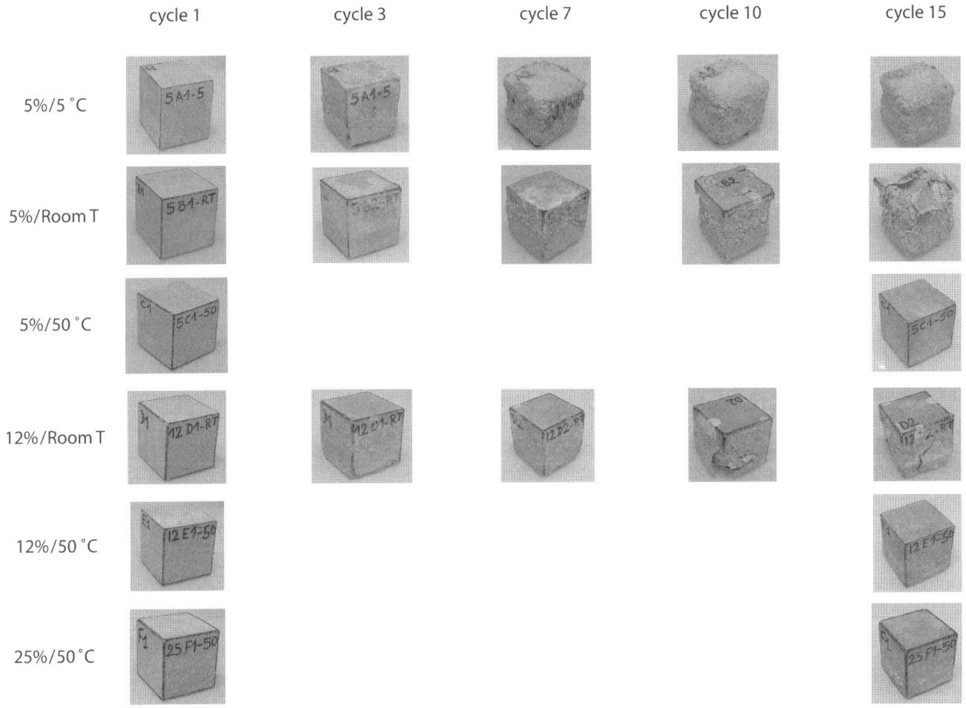

Fig. 2. Evolution of the 7 cm cubic samples in the six thermodynamic conditions during the 25 weathering cycles. The most characteristic sample is included to illustrate decay.

were immersed during the tests. The efflorescences are unexpectedly more important on the sample saturated with the 5% brine, that is, with the lowest salt concentration. At this point, no major difference is observed between the two concentrations of salts (5 and 12%). From around the 6th cycle, contour scaling starts to appear on all samples with the detached part thicker for 12% than for 5%. In fact, from this point on, the decay becomes different for the two concentrations. For 5% sodium sulphate, the contour scaling rapidly turns into crumbling from the 8th cycle. Crumbling becomes almost the only type of decay after the 11th cycle, although some residual scales are found on the top of the sample (Fig. 2; 5%/RT). The remains then turn into a very fine powder similar to the remains of the experiments at 5 °C. On the other side, at 12% sodium sulphate, contour scaling continues to be the main type of decay and the detached parts can be up to 1 cm thick. The decay in these conditions is therefore highly dependent on the heterogeneities of the stone. Although triggered by the experimental conditions, contour scaling seems to be localized by flaws in the sample (as opposed to crumbling, which seems to be regular over the whole surface exposed to evaporation). These microstructural differences also imply a slight variability in the weight loss schedule, since the scales can be separated from the sample at almost any moment after their appearance. This can explain the difference in the evolution, two samples of the weight in this condition (12%/RT): early loss of scale for one and late loss of scale for the other. More samples tested under these condition are however necessary to assess this hypothesis. It is also remarkable to note that, under very similar experimental conditions (salt, stone, cycles, temperature, RH), two samples show two different types of decay: crumbling and contour scaling. Decay only seems to depend on the quantity of salts contained in the brine.

Weight monitoring

The samples were weighed three times during the cycles: before imbibition, after imbibition and after drying. This corresponds, as mentioned earlier, to the beginning of each step of the cycle. This means that two of the weighing stages are performed on dry samples (before imbibition and after drying), and the other is performed on a fully saturated sample. Figure 3 presents an example of the weight monitoring. Figure 3a presents the full weight evolution of a sample (A1) during the 15 cycles. The parts that fell off the samples during imbibition have also been weighed.

The difference between the weights at 16 h and 18 h (before and after imbibition) corresponds to the increase due to the water and salt supply minus

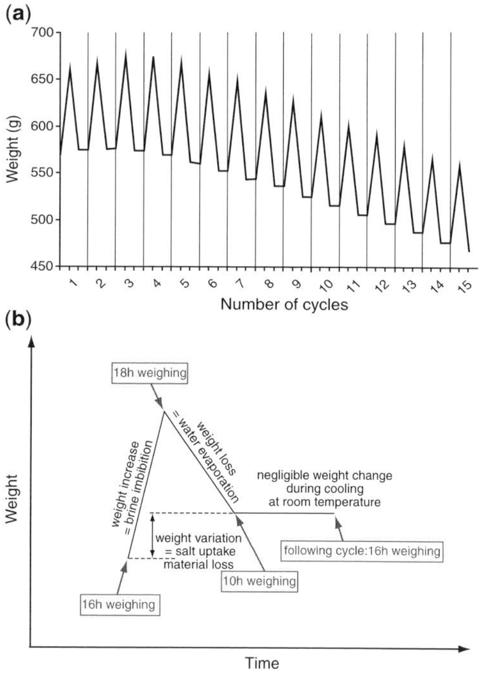

Fig. 3. (**a**) Example weighings performed during the weathering cycles for a sample at 5%/5 °C and (**b**) detail of the weighing performed during one cycle and of their experimental meaning.

the parts that fell off the sample. The difference between 18 h and 10 h (before and after drying in the oven) corresponds to the loss of water and the loss of material (negligible during this stage). The difference between the weighing at 10 h and 16 h (before and after cooling at room temperature) corresponds only to the equilibration of the water content in the sample due to the change of environmental conditions, from the oven (105 °C; RH > 90%) to room conditions (around 25 °C; 35–55%). The detailed weight variation during one cycle is presented in Figure 3b. Figure 4 depicts the weight evolution of the samples during the cycles. The noticeable curves that have been plotted are the evolution of normalized dry weight (Fig. 4a), normalized collected remains (Fig. 4b) and the sum of normalized dry weight and normalized collected remains (Fig. 4c).

Discussion

Localization of crystallization

The most striking feature of this experiment is the decay pattern of the samples. A difference in the kinetics of damage, that is, the appearance of the first signs of decay (alteration index AI, Angeli *et al.* 2007) and in the intensity was definitely

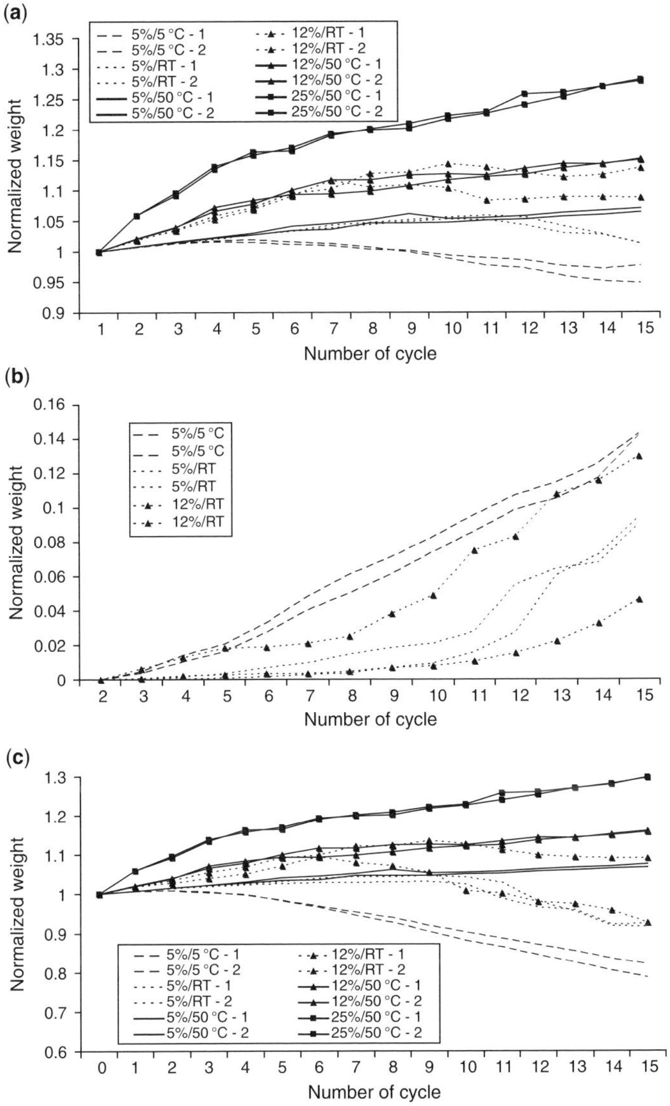

Fig. 4. (**a**) Weight evolution of the 12 samples in dry state; (**b**) weight evolution of the collected remains. The remains are collected only once per cycle, that is, it is impossible to distinguish the loss during imbition from the loss during drying. Since the samples at 50 °C were not damaged at the end of the test, they have produced absolutely no damage. Their weight evolution has therefore not been reported on the figure and (**c**) Weight evolution of the combined weight (dry state and collected remains) of the 12 samples.

suspected, since the properties of sodium sulphate, mainly saturation in water, are highly variable at these three temperatures as can be seen on the sodium sulphate phase diagram (Fig. 1). Although these features were observed during the experiments, the most unexpected feature is that the type of decay is very different from one thermodynamic condition to another. Complete values of AI and alteration velocity AV are noted in Table 2. These two parameters are introduced by Angeli *et al.* (2006, 2007): AI corresponds to the number of the cycle during which the first sign of decay can be seen on the sample; AV is the slope of the plot representing the weight loss in percent as a function of the cycle number, once the decay becomes regular. Note that the alteration velocity is not exactly the same as that defined by Angeli *et al.* (2007): it can normally only be calculated once the decay has become regular, when salt uptake becomes negligible compared to material loss. In this experiment (not

Table 2. *Alteration quantification of the samples in the six conditions (AI in cycles, AV in %/cycle)*

		5%	12%	25%
5 °C	AI	2		
	AV	2		
25 °C	AI	4	3	
	AV	2	0	
50 °C	AI	>15	>15	>15
	AV	−0.15	−0.8	−1

always the case) the AV is used to characterize the weight variations. This causes a small bias in the interpretation of the data. Since we have not reached the steady-state degradation in the case of 12%/RT, that is, past the weight increase and shape effect (phase III in Angeli *et al.* 2007), the AV is lower at 12% than at 5%. This would not have been the case with a higher number of cycles, however as suggested by the large scales starting to appear on the samples at 12%.

At 50 °C, whatever the salt concentration, the samples do not suffer any damage from salt crystallization. The weight of the sample keeps increasing through the 15 cycles of the experimental ageing test. This weight increases linearly with brines at 5 and 12% sodium sulphate. For the samples filled with 25% sodium sulphate brine, the weight evolution shows two parts: first, until the 5th cycle, the weight gain is approximately constant at around 4%/cycle. From a porosity of 37.2 vol.% and a saturation of 25% in weight, we would expect a weight increase of 9.3%/cycle since the samples are fully saturated during each cycle. This difference could be due to the relatively large size of the pores in this stone, hence a small surface area leaving very few places for the salt to crystallize. Then, after the 5th cycle, the weight gain is still grossly linear but lower (approximately 1%/cycle): this diminution in the weight increase is due to the progressive saturation of the porous network by the presence of salts from preceding imbibitions.

Tsui *et al.* (2003) concluded from their experimental ageing tests performed at 50 °C that mirabilite was the damaging phase of sodium sulphate under 32.4 °C, and that crystallization of thenardite was almost harmless for the stones. This study clearly confirms these results, since none of the samples that underwent the test at 50 °C were damaged.

At 5%/5 °C, we observe almost only crumbling or granular disintegration, except for the very beginning where a very thin contour scaling seems to be initiated (cycle 2). At 5%/RT, the pattern is very similar: decay starts with very thin contour scaling and but then changes to crumbling. The only difference is that this shift happens later at RT (6–7th cycle instead of 2nd cycle) which is quite normal; it is generally accepted that supersaturation of the solution is a mandatory condition to salt damage (Correns 1949). Sodium sulphate saturation is higher at RT (*c.* 18%) than at 5 °C (*c.* 6%), therefore more thenardite needs to be dissolved in the 5% Na_2SO_4 brine to reach supersaturation at RT than at 5 °C. This supersaturation then occurs after more cycles of imbibition and drying (Flatt 2002; Coussy 2006). In both cases, the crystallization of sodium sulphate therefore occurs close to the surface.

At 12%/RT, the decay of the samples starts after three cycles (Table 2). At first the reaction is very similar to that of the 5% brine, that is to say a few efflorescences and small damage on the corners of the immersed parts. Contour scaling also occurs after 6–7th cycle, but this time with much thicker scales. These scales can have a thickness up to 1 cm as observed in Figure 2 (12%/RT, 10th and 15th cycle). This contour scaling suggests that the crystallization of sodium sulphate takes place deep below the surface of the samples (Ioannou *et al.* 2005). Temperature and concentration of salt in the brine are therefore not only responsible for the presence or absence of damage, but also have an influence on the damage patterns.

Weight monitoring

Dry samples. Figure 4a depicts the evolution of the dry sample weights as a function of the cycles. This evolution is normalized in order to be able to compare the weight gains as a function of the concentration of salt in the brines. An important point to note is that the experiments are reproducible: for each condition, the two samples have a very similar weight evolution. The only exception is the 12%/RT condition, where we can clearly see a difference between the two samples from cycle 8. This difference is due to the type of decay; this point corresponds to the start of a thick contour scaling on both samples. The decay therefore becomes more dependent on the pre-existent heterogeneities of the samples, as explained earlier in the general observations. In this case, the first sample lost a piece close to a corner at the bottom of the sample, while the other kept all its scales during the entire experiment. Despite this variation, the alteration velocities of the two samples in this thermodynamic condition become similar from the tenth cycle.

The weight plots allow the observations made of the samples to be quantified. The most damaged samples are those at 5 °C (Fig. 4a), although the quantity of salt in them is lower than at the other temperatures due to a lower solubility of sodium sulphate in water at low temperature. The low temperature allows fast supersaturation and decay therefore appears quickly. After 15 cycles, these samples have lost 20% of their initial weight with a final AV of 2% per cycle. The samples in the 5%/RT conditions also experience crumbling from the 7th cycle: this crumbling begins later since supersaturtion is reached faster at 5 °C than at room temperature. The weight loss in this case becomes significant only after the 10th cycle. Once this crumbling has begun, the AV becomes almost the same as in the 5%/5 °C conditions.

Collected remains. Figure 4b depicts the weight of the remains that are collected for each sample during the cycles. In order to compare the weight losses, these weights have been normalized to the weight of the initial clean sample. Let us recall here that all the remains have been collected except the finest powder from the crumbling of the samples. An interesting point to note is that crumbling causes a regular loss of material at every cycle, as demonstrated by the weight of the material loss by the samples at 5%/5 °C. This is opposed to the material loss during contour scaling which is more random, demonstrated by the weight of the material lost by the 12%/RT samples.

Dry samples and collected remains. Figure 4c depicts the monitoring of the dry weight plus the collected remains, referred to as the combined remains for case. This plot is important since it allows us to have a precise idea of the evolution of the salt uptake. In fact, this weight corresponds to the weight of the initial sample plus the weight of the salt it contains minus the fine powdery remains that are not collected during the tests. Since there is no powdery remains (or very few) during the first four cycles, we can assume that the weight variation of this assembly during these cycles consists only of salt uptake. This salt uptake appears to be dependent only on the concentration of the brine. In fact, whatever the temperature, the weight gain is the same for two samples filled with the same brine: 0.8%/cycle for 5% brines, 2%/cycle for the 12% brines and 4%/cycle for the 25% brines. This uptake is proportional to the sodium sulphate weight fraction in the brine. This salt uptake is regular until the sample contains too much salt. From this moment on, the salt uptake is limited by ionic diffusion between the brine in the container and the brine in the pores. The 50 °C weight evolution will later be used as a reference for salt uptake for each concentration. For a given condition, the difference between its weight variation and the weight variation at 50 °C for the same salt concentration will give the amount of powdery remains produced by weathering.

During the first ten cycles, 5%/5 °C is the only condition under which this combined weight decreases. This means that it is the only condition where powdery remains are massively produced. This corresponds to the fact that it is the only condition that undergoes crumbling. Progressively, the combined weight of the 5%/RT samples starts to diminish with the same AV as for 5%/5 °C. The AV is now approximately 0.5%/cycle for these conditions, instead of 2%/cycle if we consider only the dry sample. The proportion of powdery remains is therefore approximately one-quarter of the total remains in both cases.

However, crumbling is not the only weathering pattern producing powdery remains. The diminution of combined weight of the 12%/RT samples suggests that, although mostly producing scales, contour scaling produces also powdery remains that have not been collected during the tests.

Variable decay at room temperature

An important fact is observed during the experiments at RT. The difference in the decay at 5% and 12% seems to be the crystallization depth. The contour scaling, starting around cycle 8, is thicker at 12% than at 5%. This means that damage takes place deeper in the stone when the concentration of salt is higher.

Since in-depth crystallization of salts causes contour scaling (Ioannou *et al.* 2005), the thickness of the scales is constrained by the depth of crystallization of the salts. This depth of crystallization depends on the evaporation speed of the brine during drying (Rijniers 2004): if drying is fast, crystallization occurs where evaporation takes place. Ruiz-Agudo *et al.* (2007) proved that for two brines with two different viscosities, salts will precipitate closer to the surface for the brine with the lower viscosity. They therefore explained the different decay mechanisms caused by sodium and magnesium sulphate. The experimental conditions are different in Ruiz-Agudo *et al.* (2007) since the imbibition is continuous, but the mechanisms at stake are the same. In our experiment, supersaturation during imbibition will occur in the zone where thenardite is available for dissolution. These zones are the in-depth locations where crystallization takes place during drying. These viscosity variations can also explain why the scales are thicker at 12%/RT than at 5%/RT. The viscosity of a brine depends on the concentration of salt and of on the temperature (Moore 1896). The higher concentration of salts at 12%/RT entails a higher viscosity of the brine than at 5%, and thus a deeper crystallization of the salts. This would explain the two decay mechanisms occurring in these two similar conditions.

The deceleration of drying can also be caused by a change in the thermal properties of the samples. Angeli (2007) have demonstrated that the thermal conductivity of a sample was reduced by the presence of salt in it. The sample is therefore slower to heat up or cool down, and the temperature of the sample is more homogeneous from its core to its surface. Evaporation becomes slower, and crystallization takes place deeper in the stone.

Conclusions

Temperature is a decisive parameter that controls the amount and pattern of alteration in the case of sodium sulphate decay. Its importance has been previously assessed since mirabilite, which is the damaging phase, only exists under certain environmental conditions. A temperature above 32.4 °C prevents the crystallization of mirabilite and prevents damage (Tsui *et al.* 2003). This study concludes that, in the case of this detritic limestone undergoing sodium sulphate exposure, the damage is higher at 5 °C than at RT. The patterns are also different since crumbling occurs at 5 °C and contour scaling at RT. This implies a difference in the loss of materials during weathering. Crumbling at 5 °C causes the loss of fine particles composed of approximately 25% of fine powder; contour scaling at RT causes the loss of scales of thickness several millimetres, but also a small proportion of powder.

The concentration of salt in the brine also has an important influence on the decay. First the salt uptake in the sample is directly dependent on this concentration, whatever the temperature. Second, this concentration can directly affect the weathering pattern of samples. The concentration change can cause a switch from crumbling to contour scaling, as was observed on the samples at room temperature. This influence of temperature and salt concentration on the weathering amplitude and patterns results from the change they cause in the evaporation velocity, hence in the crystallization depth of sodium sulphate. This change is expected to be due to the variable viscosity of the brine and to the modification of thermal conductivity due to the salt content in the sample.

References

ANGELI, M., BIGAS, J.-P., MENÉNDEZ, B., HÉBERT, R. & DAVID, C. 2006. Influence of capillary properties and evaporation on salt weathering of sedimentary rocks. *In:* FORT, R., ALVAREZ DE BUERGO, M., GOMEZ-HERAS, M. & VAZQUEZ-CALVO, C. (eds) *Heritage, Weathering and Conservation.* Taylor & Francis/Balkema, AK Leiden, The Netherlands, 253–259.

ANGELI, M. 2007. *Multiscale study of stone decay by salt crystallization in porous networks.* PhD Thesis, Université de Cergy-Pontoise, France.

ANGELI, M., BIGAS, J.-P., BENAVENTE, D., MENÉNDEZ, B., HÉBERT, R. & DAVID, C. 2007. Salt crystallization in pores: quantification and estimation of damage. *Environmental Geology*, **52**, 205–214.

BENAVENTE, D., GARCIA DEL CURA, M. A., FORT, R. & ORDOÑEZ, S. 1999. Thermodynamic modelling of changes induced by salt pressure crystallization in porous media of stone. *Journal of Crystal Growth*, **204**, 168–178.

CORRENS, C. W. 1949. Growth and dissolution of crystals under linear pressure. *Discussions of the Faraday Society*, **5**, 267–271.

COUSSY, O. 2006. Deformation and stress from in-pore drying-induced crystallization of salt. *Journal of the Mechanics and Physics of Solids*, **54**, 1517–1547.

FLATT, R. J. 2002. Salt damage in porous materials: how high supersaturations are generated. *Journal of Crystal Growth*, **242**, 435–454.

GOUDIE, A. S. 1986. Laboratory simulation of "the wick effect" in salt weathering of rock. *Earth Surface Processes and Landforms*, **11**, 275–285.

GOUDIE, A. S. 1993. Salt weathering simulation using a single-immersion technique. *Earth Surface Processes and Landforms*, **18**, 369–376.

HOUGEN, O. A., WATSON, K. W. & RAGATZ, R. A. 1954. *Chemical Process Principles. Part I.* 2nd edn. Wiley, New York.

IOANNOU, I., HALL, C., HOFF, W. D., PUGSLEY, V. A. & JACQUES, S. D. M. 2005. Synchrotron radiation energy-dispersive X-ray analysis of salt distribution in Lépine limestone. *Analyst*, **130**, 1006–1008.

MOORE, B. E. 1896. On the viscosity of certain salt solution. *The Physical Review*, **5**, 312–334.

RIJNIERS, L. A. 2004. *Salt crystallization in porous materials: A NMR study.* PhD Thesis, Techniche Universiteit Eindhoven, The Netherlands.

RODRIGUEZ-NAVARRO, C. & DOEHNE, E. 1999. Salt weathering: influence of evaporation rate, supersaturation and crystallization pattern. *Earth Surface Processes and Landforms*, **24**, 191–209.

RUIZ-AGUDO, E., MEES, F., JACOBS, P. & RODRIGUEZ-NAVARRO, C. 2007. The role of saline solution properties on porous limestone salt weathering by magnesium and sodium sulfates. *Environmental Geology*, **52**, 269–281.

SCHERER, G. 2004. Stress from crystallization of salt. *Cement and Concrete Research*, **34**, 1613–1624.

STEIGER, M. 2005*a*. Crystal growth in porous materials – I: the crystallization pressure of large crystals. *Journal of Crystal Growth*, **282**, 455–469.

STEIGER, M. 2005*b*. Crystal growth in porous materials – II: Influence of crystal size on the crystallization pressure. *Journal of Crystal Growth*, **282**, 470–481.

TSUI, N., FLATT, R. J. & SCHERER, G. W. 2003. Crystallization damage by sodium sulfate. *Journal of Cultural Heritage*, **4**, 109–115.

Is sodium sulphate invariably effective in destroying any type of rock?

SWE YU* & CHIAKI T. OGUCHI

Geosphere Research Institute, Saitama University, Saitama 338-8570, Japan

**Corresponding author (e-mail: sweyu76@gmail.com)*

Abstract: Sodium sulphate has been implicated as one of the most destructive weathering agents in many field observations and numerous laboratory studies. We hypothesize however, that sodium sulphate would not be invariably effective on any type of rock. To verify the supposition, a laboratory cyclic impregnation–drying experiment was undertaken. In addition to sodium sulphate, two other destructive hydratable salts, magnesium sulphate and sodium carbonate, were used to attack eight types of rock. In all three salt attacks, rock breakdown occurred only during immersion due to the exertion of higher crystallization pressure driven by the greater supersaturation reached after dissolution of the crystals precipitated during drying. Sodium sulphate was the most destructive salt in six out of the eight rocks tested, and even granite was substantially disintegrated. However, although probability is small, sodium sulphate indeed manifested its impotency against a relatively weak rock (Tago Sandstone). Contrary to its modest damaging power on other rocks, magnesium sulphate destroyed Tago Sandstone which could resist sodium sulphate attack. Sodium carbonate was the least destructive of the three hydratable salts. The general damage mechanism of hydratable salts, the process of damage of Tago Sandstone by magnesium sulphate and the possible reasons behind the impotency of sodium sulphate against Tago Sandstone are all investigated.

Salt weathering is known to be one of the most destructive mechanisms in the deterioration of natural rocks and porous building materials such as stone, concrete, mortar and brick (Evans 1970; Amoroso & Fassina 1983; Winkler 1994; Goudie & Viles 1997). It is therefore important to fully understand the various aspects of salt weathering in order to find efficient ways to preserve not only our priceless architectural and archaeological monuments but also our modern buildings and engineering structures. Between the 1950s and 1980s, numerous laboratory salt-efficacy and/or rock-durability tests were performed. In almost all of the various salt-efficacy tests, sodium sulphate emerged as the most destructive salt (Table 1). Moreover, sodium sulphate weathering has been prove to be more effective than most other types of simulated physical weathering (Luquer 1895; Goudie 1974). The unmatchable damage potential of sodium sulphate has therefore led to the widespread use of this salt in many standard durability tests, for example, the American standard test ASTM (C-88) and the European standard test EN (12370), as a measure of the weathering resistance of stones and porous building materials (Price 1978; Benavente *et al.* 2001). It has also been used in most laboratory evaluations of the relative resistance of various rocks (Luquer 1895; Schaffer 1955; Cooke 1979; Goudie 1999). Furthermore, in recent decades, sodium sulphate has most frequently been employed in in-depth investigations of various aspects of salt weathering (e.g. Rodriguez-Navarro & Doehne 1999; Flatt 2002; Benavente *et al.* 2004*a*).

Sodium sulphate occurs widely in deserts and arid environments (Goudie & Cooke 1984). The pre-eminent potency of sodium sulphate and its damage mechanisms are the most cited and most stressed issues in salt–weathering studies (e.g. Schaffer 1932; Birot 1954; Kwaad 1970; Evans 1970; Goudie 1977; Cooke 1979; Sperling & Cooke 1985). Goudie & Viles (1997) compiled the reasons behind this efficacy; some of them have been verified, whereas others have been ruled out or are still under debate. The solubility of sodium sulphate is more temperature sensitive than those of other common salts. This enables a sodium sulphate solution to supersaturate and induce crystallization damage both on nocturnal cooling (with marked solubility reduction below 32.3 °C) and under diurnal heating (with a decrease in solubility plus evaporation above 32.3 °C). Moreover, as mirabilite crystals (monoclinic) have a very long prismatic (Wells 1923) to acicular (010) habit and thenardite crystals (orthorhombic) have a dipyramidal (111) or tabular (010) habit, sodium sulphate tends to increase its rock-damaging power by concentrating crystal growth pressure along one axis.

In addition, volume expansion pressure caused by the rapid hydration processes occurring above and below its transition temperatures, that is,

From: PŘIKRYL, R. & TÖRÖK, Á. (eds) *Natural Stone Resources for Historical Monuments.* Geological Society, London, Special Publications, **333**, 43–58.
DOI: 10.1144/SP333.5 0305-8719/10/$15.00

Table 1. *Results of past salt-efficacy ranking tests*

Researcher	Specimen	Salt efficacy ranking (in decreasing order)
Birot (1954)	Granite	Sodium hyposulphite, Na_2CO_3, Na_2SO_4, NaCl
Kwaad (1970)	Granite	Na_2SO_4, Na_2CO_3, $MgSO_4$, NaCl, $CaSO_4$
Goudie *et al.* (1970)	Arden sandstone	Na_2SO_4, $MgSO_4$, $CaCl_2$, Na_2CO_3, NaCl
Goudie (1974)	Chalk and silica-cemented sandstone	Na_2SO_4, $MgSO_4$, $CaCl_2$, Na_2CO_3 (Sandstone)
		Na_2SO_4, Na_2CO_3, $NaNO_3$, $CaCl_2$ (Chalk)
Cooke (1979)	Two sandstones and slate	$Na_2SO_4 \ggg$ NaCl, $CaSO_4$, $NaNO_3$
Smith & McGreevy (1983)	Sandstone	Na_2SO_4, $MgSO_4$, NaCl
Goudie (1986)	York stone	Na_2CO_3, $MgSO_4$, Na_2SO_4, NaCl, $NaNO_3$, $CaSO_4$
Smith & McGreevy (1988)	Darney stone	$MgSO_4$, 10% Na_2SO_4, 10% $MgSO_4$, saturated NaCl, 10% NaCl
Goudie (1993)	Mortar	Na_2SO_4, Na_2CO_3, $NaNO_3$, $MgSO_4$, NaCl

Note: The literature cited employed different experimental setups, salt supply techniques and/or environmental conditions, indicating consistent destructiveness of sodium sulphate under various laboratory conditions.

32.383 °C (thenardite–mirabilite) and 23.465 °C (thenardite–heptahydrate), has been attributed to observed sodium sulphate damage (Mortensen 1933; Winkler & Wilhelm 1970; Goudie & Viles 1997). Some workers have claimed that hydration pressure alone is ample to disrupt various rocks (Evans 1970; Goudie 1977; Sperling & Cooke 1980). However, recent investigations, including Environmental Scanning Electron Microscopy (ESEM) studies, have revealed that the transition of anhydrite ($CaSO_4$) to gypsum ($CaSO_4 \cdot 2H_2O$) and of thenardite to mirabilite occurs through a process of complete dissolution with subsequent recrystallization, rather than by solid-state hydration (Charola & Weber 1992; Doehne 1994; Rodriguez-Navarro *et al.* 2000; Scherer 2000; Flatt 2002; Benavente *et al.* 2004*a*; Thaulow & Sahu 2004). Moreover, the crystal volume change resulting from the hydration process is insignificant because the total volume of the system shrinks: for example 14% in sodium carbonate to soda ($Na_2CO_3 \cdot 10H_2O$) transition; 5.6% in thenardite to mirabilite conversion; and 10.9% in kieserite ($MgSO_4 \cdot H_2O$) to epsomite ($MgSO_4 \cdot 7H_2O$) transformation (Correns & Steinborn 1939; Chatterji & Jensen 1989; Haynes 2005). Therefore, the hydration pressure hypothesis has been ruled out as a credible cause of salt damage (Thaulow & Sahu 2004). It seems that the damage attributed in the literature to hydration pressure of hydratable salts actually corresponds to the crystal growth pressure of the hydrated phases (Rodriguez-Navarro *et al.* 2000).

Regardless of the mechanism involved, many studies have proved the destructiveness of sodium sulphate. Price (1978), however, revealed that the results of various sodium sulphate crystallization tests using the same standards and materials were inconsistent. Although various studies have been undertaken in an effort to find out why sodium sulphate is so destructive, no studies have tested the invariability of this salt's potency against rock type. Furthermore, some other hydratable salts such as sodium carbonate and magnesium sulphate have outranked sodium sulphate under some environmental conditions (Goudie 1993) and also some laboratory salt-efficacy tests (e.g. Goudie 1986; Smith & McGreevy 1988). Accordingly, the accuracy of prediction of the resistance of a rock against salt weathering is in question, as is the reliability of durability tests, if the rock(s) is treated only with sodium sulphate. The primary purpose of the present study, therefore, is to investigate whether sodium sulphate is invariably effective on all types of rocks. The secondary objective is to compare the destructive power of sodium sulphate with those of two other aggressive hydratable salts on rocks of various types.

Experimental design

Rock types studied

Eight types of rock, namely Oya Tuff, Ashino Tuff, Koga Rhyolite, Indian Sandstone, Tago Sandstone, Italian Travertine, Kuzuu Dolomite and Makabe Granite were used. These rocks have been widely used as building and/or decorative stones in Japan. With the exception of Italian Travertine and Indian Sandstone, the rocks are quarried domestically. Their mineralogical and lithological characteristics are described below.

Oya Tuff (OT) was formed from submerged volcanic eruption during the Miocene. It is a green, fine-textured tuff that contains quartz, albite, clinoptirolite and clay-rich patches composed mainly of Fe-rich montmorillonite and chlorite; the clay patches change from dark green to brown immediately after the rock is mined.

Ashino Tuff (AT) was formed from Ashino pyroclastic flow during the Pleistocene. It is a grey, pyroclastic dacite tuff with a very small amount of amphibole and monoclinic pyroxene.

Koga Rhyolite (KR) is a vesicular, highly porous rhyolite which was formed by an eruption during the Holocene on a small island about 180 km south of Tokyo in the Pacific Ocean. It is composed of phenocrysts (quartz, plagioclase and biotite) making 15 vol. % and about 85 vol. % glassy groundmass with a few traces of mica and kaolin clay minerals (Oguchi *et al.* 1999).

Indian Sandstone (IS) is a member of upper Cretaceous era rock. It is a very smooth, homogeneous, fine-grained quartz sandstone with ferruginous cement, which makes it red.

Tago Sandstone (TS) is a yellowish-brown, coarse-grained, early Miocene sandstone. It is composed mainly of monocrystalline quartz, orthoclase and plagioclase. Kaolinite cement partly fills the primary interparticle porosity, thereby reducing the size of the pores.

Italian Travertine (IT) is Jurassic in age and yellowish white, with very low porosity. It is composed mainly of calcite (94 vol. %) recrystallized from aragonite.

Kuzuu Dolomite (KD) is Permian in age, dark grey and usually microcrystalline, massive and homogeneous. It is a hard, brittle and Mg-rich rock of 60% $CaCO_3$ and nearly 40% $MgCO_3$ (Fukui *et al.* 2005).

Makabe Granite (MG) is a Cretaceous-period fine-grained granite consisting of quartz, potassium feldspar, biotite and plagioclase.

Petrophysical properties

The rocks studied exhibit a range of different petrophysical properties (Table 2). The porosity values (investigated with an AutoPore IV 9500 Micromeritics mercury porosimeter) lie in a wide range, from less than 1 vol. % to nearly 40 vol. %. Microporosity, which is associated with the generation of high crystallization pressure (Scherer 2004; Steiger 2005*a*), is defined here as the volume percentage of pores less than 1 μm in radius (Benavente *et al.* 2004*b*); the values vary from 0.03 to 28.6 vol. %.

Water absorption capacity (WAC) is a measure of the amount of water absorbed under atmospheric pressure within 24 h. It is expressed as the percentage of the initial dry weight of the specimen and ranges from as low as 0.1 wt % to as high as 25 wt %. For each rock, the saturation coefficient (C_s), which is defined as the ratio of the water-filled porosity under atmospheric conditions (i.e. WAC) to the total accessible porosity, was also calculated. These petrophysical properties are presented solely to provide a basic understanding of the eight rock types. More comprehensive properties and their influence on rock durability and salt weathering are investigated and thoroughly discussed elsewhere.

Salt types

Apart from the particularly destructive sodium sulphate, two other commonly found hydratable salts, magnesium sulphate and sodium carbonate, were used as controls to check for any anomalies in the destructiveness of sodium sulphate. If a rock resisted sodium sulphate but was prone to the attack of control salts, and the discrepancy was too large, the impotency of sodium sulphate could be noted. If sodium sulphate alone was employed, this would not be realized; in this case, the rock properties would likely be attributed to the resistance of the rock. To set a standard for comparison among different salt treatments, saturated solutions were prepared at 20 °C with super-distilled water. Initial salt uptake was also evaluated by individually soaking 5 cm cube specimens of each rock type in the appropriate saturated salt solutions at 20 °C for 24 h. The specimens were then dried in an oven at 105 °C until constant weight was reached. Salt

Table 2. *Petrophysical properties of the eight types of rock*

Rock	Porosity (vol. %)	Microporosity (vol. %)	Density (g cm^{-3})	Surface hardness[1]	Water absorption capacity (wt. %)	Saturation coefficient, C_s
OT	39.59	28.60	2.05	238	24.41	0.91
AT	14.84	13.57	2.58	625	5.78	0.90
KR	39.27	2.95	2.37	128	14.97	0.36
IS	14.38	2.28	2.61	569	3.64	0.61
TS	27.98	20.68	2.58	273	10.25	0.72
IT	1.24	0.17	2.68	630	0.20	0.46
KD	0.69	0.03	2.69	798	0.10	0.39
MG	1.39	0.74	2.65	826	0.36	0.70

[1]Rebound readings (L-values) of Equotip hardness tester.

uptake is expressed as the percentage of the initial dry weight of the specimen (Table 3).

Salt weathering experiment

Seven unpolished 5 cm cubes of each rock type were oven dried initially at 105 °C for 24 h and their initial dry weights were recorded. Three specimens (plus three more specimens for replication of the test) of each rock type were immersed at 20 °C for 2 h in each of three transparent containers, each containing one of three saturated salt solutions. The remaining specimen of each rock type was treated with super-distilled water as an experimental control. All cubes that remained as coherent stone (weighing more than 1 g) were lifted out from the containers, placed in separate trays and dried in the oven at 50 °C for 20 h. It is believed that this drying temperature is a fair representation of the blistering heat of most deserts (Goudie 1974). At the end of each drying phase, the weights of the coherent specimens were recorded. The still-hot specimens were then allowed to cool at 20 °C for 2 h. This 24 h immersion–drying–cooling cycle was performed a total of 70 times to gain a clear salt efficacy ranking, even for a highly durable rock such as granite.

At the end of the experiment, any remaining specimens were thoroughly leached with distilled water. They were oven dried at 105 °C for a final time and weighed (expressed as percentage of initial dry weight). Some crucial results were further investigated with SEM-EDS. The experimental design adopted in this study was not intended to reproduce or be comparable to the results of previous tests, but to ensure that all eight types of rock underwent an analysable degree of weathering in a limited time interval.

Results

Relative efficacy of three hydratable salts v. relative durability of eight rocks

Figure 1 shows the weathering progress (weight loss and/or gain) of the specimens during the experiment. There were generally four distinct weathering trends: a very rapid disintegration (OT: completely destroyed); a steady, continuous disintegration (AT, IS and MG); an intermittent, slow disintegration (IT and KD); and an impotent weathering that promoted only encrustation/efflorescing (KR). Throughout the 70 cycles of the test, specimens that were disintegrating well (excluding those being intermittently destroyed) showed no retardation or culmination in debris liberation, regardless of the salt type. If the test had been continued, these specimens (along with the sporadically disintegrating specimens) would have continued liberating debris.

All damage, regardless of both rock type and salt type, occurred only during the immersion phase; the use of transparent containers allowed us to witness the falling of particles from their parent stones. As far as salt efficacy is concerned, sodium sulphate manifested itself as the most destructive salt in six out of the eight types of rock (Table 4). OT suffered total destruction at the 10th cycle. AT and IS were also severely destroyed and even the hardest rocks (IT, MG and KD) were noticeably damaged by its attack.

Magnesium sulphate also disintegrated AT and IS considerably, although not to the same extent as sodium sulphate damage. It has a tendency to cause encrustation (Keller *et al.* 1986) especially on the surfaces of highly porous rocks such as KR and TS. Although OT had the highest porosity, the weathering was so fast that specimens were totally destroyed before becoming encrusted. The crusts formed on KR not only sealed the surface but continued to grow enormously throughout the test, without damaging the rock structure. On the other hand, TS suffered significant disruption by repeated coating and detaching of crusts, in addition to crystallization stress.

Of the three hydratable salts, sodium carbonate caused a less marked and rapid change in weight for all the specimens except OT. However, there was a very close competition between magnesium sulphate and sodium carbonate (Table 4). The extraordinary potency of magnesium sulphate in

Table 3. *Initial salt uptake (SU) values of the eight types of rock*

Salt	Concentration* (wt. %)	SU** (wt. %)							
		OT	AT	KR	IS	TS	IT	KD	MG
Na_2CO_3	18.1	5.69	1.08	2.71	0.61	2.67	0.05	0.01	0.05
Na_2SO_4	16.1	5.93	1.77	4.64	0.56	2.36	0.03	0.01	0.06
$MgSO_4$	25.2	9.92	2.55	5.77	1.47	6.58	0.08	0.02	0.15

*Values at 20 °C.
**Values are average for two specimens of each type of rock.

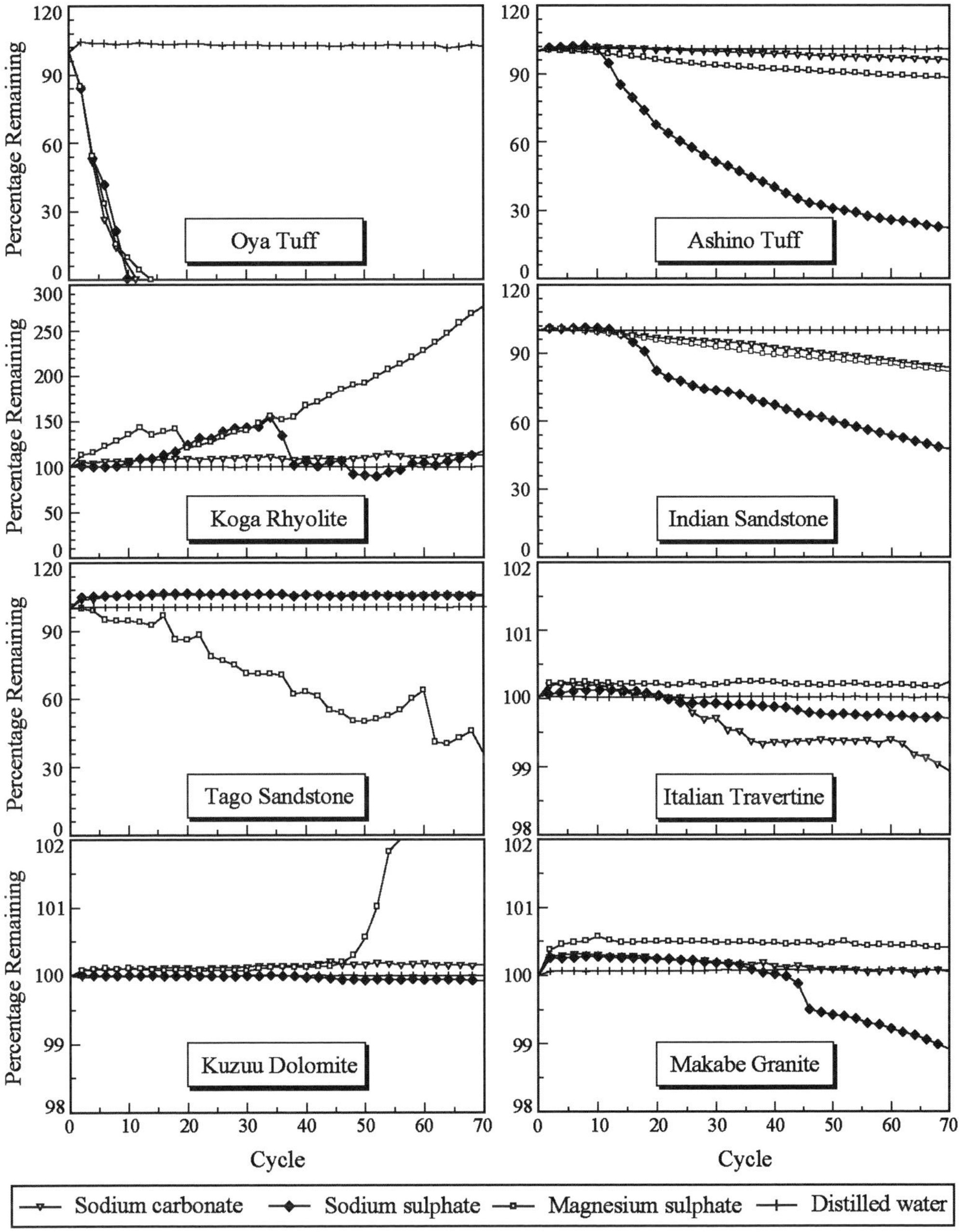

Fig. 1. Weathering progression curves of the eight types of rock against the three hydratable salts. Note: the vertical scales are different.

attacking TS, however, left sodium carbonate as the least effective salt. Whether its efficacy could be substantially enhanced under different environmental conditions (cf. Goudie 1986, 1993) is questionable.

The weathering degree of each type of rock was not ranked in the same order for the three salt treatments. The relative durability rankings of the eight rock types differed with the different salt treatments (Table 5). Although rock physical properties can

Table 4. *Relative effectiveness rankings of the three hydratable salts*

Rock Type	Sodium carbonate		Sodium sulphate		Magnesium sulphate	
	% remaining	Ranking	% remaining	Ranking	% remaining	Ranking
OT*	0.00	2	0.00	1	0.00	3
AT	95.24	3	22.50	1	87.91	2
KR	99.96	3	80.57	1	99.86	2
IS	82.58	3	47.02	1	80.36	2
TS	98.07	2	99.11	3	35.96	1
IT	98.94	1	99.79	2	100.02	3
KD	100.01	3	99.92	1	99.99	2
MG	99.71	2	98.63	1	99.79	3

*Totally destroyed at 11th cycle by sodium carbonate, at 10th cycle by sodium sulphate, and at 14th cycle by magnesium sulphate.
Note: Values are averages of the results for two specimens.

usually be attributed to the different resistances of different rocks against one type of salt, the variation in rock durability rankings with salt type indicates the influence of a salt–solution–rock interaction in terms of such factors as contact angle or disjoining forces (Scherer 2004). The same salt may therefore act differently in different rocks, even if their petrophysical properties are similar. Likewise, different salts with the same efficacy could show different degrees of damage against the same type of rock.

Is sodium sulphate invariably effective?

In effect, magnesium sulphate and sodium carbonate were employed as the controls to indirectly check whether the degree of damage caused by sodium sulphate was consistent with the durability of each rock. The negligible sodium sulphate damage associated with the very durable KD (dolomite) was normal, as were the large degrees of disintegration in less resistant rocks such as OT, AT and IS. Comparison of the degrees of damage of OT, AT, KR, IS, MG and KD treated with sodium sulphate and the control salts showed nothing abnormal; the salts were reasonably competitive, although sodium sulphate was still the top-ranked salt in these cases.

Sodium sulphate, however, was outranked in two cases (IT and TS). Although sodium carbonate dominated in damage to IT because of the plucking out of large fragments by crystallization pressure, sodium sulphate also demonstrated its potency by intermittently causing substantial fragmentations throughout the test. Furthermore, the difference in the degrees of IT weathering between these two salts was not notable. This can also be regarded as normal.

On the other hand, the degree of damage caused to TS by sodium sulphate was improbably less than that caused by magnesium sulphate; the latter showed incredible power in disrupting TS, although it was second to sodium sulphate in all other treatments in this study (Table 4). The TS specimen treated with sodium sulphate retained 99.11% of its initial dry weight, whereas the specimen treated with magnesium sulphate retained only 35.96% (see the size difference in Fig. 2a). The domination of sodium sulphate in six out of the eight types of rock indicated that the present experimental set-up favoured the destructiveness of this salt. Thus, even under the same experimental conditions, its pronounced ineffectiveness on TS, which possesses salt-damage-prone properties such as high porosity (27.98 vol. %), high microporosity (20.68 vol. %), a fairly high saturation coefficient (0.72) and low surface hardness (L-value = 273), was surprising. Moreover, the sodium sulphate damage to TS was even smaller than that to the apparently more durable MG (granite). The breakdown of MG was 4.62 g due to progressive salt damage until the end of the test, whereas the breakdown of TS was only 2.18 g due to the detachment of grains (loosened

Table 5. *Relative durability rankings of the eight types of rock*

Rock	Rock durability ranking (for each different treatment)		
	Sodium carbonate	Sodium sulphate	Magnesium sulphate
OT	8	8	8
AT	6	7	5
KR	2	5	4
IS	7	6	6
TS	5	3	7
IT	4	2	1
KD	1	1	2
MG	3	4	3

Fig. 2. Post-experiment specimens of (**a**) Tago Sandstone and (**b**) Indian Sandstone.

by specimen cutting) from the edges during only the early cycles.

The exceptionally large discrepancy between the efficacies of two well-known destructive salts, that should have shown relatively close efficacy, strikingly illustrates the unnatural behaviour of sodium sulphate; if such a large divergence was observed between the efficacies of sodium chloride and magnesium sulphate, it could be considered as an ordinary case. The result suggests that there are rocks—although possibly only a few—that are extraordinarily resistant to sodium sulphate even though they have weathering-prone petrophysical properties. The reasons why TS totally resisted the most effective salt on the one hand and was massively destroyed by a relatively less effective salt on the other hand are discussed in detail below.

Discussion

Salt uptake and salt efficacy

In terms of their rankings, the initial salt uptakes (SU) of the rocks perfectly mirrored their porosity (Table 3): that is, more porous rocks absorb more salts. A closer look at the amount of salt absorbed, however, reveals that microporosity plays a more important role in salt uptake. For instance, the salt uptake of OT was significantly greater than that of KR, whereas their porosities were almost identical. Moreover, although TS had markedly lower total porosity than KR, it absorbed almost the same amount of salts. The reason behind this is the different microporosities that these rocks possess. Despite having approximately the same total porosity, AT, which has more micropores (13.57 vol. %), absorbed more salts than IS (2.28 vol. %).

It was also observed that solubility and salt uptake are not perfectly correlated (McGreevy 1982; Goudie 1993). Magnesium sulphate was absorbed the most by all eight types of rock (Table 3), reflecting its highest solubility among the three salts. The uptake of sodium carbonate, however, was less than might be expected from its solubility. Sodium carbonate uptake was even lower than, or nearly the same as, that of sodium sulphate which has a lower solubility. In effect, the physical properties of the salt solutions, such as surface tension, vapour pressure and viscosity, as well as the rock properties such as microporosity, play a coupled role in the salt uptake process (cf. Rodriguez-Navarro & Doehne 1999; Benavente *et al.* 2001).

The amount of individual salt absorbed by each rock was fairly well correlated with the degree of rock breakdown (Spearman's rank correlation coefficient: 0.76 to 0.79). The differences in uptakes of these three salts, however, did not seem to have a simple linear effect on the different weathering degree of the individual rocks. Although magnesium sulphate was absorbed the most by all types of rock, it did not emerge as the most destructive salt (except in the case of TS). Still, magnesium sulphate was more aggressive than the two other salts in the early cycles. It seems that there is a threshold amount of salt that must be accumulated within the pores in order for rock disruption to commence (Tsui *et al.* 2003). For crystallization stresses to be generated inside the pore network, the crystals must be confined but still able to grow (Correns 1949; Scherer 1999, 2004; Steiger 2005*a*, *b*). A crystal, however, most likely tends to grow into the spaces without constraint. To overcome this, salt crystals themselves may take part as a confining barrier by creating smaller pores, thereby raising the potential crystallization pressure. Damage can therefore occur once enough salt has accumulated to prevent unrestrained crystal growth. This mechanism works particularly with hydrated salts, as their massive and rapid growth permits high stresses under non-equilibrium conditions that are not correctly captured by the equilibrium thermodynamic equations.

After the first immersion–drying cycle, magnesium sulphate precipitated out to a greater extent than sodium sulphate and sodium carbonate because of its higher concentration. It could therefore accumulate more rapidly to reach the threshold-salt amount and its destruction speed was faster in the early cycles. With an increasing number of cycles, its aggressiveness was stabilized. On the other hand, sodium sulphate, which had a slow weathering rate initially, accelerated its destructive work with an increasing number of cycles as the threshold-salt amount accumulated. It eventually outperformed magnesium sulphate and was continuing its destructive pace until the conclusion of the test, thus proving the pre-eminence of its rock-destroying power.

Damage mechanism of hydratable salts in the cyclic impregnation–drying test

Although Steiger *et al.* (2008) observed that true solid-state kieserite–hexahydrite hydration could occur below the deliquescence humidity of kieserite, the process was found to be kinetically hindered. Crystal growth pressure is therefore considered to be the sole mechanism of damage by the three hydratable salts during both the drying and the immersion regimes. Apart from the very complex salt-hydrate systems shown in Table 6, new metastable phases of sodium sulphate (e.g. Genkinger & Putnis 2007; Hamilton & Hall 2008) and magnesium sulphate (e.g. Steiger *et al.* 2008) are still being reported. Instead of verifying the exact phase, the thermodynamically most possible phases of each salt (relevant to the temperatures used) are predicted to precipitate: thermonatrite ($Na_2CO_3 \cdot H_2O$), thenardite (Na_2SO_4) and kieserite ($MgSO_4 \cdot H_2O$) or more likely starkeyite ($MgSO_4 \cdot 4H_2O$) during the drying regime; and natron ($Na_2CO_3 \cdot 10H_2O$), mirabilite ($Na_2SO_4 \cdot 10H_2O$) and hexahydrite ($MgSO_4 \cdot 6H_2O$) or epsomite ($MgSO_4 \cdot 7H_2O$) during the wetting regime. Kinetics and crystallization characteristics, however, may be more important. Thermodynamically less favourable metastable phases are therefore also expected to crystallize and coexist with stable phases (Rodriguez-Navarro *et al.* 2000; Genkinger & Putnis 2007; Steiger *et al.* 2008).

The fact that all damage occurred only during the immersion regime is in agreement with the results of previous cyclic impregnation–drying tests (Chatterji & Jensen 1989; Tsui *et al.* 2003; Benavente *et al.* 2001). A possible reason is the difference between the degree of supersaturation reached during the drying and the immersion phases: supersaturation determines the magnitude of crystallization pressure (Correns 1949; Steiger 2005*a*, *b*). When the specimens are submerged into the solutions at 20 °C the crystals, directly precipitated and/or converted from the dehydration of higher hydrate(s) during the drying phase (e.g. thenardite for sodium sulphate), dissolve. This solution, saturated with thenardite, is supersaturated with respect to mirabilite because of the difference in solubility of these two minerals at 20 °C, providing an effective driving force for crystallization (Scherer 1999; Flatt 2002). Furthermore, the higher hydrates formed during the immersion regime occupy larger pore spaces as they have a greater molar volume. This increases the contact area between the growing crystal and the pore wall, thereby creating a narrower confining barrier and a greater stress field. We do not rule out the likely crystallization stresses induced by the

Table 6. *Salt-hydrate systems of the three hydratable salts (Goudie & Viles 1997)*

Salt-hydrate system					
Sodium carbonate		Sodium sulphate		Magnesium sulphate	
Thermonatrite	$Na_2CO_3 \cdot H_2O$	Thenardite	Na_2SO_4	Kieserite	$MgSO_4 \cdot H_2O$
Heptahydrite	$Na_2CO_3 \cdot 7H_2O$	Heptahydrate	$Na_2SO_4 \cdot 7H_2O$	Sanderite	$MgSO_4 \cdot 2H_2O$
Natron/natrite	$Na_2CO_3 \cdot 10H_2O$	Mirabilite	$Na_2SO_4 \cdot 10H_2O$	Starkeyite	$MgSO_4 \cdot 4H_2O$
				Pentahydrite	$MgSO_4 \cdot 5H_2O$
				Hexahydrite	$MgSO_4 \cdot 6H_2O$
				Epsomite	$MgSO_4 \cdot 7H_2O$
				Dodekahydrate	$MgSO_4 \cdot 12H_2O$

growth of anhydrous or lower hydrate crystals during drying, however. The mechanism may conceivably be involved in the similar processes of damage by magnesium sulphate (kieserite or starkeyite dissolution – hexahydrite or epsomite reprecipitation) and sodium carbonate (thermonatrite dissolution – natron recrystallization). It seems, however, that the crystallization speed (which defines the severity of weathering) of mirabilite during immersion must have been faster than epsomite and natron in most cases (Correns & Steinborn 1939).

Mechanism of magnesium sulphate damage of Tago Sandstone

The extraordinary efficacy of magnesium sulphate in attacking TS can partly be explained by its destructive encrustation effect in addition to its crystallization pressure. During drying, the absorbed magnesium sulphate solution moved towards the surface. With evaporation, the solution became supersaturated and crystallization began in the near-surface pores. The crystals firmly plugged the surface pores while forming a thin, indurated crust. Two hours of immersion could not dissolve the crust-forming crystals. With an increasing number of cycles, the crusts became rapidly thickened (to 1–1.5 mm) and they eventually bulged outwards by growing in irregular shapes. Meanwhile, second-generation crystals grew underneath the surface layer. As there were still enough spaces between the crusts and the surfaces of the engulfed specimen, surface sealing was negligible. The solution could therefore directly move into the specimen during immersion, and evaporation was undisturbed during drying. Crystal growth and granular disintegration proceeded underneath the surface layer, while the specimen weight continued to increase as long as the crusts grew and still covered the specimen.

Eventually, the crusts lost their support and started peeling off because the grains on which they rested had been expelled by the subflorescent crystal growth. The crusts were sometimes detached as a whole without losing their integrity (Fig. 3). Since the crusts firmly plugged the surface pores, the sand grains, which were coalesced with the crusts, were plucked away when the crusts were detached. After detachment, the weakened surfaces (covered with loosened sand grains) were ready for the formation of the next generation of crusts. The subflorescent crystallization damage was exacerbated by the cyclic engulfing–disjoining nature of the magnesium sulphate crusts. The process appeared to depend on the rock fabric characteristics, porosity and pore size of the rock; of the eight rocks treated with magnesium sulphate, the process was only associated with TS.

Fig. 3. Detached magnesium sulphate crust (with coalesced sand grains inside) and a specimen of Tago Sandstone destroyed by the subfloresent magnesium sulphate and its crust.

SEM micrographs showed that magnesium sulphate formed inside the subsurface pores of TS was needle-shaped radiating crystals (Fig. 4a, b). The crystals filled both small and large pores. This highly anisotropic (non-equilibrium) crystal morphology indicated that a high degree of supersaturation could be reached before nucleation; this in turn points out the likely generation of high crystallization pressures during drying. Dissolution of these acicular crystals during immersion could have created a much higher supersaturation with respect to hexahydrite or epsomite (even higher than that of mirabilite nucleation), thereby exerting greater crystallization pressure (Dr Rosa Espinosa, personal communication 2008). These acicular crystals could have generated high crystallization stresses during both the drying and the immersion regimes leading to massive devastation of TS.

Scherer (1999) stated that if acicular (needle-shaped) crystals are not extremely long, they will be able to exert a stress at each of their ends limited only by the driving force and interfacial energy. Moreover, since the crystallization pressure is unlikely to exceed the yield stress of a crystal, shear or buckling of the observed short needles is unlikely. As the area of contact with the wall is small for this type of crystal, the stress they exert on the wall will not be great if they touch only in a few points. The substantial quantity of acicular crystals in TS, however, seems most likely to have generated ample stresses to disrupt the rock structure.

Possible reasons for sodium sulphate inefficacy

Having elucidated the extraordinary efficacy of magnesium sulphate, we then attempted to

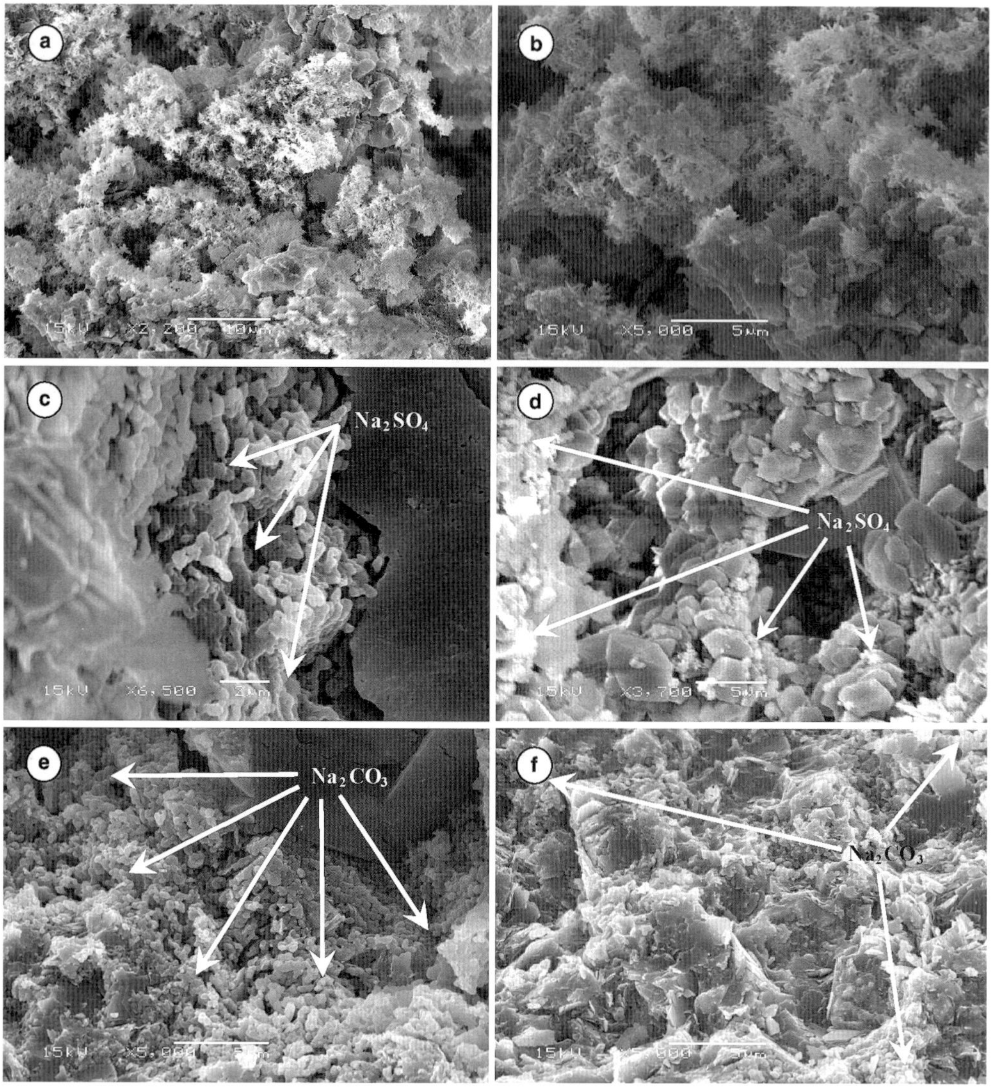

Fig. 4. SEM micrographs of salt-weathered Tago Sandstone specimens: (**a**, **b**) radiating needle-shaped crystals of magnesium sulphate inside the pore network of Tago Sandstone; (**c**, **e**) mass precipitation of prismatic sodium sulphate and sodium carbonate crystals within 3 mm of the specimen surface and (**d**, **f**) relatively few sodium sulphate and sodium carbonate crystals observed at a location beyond 3 mm from the specimen surface.

determine the reasons behind the impotency of sodium sulphate (and of sodium carbonate) in the weathering of TS (Fig. 1). As a comparison, specimens of TS but also of IS (which were significantly damaged by all three salt attacks; Fig. 2b) were investigated with SEM-EDS. In IS, the sodium sulphate crystals (Fig. 5a) were very similar to the thenardite crystals observed by Ruiz-Agudo *et al.* (2007). Dehydrated magnesium sulphate crystals (Fig. 5c) and prismatic crystals (Fig. 5d), indicative of the higher supersaturation reached before nucleation, were also observed. Dehydrated whiskers of sodium carbonate were found inside the pores of IS (Fig. 5b). Except for the protruding whiskers at the pore entrance (which would have consumed supersaturation), the columnar whiskers inside were found growing diagonally across the pore walls. The latter could ultimately disrupt the pore without necessarily filling the whole pore if the solution films were present at each end (Arnold & Zehnder 1990; Scherer 2004). Although the small quantities of crystals of all three salts

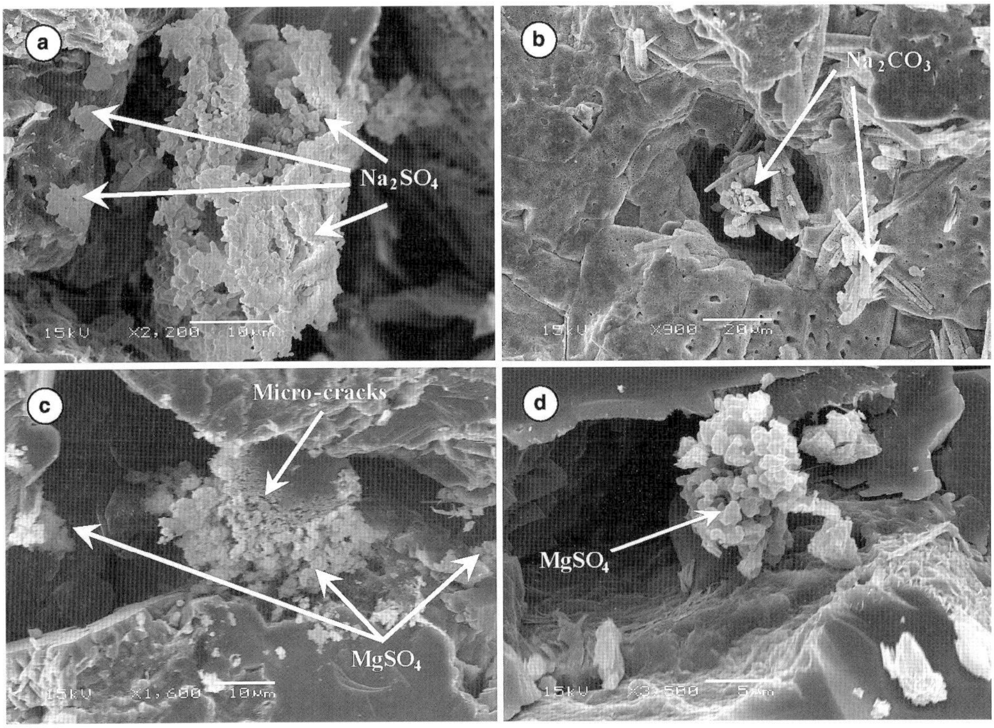

Fig. 5. SEM micrographs of salt-weathered Indian Sandstone specimens: (**a**) thenardite crystals; (**b**) dehydrated sodium carbonate whiskers; (**c**) dehydrated magnesium sulphate crystals and (**d**) prismatic magnesium sulphate crystals formed during drying.

found in IS did not totally occupy the pores, they tended to grow against constraints after they had filled the threshold-pore-spaces, thereby generating tensile stresses (Tsui *et al.* 2003).

Unlike for IS, the pore networks within 2–3 mm of the surface of TS were massively occupied by the crystals of all three salts (Fig. 4a, b, c, e). This is in agreement with the results of Goudie & Viles (1995) who observed an impressive amount of prismatic sodium carbonate crystals in pores of all sizes in their specimens. Beyond this depth, crystals were scarce (Fig. 4d, f).

An examination of the petrophysical properties relevant to salt weathering gives no hint as to why TS could resist the attack of a proven destructive salt. It has much higher porosity, microporosity, water absorption capacity, saturation coefficient and salt uptake than IS. Moreover, its hardness value is lower than that of IS. Nevertheless, IS was significantly destroyed by all three salts, whereas TS was susceptible only to magnesium sulphate. This discrepant result (sodium sulphate inefficacy on TS) is therefore difficult to interpret in terms of physical rock properties. The possible reasons behind it can only be surmised from already established theoretical information and empirical evidence, as (1) pore clogging, (2) supersaturation, (3) nucelation and/or (4) disjoining pressure.

Pore clogging. This is the most obvious reason and was supported by evidence from the weathering progress curves and SEM images. The curve of TS v. sodium sulphate (Fig. 1) showed a continuous increase in specimen weight up to 10 cycles. After that point, a constant weight was sustained for the remainder of the test, that is, during the next 60 cycles. Examination of the daily record of damage visualization revealed that breakdown occurred only during the weight-increase period, that is, within the first 10 cycles, indicating the prevalence of salt uptake over breakdown. No damage was induced after the 10th cycle and weight increase also stopped. SEM micrographs indicated mass precipitation of salts in pores within 3 mm of the surface (Fig. 4c), whereas salts were scarce beyond this (Fig. 4d). It is likely that a steady state between evaporation and solution flow was established at 2–3 mm from the surface. Initial crystallization was therefore concentrated at this drying front. The crystals appeared to propagate (from pore to pore) wherever they did not encounter constraint. They therefore created no crystallization pressure; this is confirmed by the negligible breakdown. After filling of the pores within 3 mm of the surface, pore clogging occurred and consequently the cyclic supply of solution seemed to reach a plateau.

In effect, pore clogging could have been overcome if precipitated crystals had disrupted the near-surface pores so that the blocking salts could fall down, along with the detached grains. This surface disintegration could have created new, non-salt-blocked surfaces that allowed easy penetration and evaporation of the solution, thereby enhancing effective crystallization. This process is usually typical of sodium sulphate damage, but it was not the case with TS. Espinosa & Scherer (2008*b*) observed that the more viscous magnesium sulphate caused more effective pore clogging than sodium sulphate. They suggested that the material properties also affect the intensity of pore clogging. The present result indicates that pore blocking depends on the mutual action of the material and the salt/solution. For example, sodium sulphate kept blocking the pores of TS throughout the test whereas magnesium sulphate overcame the pore clogging by the mechanism discussed above. Magnesium sulphate, however, could not overcome this blocking effect in KR because it was unable to induce damage in this rock. It is clear that the actual cause of pore clogging, and thus sodium sulphate inefficacy in damaging TS, was due to the non-existence, or insufficient exertion of crystallization pressure.

Generation of a high crystallization pressure during drying requires rapid evaporation. Vapour diffusion rather than capillary transport through the blocked region reduced the evaporation rate. Consequently, crystallization may have occurred at very low supersaturation, without exerting pressure. During the immersion phase, because of surface pore blockage, the solution penetration became difficult and only a limited amount of thenardite would have dissolved. Accordingly, less precipitated mirabilite would have led to the induction of a lower crystallization pressure (because of action on only a small area) which would not have been sufficient to disrupt TS. Crystallization pressure and its magnitude are influenced by supersaturation, nucleation and the magnitude of the disjoining forces (Correns 1926, 1949; Correns & Steinborn 1939; Scherer 1999, 2004; Benavente *et al.* 2004*a*).

Supersaturation. The novel ideas of Correns (1926, 1949) about the influence of supersaturation and interfacial energies on crystallization pressure seem to be useful to answer the questions 'Why are certain salts much more damaging than others?', 'Why are certain types of stone much more vulnerable than other types to salt damage?' (Doehne 2002) and 'Why can even the destructive salts sometimes be ineffective on certain types of rock?' Supersaturation controls nucleation and the magnitude of crystallization pressure, which is directly proportional to the degree of supersaturation. Sodium sulphate solution, whenever conditions are favourable, can more easily reach very high supersaturations than other salts (Rodriguez-Navarro *et al.* 2000; Steiger & Asmussen 2008). From the breakdown data (Table 4), we can predict that sodium sulphate could have reached such high supersaturations in AT and IS, although these rocks had less weathering-prone physical properties compared to TS. It was magnesium sulphate—not sodium sulphate—that reached very high supersaturation in TS. This indicates that the degree of supersaturation depends on both the salt type and the rock type (in a controlled environment); that is, it can differ with the interactions between different minerals and different salt solutions.

Another possible reason for the inefficacy of sodium sulphate is the presence of unloaded crystals/crystal faces that consume the supersaturation required for loaded crystals to grow against a load (Steiger 2005*b*). Since a loaded crystal is slightly more soluble than an unloaded crystal, its growth against pressure is determined by the competition between the available supersaturation and the quantity and location of unloaded crystals (Steiger 2005*b*). If substantial unloaded crystals are close enough to the loaded crystals, their consumption of the supersaturation is sufficient to stop the loaded crystals from growing against the constraint (Becker & Day 1916; Taber 1916). It is possible that the availability and growth rate of unloaded sodium sulphate crystals in TS overtook the kinetics involved in this experiment. Sufficient supersaturation would therefore not have been available for the loaded crystals to grow and exert pressure. The massive amount of crystals apparent in Figure 4c manifests the possibility of non-pressure crystallization.

Nucleation. Only a few mirabilite could have nucleated since only a limited amount of oven-dried thenardite dissolved during the destructive immersion regime. This is because a shell of mirabilite encapsulates thenardite, thereby preventing rapid dissolution and further hydration of thenardite if the latter is large enough (Haynes 2005; Linnow *et al.* 2006). Another possible reason is that the metastable heptahydrate might have precipitated in TS in preference to the more destructive mirabilite. The latter seems to be unfavourable to nucleate in

TS because of its larger crystal–solution interfacial energy (γ_{cl}), which is a barrier to the nucleation of crystals. The nucleation of mirabilite requires a much higher supersaturation; if fulfilled, it can exert a higher crystallization pressure (Espinosa & Scherer 2008*a*). On the other hand, a lower degree of supersaturation would reach heptahydrate nucleation at a given concentration (Steiger & Asmussen 2008). Consequently, the slower rate of crystallization may have hindered the generation of enough pressure to damage TS. In this context, it seems that magnesium sulphate was extraordinarily effective in destroying TS because of the nucleation of epsomite (the crystal–solution interfacial energy of which is even higher than that of mirabilite), needed to overcome the higher energetic barrier. A very high supersaturation must have been reached before epsomite nucleation, thereby exerting the greater crystallization pressure. It is therefore clear that the ease of nucleation also depends on both the salt and the rock type. The mineral/structural-dependent nature of nucleation and supersaturation has been observed elsewhere (Linnow *et al.* 2006).

Nucleation mode also controls salt efficacy. Homogeneous nucleation requires a high supersaturation. The difficulty is reduced by heterogeneously nucleating with favourable nucleation sites such as a variety of minerals or crystal phases, impurities and defects that are exposed to the pore walls. If the minerals on the pore wall themselves were the effective nucleating agent, then heterogeneous nucleation, which excludes the solution film and therefore crystal growth against the pore wall, would induce propagation of salts throughout the pore network (Scherer 1999). An ESEM study by Benavente *et al.* (2004*a*) suggested that sodium sulphate has a stronger tendency to preserve a liquid film at its surface and, therefore, a high disjoining pressure that produces serious damage in porous materials. The present results, however, indicate that the nucleation mode also depends on the interaction between the solution and the rock minerals; sodium sulphate did not seem to nucleate homogeneously inside the pore network of TS and we attribute this to the influence of interfacial energies, as advocated by Correns (1949).

Disjoining pressure. A crystal can grow against pressure only if a supersaturated solution film, which acts as a diffusion path for ion exchange, exists between its loaded face and the constraint. Although the existence of the solution film has been proved by many studies (see Henniker 1949), the origin of the disjoining forces that create a nanometer gap between the pore wall and salt crystals for solution penetration is still obscure. The most cited origin is that proposed by Correns (1926, 1949). A solution film can penetrate between the crystal and the pore wall if the interfacial tension between crystal and rock, γ_{cs}, is larger than the sum of those between salt crystal and solution and between solution and rock ($\gamma_{cl} + \gamma_{sl}$). It will disappear if the load is too large or if the interfacial energy difference is negligible, that is, $\gamma_{cs} < (\gamma_{cl} + \gamma_{sl})$ or $\gamma_{cs} \approx (\gamma_{cl} + \gamma_{sl})$. The interfaces between crystal and solution and between solution and pore wall will therefore be replaced by an interface between crystal and pore wall. Crystallization pressure will not increase with further increases in supersaturation (Scherer 1999; Flatt 2002; Steiger 2005*a*, *b*; Flatt *et al.* 2006). The disjoining force is thus proportional to $\gamma_{cs} - (\gamma_{cl} + \gamma_{sl})$ and its magnitude depends on the variability of energies between these three interfaces. As long as the effect of the interfacial forces does not come into play, supersaturation can proportionally set the upper bound of the crystallization pressure. The relative magnitudes of the interfacial energy, however, define the maximum crystallization pressure that a loaded crystal can exert while maintaining a solution film between itself and the pore wall. They can also prevent the crystallization pressure from developing.

In the case of TS, the sodium sulphate crystal–pore wall interfacial tensions (γ_{cs}) must not have been larger than the sum of the other two terms (i.e. $\gamma_{cl} + \gamma_{sl}$). The crystallization pressure of sodium sulphate therefore might have been non-existent or very low even if the supersaturation is high. Additionally, the surface energies are different for different crystalline planes as well as different confining minerals (Correns & Steinborn 1939). This behaviour creates a different degree of repulsion between the crystal face and the confining surface. In some cases, as in TS v. sodium sulphate, there would be no repulsion and therefore no crystal growth pressure against constraint. The different interfacial tensions of the surfaces may arise from differences in the molecular structures of the respective surfaces.

Apart from differences in the interfacial energies, the repulsive forces may also originate from the electrostatic forces of diffuse-layer or direct-surface charges and from the hydration/solvation forces created by ordered layers of the adsorbed water molecules against the solid surfaces (Scherer 2004). For example, an electrostatic attraction between NaCl and quartz leads to a direct contact between their surfaces (G. W. Scherer, pers. comm. 2008). In a different approach, Scherer (1999) suggested that a solution film can exist between the crystal and the wall only when the contact angle (θ) is 180°. For any contact angle less than 180° there will be a direct contact of the crystal with the wall. If $\theta < 90°$, that is, the energy of the system decreases when salt touches the wall, the crystal will behave like a wetting liquid and will spontaneously penetrate pores of all sizes. These newer approaches may also help to explain our discrepant results. For instance, there might have been an electrostatic attraction and/or a contact angle of less than 180° between the sodium sulphate crystals and minerals in TS.

All the above-discussed factors may act independently or synergistically in the observed impotency of sodium sulphate in destroying TS. Empirical verification of the influence of these factors needs further in-depth investigations with more sophisticated techniques.

Conclusions

The striking conclusions that can be drawn are as follows.

(1) Rock breakdown occurred only during immersion because of the exertion of the higher crystallization pressure driven by the greater supersaturation reached (with respect to the higher hydrates) after dissolution of the anhydrous or lower hydrate crystals precipitated during drying.
(2) Of three destructive hydratable salts, sodium sulphate was the most destructive in rock disintegration. However, it may not be invariably effective in destroying all types of rock. There might be some types of rock that can resist sodium sulphate attack, even under efficacy-favouring conditions; Tago Sandstone exhibited this possibility.
(3) Rock decay seems to depend on both the rock properties and the inherent effectiveness of salts. The efficacy of salts and the durability of rocks may be influenced by subtle rock mineral–solution–salt crystal interactions, which differ individually. This needs further investigation.
(4) The present results predict that in nature, some salt mixtures for example, magnesium-sulphate-rich groundwater, may destroy stones whose durability is confirmed by the sodium sulphate test. It is therefore recommended that the selection/testing of stones, especially for monument restoration, should not be performed only with sodium sulphate.

We express our heartfelt gratitude to Professor George W. Scherer, Professor Michael Steiger, Professor Andrew S. Goudie and Dr Rosa Espinosa for their valuable discussions and helpful critical comments on the manuscript. The authors gratefully acknowledge Dr Tamao Hatta (JIRCAS) for kindly allowing us to use XRD and SEM-EDS and Dr Yasuhiko Takaya for providing rock samples. This study was supported by the Science Research Fund of the JSPS (No.18680054).

References

AMOROSO, G. G. & FASSINA, V. 1983. *Stone Decay and Conservation: Atmospheric Pollution, Cleaning, Consolidation and Protection.* Material Science Monographs, 11, Elsevier, Amsterdam.

ARNOLD, A. & ZEHNDER, K. 1990. Salt weathering on monuments. *In*: ZEZZA, F. (ed.) *The Conservation of Monuments in the Mediterranean Basin.* Brescia: Grafo, 31–58.

BECKER, G. F. & DAY, A. L. 1916. Note on the linear force of growing crystals. *Journal of Geology*, **4**, 313–333.

BENAVENTE, D., GARCÍA DEL CURA, M. A., BERNABÉU, A. & ORDÓÑEZ, S. 2001. Quantification of salt weathering in porous stones using an experimental continuous partial immersion method. *Engineering Geology*, **59**, 313–325.

BENAVENTE, D., GARCÍA DEL CURA, M. A., GARCÍA-GUINEA, J., SÁNCHEZ-MORAL, S. & ORDÓÑEZ, S. 2004*a*. Role of pore structure in salt crystallization in unsaturated porous stone. *Journal of Crystal Growth*, **260**, 532–544.

BENAVENTE, D., GARCÍA DEL CURA, M. A., FORT, R. & ORDÓÑEZ, S. 2004*b*. Durability estimation of porous building stones from pore structure and strength. *Engineering Geology*, **74**, 113–127.

BIROT, P. 1954. Dèsagrègation des roches cristallines sous l'action des sels. *Comptes Rendus de l'Academie des Sciences Paris*, **28**, 1145–1146.

CHAROLA, A. E. & WEBER, J. 1992. The hydration-dehydration mechanism of sodium sulfate. *In*: RODRIGUES, J. D., HENRIQUES, F. & JEREMISAS, F. T. (eds) *7th International Congress on the Deterioration and Conservation of Stone, Lisbon. Laboratorio Nacional de Engenharia Civil, Lisbon*, 581–590.

CHATTERJI, S. & JENSEN, A. D. 1989. Efflorescence and breakdown of building materials. *Nordic Concrete Research*, **8**, 56–61.

COOKE, R. U. 1979. Laboratory simulation of salt weathering process in arid environments. *Earth Surface Processes*, **4**, 347–359.

CORRENS, C. W. 1926. Über die erklärung der sogennanten kristallisationskraft. *Berichte der Preussischen Akademie der Wissenschaft*, **11**, 81–88.

CORRENS, C. W. 1949. Growth and dissolution of crystals under linear pressure. *Discussions of the Faraday Society*, **5**, 267–271.

CORRENS, C. W. & STEINBORN, W. 1939. Experimente zur messung und erklärung der sogennanten kristallisationskraft. *Zeitschrift für Kristallographie*, **101**, 117–133.

DOEHNE, E. 1994. In situ dynamics of sodium sulfate hydration and dehydration in stone pores: Observations at high magnification using the environmental scanning electron microscope. *In*: FASSINA, V., OTT, H. & ZEZZA, F. (eds) *3rd International Symposium on the Conservation of Monuments in the Mediterranean Basin, Venice.* Soprintendenza ai Beni Artistici e Storici di Venezia, 143–50.

DOEHNE, E. 2002. Salt weathering: a selective review. *In*: SIEGESMUND, S., WEISS, T. & VOLLBRECHT, A. (eds) *Natural Stone, Weathering Phenomena, Conservation Strategies and Case Studies.* Geological Society, London, Special Publications, **205**, 51–64.

ESPINOSA, R. & SCHERER, G. W. 2008*a*. Crystallization of sodium sulfate salts in limestone. *Environmental Geology*, **56**, 605–621.
ESPINOSA, R. & SCHERER, G. W. 2008*b*. Study of the pore clogging induced by salt crystallization in Indiana Limestone. *Proceedings of the 11th International Congress on the Deterioration and Conservation of Stone*. Nicholas Copernicus University Press, Torun, **1**, 81–88.
EVANS, I. S. 1970. Salt crystallization and rock weathering: a review. *Revue de Géomorphologie Dynamique*, **19**, 153–177.
FLATT, R. J. 2002. Salt damage in porous materials: How high supersaturations are generated. *Journal of Crystal Growth*, **242**, 435–454.
FLATT, R. J., STEIGER, M. & SCHERER, G. W. 2006. A commented translation of the paper by C. W. Correns and W. Steinborn on crystallization pressure. *Environmental Geology*, **52**, 187–203.
FUKUI, K., OKUBO, S. & TERASHIMA, T. 2005. Electromagnetic radiation from rock during uniaxial compression testing: The effects of rock characteristics and test conditions. *Rock Mechanics and Rock Engineering*, **38**(5), 411–423.
GENKINGER, S. & PUTNIS, A. 2007. Crystallization of sodium sulfate: supersaturation and metastable phases. *Environmental Geology*, **52**, 295–303.
GOUDIE, A. S., COOKE, R. U. & EVANS, I. S. 1970. Experimental investigation of rock weathering by salts. *Area*, **2**, 42–48.
GOUDIE, A. S. 1974. Further experimental investigation of rock weathering by salt and other mechanical processes. *Zeitschift für Geomorphologie*, Supplementband, **21**, 1–12.
GOUDIE, A. S. 1977. Sodium sulfate weathering and the disintegration of Mohenjo-Daro, Pakistan. *Earth Surface Processes*, **2**, 75–86.
GOUDIE, A. S. 1986. Laboratory simulation of 'the wick effect' in salt weathering of rock. *Earth Surface Processes and Landforms*, **11**, 275–285.
GOUDIE, A. S. 1993. Salt weathering simulation using a single-immersion technique. *Earth Surface Processes and Landforms*, **19**, 369–376.
GOUDIE, A. S. 1999. A comparison of the relative resistance of limestones to frost and salt weathering. *Permafrost and Periglacial Processes*, **10**, 309–316.
GOUDIE, A. S. & COOKE, R. U. 1984. Salt efflorescences and salt lakes: a distributional analysis. *Geoforum*, **15**, 563–582.
GOUDIE, A. S. & VILES, H. A. 1995. The nature and pattern of debris liberation by salt weathering: a laboratory study. *Earth Surface Processes and Landforms*, **20**, 437–449.
GOUDIE, A. S. & VILES, H. A. 1997. *Salt Weathering Hazards*. Wiley, Chichester.
HAMILTON, A. & HALL, C. 2008. Sodium sulfate heptahydrate: a synchrotron energy-dispersive diffraction study of an elusive metastable hydrated salt. *Journal of Analytical Atomic Spectrometry*, **23**, 840–844.
HAYNES, H. 2005. ASTM C 88 test on soundness of aggregate using sodium sulfate or magnesium sulfate: A study of the mechanisms of damage. *Journal of ASTM International*, doi: 10.1520/JAI12517.
HENNIKER, J. C. 1949. Depth of the surface zone of a liquid. *Reviews of Modern Physics*, **21**, 322–342.
KELLER, L. P., MCCARTHY, G. J. & RICHARDSON, J. L. 1986. Laboratory modeling of northern Great Plains salt efflorescence mineralogy. *Soil Science Society of America Journal*, **50**, 1363–1367.
KWAAD, F. J. P. M. 1970. Experiments on the granular disintegration of granite by salt action. *University Amsterdam Fysisch Geografisch en Bodemkundig Laboratorium Publicatie*, **16**, 67–80.
LINNOW, K., ZEUNERT, A. & STEIGER, M. 2006. Investigation of sodium sulfate phase transitions in a porous material using humidity and temperature controlled X-ray diffraction. *Analytical Chemistry*, **78**, 4683–4689.
LUQUER, L. McI. 1895. The relative effects of frost and the sulfate of soda efflorescence tests on building stones. *Transactions, American Society of Civil Engineers*, **33**, 235–256.
MCGREEVY, J. P. 1982. Frost and salt weathering: further experimental results. *Earth Surface Processes and Landforms*, **7**, 475–488.
MORTENSEN, H. 1933. Die "Salzsprengung" und ihre Bedeutung für die Regionalklimatische Gliederung der Wüsten. *Dr. A. Petermanns Mitteilungen*. Justus Perthes, Gotha, **79**, 130–35.
OGUCHI, C. T., HATTA, T. & MATSUKURA, Y. 1999. Weathering rates over 40,000 years based on changes in rock properties of porous rhyolite. *Physics and Chemistry of the Earth*, **24**, 861–870.
PRICE, C. A. 1978. The use of sodium sulfate crystallization test for determining the weathering resistance of untreated stone. *Deterioration and Protection of Stone Monuments, International Symposium*, Paris, June 5–9 1978. Unesco/Rilem, Paris.
RODRIGUEZ-NAVARRO, C. & DOEHNE, E. 1999. Salt weathering: influence of evaporation rate, supersaturation and crystallization pattern. *Earth Surface Processes and Landforms*, **24**, 191–209.
RODRIGUEZ-NAVARRO, C., DOEHNE, E. & SEBASTIAN, E. 2000. How does sodium sulfate crystallize? Implications for the decay and testing of building materials. *Cement and Concrete Research*, **30**, 1527–1534.
RUIZ-AGUDO, E., MEES, F., JACOBS, P. & RODRIGUEZ-NAVARRO, C. 2007. The role of saline solution properties on porous limestone salt weathering by magnesium and sodium sulfates. *Environmental Geology*, **52**, 269–281.
SCHAFFER, R. J. 1932. *The Weathering of Natural Building Stones*. Special Report No.18, Building Research Establishment, Garston.
SCHAFFER, R. J. 1955. Stone in architecture: I. Stone as a building material. *Journal of the Royal Society of Arts*, 837–867.
SCHERER, G. W. 1999. Crystallization in pores. *Cement and Concrete Research*, **29**, 1347–1358.
SCHERER, G. W. 2000. Stress from crystallization of salt in pores. *In*: FASSINA, V. (ed.) *Proceedings of the 9th International Congress on Deterioration and Conservation of Stone*, Venice, 19–25 June, Elsevier, Amsterdam, 187–194.
SCHERER, G.W. 2004. Stress from crystallization of salt. *Cement and Concrete Research*, **34**, 1613–1624.

SMITH, B. J. & MCGREEVY, J. P. 1983. A simulation study of salt weathering in hot deserts. *Geografiska Annaler*, **65** (A), 127–33.

SMITH, B. J. & MCGREEVY, J. P. 1988. Contour scaling of a sandstone by salt weathering under simulated hot desert conditions. *Earth Surface Processes and Landforms*, **13**, 697–705.

SPERLING, C. H. B. & COOKE, R. U. 1980. Salt weathering in arid environments: experimental investigations of the relative importance of hydration and crystallization processes. II Laboratory studies. *Bedford College London Papers in Geography*, **9**, 45.

SPERLING, C. H. B. & COOKE, R. U. 1985. Laboratory simulation of rock weathering by salt crystallization and hydration processes in hot, arid environments. *Earth Surface Process and Landforms*, **10**, 541–555.

STEIGER, M. 2005*a*. Crystal growth in porous materials-II: Influence of crystal size on the crystallization pressure. *Journal of Crystal Growth*, **282**, 470–481.

STEIGER, M. 2005*b*. Crystal growth in porous materials-I: The crystallization pressure of large crystals. *Journal of Crystal Growth*, **282**, 455–469.

STEIGER, M. & ASMUSSEN, S. 2008. Crystallization of sodium sulfate phases in porous materials: The phase diagram Na_2SO_4–H_2O and the generation of stress. *Geochimica et Cosmochimica Acta*, **72**, 4291–4306.

STEIGER, M., LINNOW, K., JULING, H., GÜLKER, G., JARAD, A. E., BRÜGGERHOFF, S. & KIRCHNER, D. 2008. Hydration of $MgSO_4 \cdot H_2O$ and generation of stress in porous materials. *Crystal Growth and Design*, **8**(1), 336–343.

TABER, S. 1916. The growth of crystals under external pressure. *American Journal of Science*, **41**, 532–556.

THAULOW, N. & SAHU, S. 2004. Mechanism of concrete deterioration due to salt crystallization. *Materials Characterization*, **53**, 123–127.

TSUI, N., FLATT, R. J. & SCHERER, G. W. 2003. Crystallization damage by sodium sulfate. *Journal of Culture Heritage*, **4**, 109–115.

WELLS, R. C. 1923. Sodium sulfate: its sources and uses. *United States Geological Survey*, Bulletin **717**, 43.

WINKLER, E. M. 1994. *Stone in Architecture*. Springer-Verlag, Berlin.

WINKLER, E. M. & WILHELM, E. J. 1970. Saltburst by hydration pressures in architectural stone in urban atmosphere. *Bulletin Geological Society of America*, **81**, 567–572.

Simultaneous wetting/drying, freeze/thaw and salt crystallization experiments of three types of Oya tuff

CHIAKI T. OGUCHI[1]* & HAYATO YUASA[2]

[1]*Geosphere Research Institute, Saitama University, 255 Shimo-Okubo, Sakura-ku, Saitama, 338-8570, Japan*

[2]*Faculty of Civil Engineering, Saitama University, 255 Shimo-Okubo, Sakura-ku, Saitama, 338-8570, Japan*

**Corresponding author (e-mail: ogchiaki@mail.saitama-u.ac.jp)*

Abstract: An abandoned subsurface Oya tuff quarry in Japan had abundant salt efflorescence in winter. Besides salt weathering, freeze-thaw weathering and slaking were likely to occur because of winter temperatures below 0 °C and the presence of swelling clays in the rock. Field surveys were performed to collect salts. Thenardite and gypsum were detected by X-ray diffraction (XRD) as the main salts, along with zeolites as secondary minerals. Oya tuff is categorized into three types for practical usage. To investigate petrophysical differences among the three types of Oya tuff, mercury intrusion porosimetry and tensile strength tests were performed. To determine the influence of petrophysical properties on salt weathering, freeze-thaw weathering and slaking (wet-dry weathering), all three types of Oya tuff were used for experiments. Prismatic specimens, the bases of which were sunk into distilled water, were used for the freeze-thaw and slaking experiments and Na_2SO_4 saturated solution was used for the salt-weathering experiment. The results show that the specimens subjected to salt weathering were the most severely damaged. The coarse-type Oya tuff sustained the most severe damage, whereas the fine type received the least. There was a large amount of debris in the coarse type, but less in the fine type. The weathering susceptibility index WSI was also calculated from the results of the pore size analyses and tensile strength. The index decreases with increasing weathering cycles representing resistant rocks. The phenomena of weathering of Oya tuff were explained by three weathering experiments on three kinds of tuff. The WSI may be useful as a practical indicator of rock weathering.

Japan has many volcanoes and many kinds of tuff are used as building stones. Oya tuff is one of the most famous building stones because it is easy to cut and has a beautiful rock fabric. However, famous sculptures and buildings made from Oya tuff are often severely weathered. In addition, weathering of Oya tuff often causes rock failure in the Oya area (e.g. Aoki *et al.* 2005). This rock has three types of petrophysical properties (Nakamura *et al.* 1981), and its quality as a building stone differs with the type. However, these rock properties have not been thoroughly investigated in relation to weathering mechanisms. We therefore performed three kinds of weathering experiment using three types of Oya tuff to examine the relationships between rock properties and weathering mechanisms.

There is an abundance of studies on the weathering of building stones. Most of these have examined salt weathering (e.g. Zehnder & Arnold 1989; Rodriguez-Navarro & Doehne 1999; Doehne 2002; Angeli *et al.* 2007) and freeze-thaw weathering (e.g. Nicholson & Nicholson 2000; Thomachot & Jeannette 2002; Thomachot *et al.* 2006). Slaking is not as well reported in the research field of building stone weathering, although it is widely studied in geomorphological research. However, few have compared these three weathering mechanisms (e.g. Ruedrich & Siegesmund 2007). We therefore attempt to understand the types of weathering occurring in the Oya tuff distributed area and perform three weathering experiments using three kinds of Oya tuff to examine petrophysical influences on weathering.

Oya tuff building stone and its weathering conditions

Oya tuff is a famous Japanese building stone showing plausible macroscopic appearance. It is quarried in the town of Oya, located about 8 km NW of the city of Utsunomiya in Central Japan (Fig. 1a). In Japan it is categorized as a green tuff. It was originally deposited as a rhyolitic ash in the old Sea of Japan during the Miocene. It belongs to the Oya geological formation and occurs in an area measuring about 4 km east–west and 6 km north–south. It is divided into three parts: upper,

From: PŘIKRYL, R. & TÖRÖK, Á. (eds) *Natural Stone Resources for Historical Monuments.*
Geological Society, London, Special Publications, **333**, 59–72.
DOI: 10.1144/SP333.6 0305-8719/10/$15.00

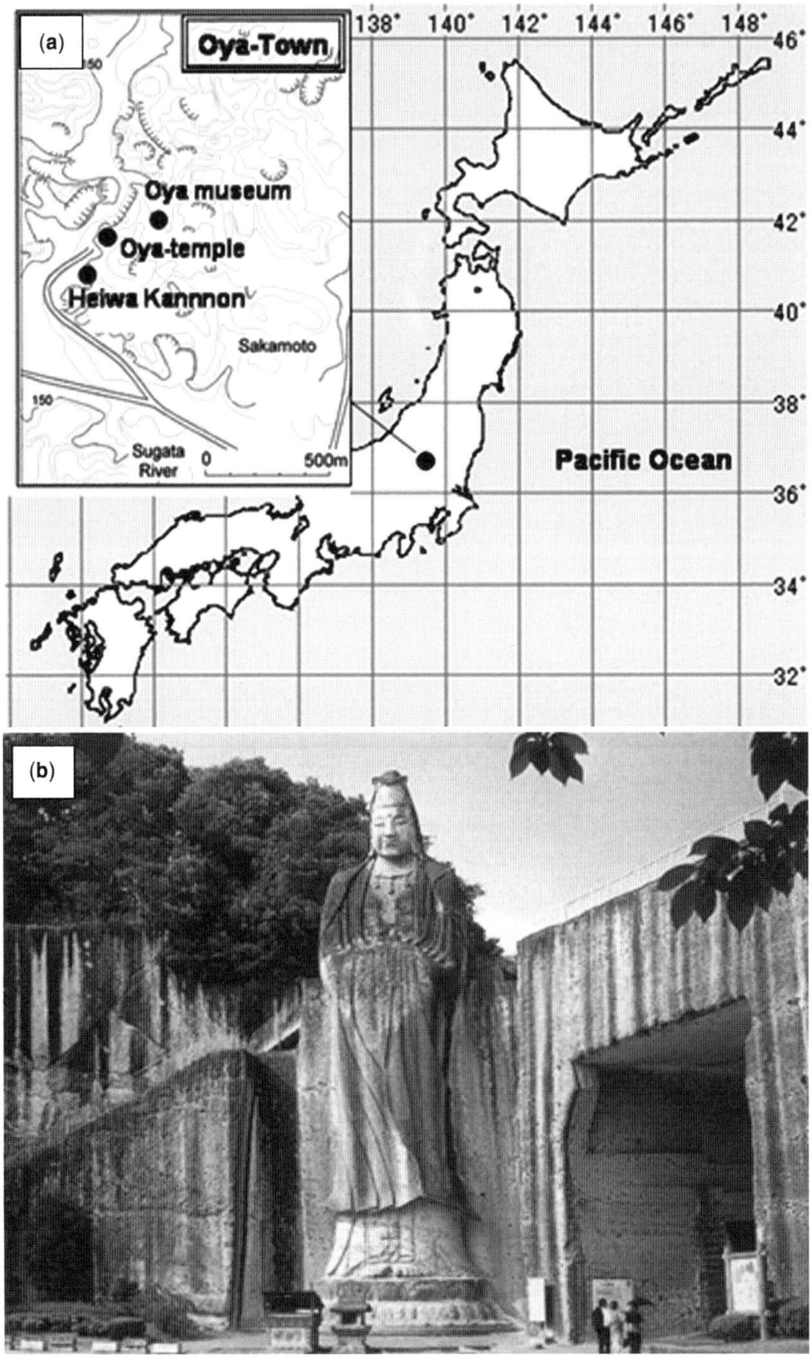

Fig. 1. Study area and Oya tuff: (**a**) location of Oya tuff formation, (**b**) Heiwa-Kannon statue.

middle and lower. The strike of the bed is N20°E on average and the dip is 5–10°E. The middle layer and the lower part of the upper layer, which provides higher quality building stones, have been exploited. For commercial reasons, the stone is divided into three categories on the basis of its fabric (Nakamura *et al.* 1981). The most valuable is the fine type with small clay patches; the other two have coarse fabrics with large clay patches and are differentiated from each other by strength. The high-quality type is now difficult to find in quarries and is now being mined at increasing depths because of its limited availability. Over 200 mining shafts, including abandoned shafts, are found in the Oya area. The clay patches in Oya tuff contain large amounts of swelling-type clay minerals such as montmorillonite, which is Fe-rich and tends to change from dark green to brown immediately after being mined (Nakamura *et al.* 1981). The existence of montmorillonite in Oya tuff suggests that slaking, one of the weathering processes of alternate swelling and shrinking that occurs in rocks containing clay minerals, might take place in this rock. Other minerals contained in this tuff are quartz, albite and clinoptilolite.

Oya tuff has the advantages of being easily cut and processed, which is why it has been used as a building stone or engineering material for over 1000 years. Gravestones were the earliest recorded use, then as cornerstones and base-stones for traditional castles in the Middle Ages and finally in large buildings such as schools, churches and factories during the 20th century. Famous buildings made of Oya tuff are the Tokyo Imperial Hotel designed by Frank Lloyd Wright and built in 1922 and the Matsugamine Catholic Church designed by Max Hinderand and built in 1932. Famous sculptures in this area are the Senju-Kannon Buddha (classified as an Important Cultural Property of Japan) in the Oyaji Temple, which was sculptured in 810 AD and the Heiwa-Kannon Buddha which was carved in 1954 (Fig. 1b). The former Buddha has been suffering from weathering, and reinforcement work to prevent destruction was carried out in 1965. Black varnish is observed on the latter Buddha and exfoliation has occurred on the surface. On the cut-off walls of abandoned Oya tuff quarries there are sulphate efflorescences mainly of thenardite (Na_2SO_4) (Imogawa *et al.* 2000).

There also many underground mines in the Oya area. One of them was mined from 1919–1986 and has since been used as the Oya museum (Fig. 2a). Large amounts of salt are observed on the walls (Fig. 2b, c). Examination of salt materials and debris from several locations by X-ray diffraction (XRD) revealed that the main salt detected was thenardite (Fig. 3). Quartz, albite, zeolite and gypsum were also detected in debris samples.

According to the Utsunomiya Meteorological Station, which is the station nearest to the Oya area, the highest and lowest monthly mean air temperatures from 1971–2000 were 30.1 °C (in August) and −3.5 °C (in January). The maximum and minimum monthly mean rainfalls were 234.5 mm (in September) and 25.5 mm (in December) over the same period. The mean annual temperature in the underground quarry surveyed is about 8 °C. This climate information indicates that not only salt weathering but also freeze-thaw weathering occurs in this area.

Petrophysical properties of Oya tuff

To determine the differences in the rock properties of the coarse, medium and fine types of Oya tuff, we investigated bulk density, tensile strength, mineralogical density (specific gravity) and pore size distribution. Bulk density ρ_{bulk} was calculated for five cylindrical specimens with 35 mm diameter and 35 mm height of each rock type by dividing the dry weight by the volume (Table 1). The specimens were oven dried at 110 °C for 48 h and then weighed.

The bulk density was 1370 kg m^{-3}, 1400 kg m^{-3} and 1430 kg m^{-3} for the coarse, medium and fine types, respectively. Mineralogical density ρ_m was measured with a 100 mL Gay-Lussac type pyknometer. Three specimens of each rock type were ground with an agate mortar into fine powder. The true densities were 2500 kg m^{-3} (coarse type), 2460 kg m^{-3} (medium type) and 2500 kg m^{-3} (fine type) (Table 2). From the results for bulk density and true density, total porosity n was calculated for the coarse, medium and fine types as 45.2%, 43.1% and 43.2%, respectively (Table 2).

Tensile strength S_t was measured on cylindrical specimens 35 mm in diameter and 35 mm height. The test was performed under both dry and wet conditions on five specimens of each rock for each condition; the results are listed in Table 1. Under dry conditions, tensile strength was 2.48 MPa (24.3 kgf cm^{-3}), 2.51 MPa (24.6 kgf cm^{-3}) and 2.89 MPa (28.3 kgf cm^{-3}) for the coarse, medium and fine types, respectively.

Pore size distributions were measured with an Autopore IV mercury intrusion porosimeter (Micromeritics Corporation). The connected pore volume V_{total} for pore diameter ranges from 0.003 to 200 μm. Three specimens of each rock type were tested. Figure 4 shows histograms of the average pore volume for each type of Oya tuff. The coarse-grained type had a wide pore volume range, with diameters between 10^{-2} and 10^{2} μm. The medium-grained type also had a wide range of pore diameters, but their volumes were generally less than those of the coarse-grained type. The fine-grained type was characterized by a small proportion of large diameter pores (diameter more than $10^{-0.5}$ μm) and large numbers of small diameter pores (diameter less than $10^{-0.5}$ μm). Table 3 lists the pore volumes of a range of pores with different diameters. The large pore (V_1), medium pore (V_2) and small pore volumes (V_3) represented the total

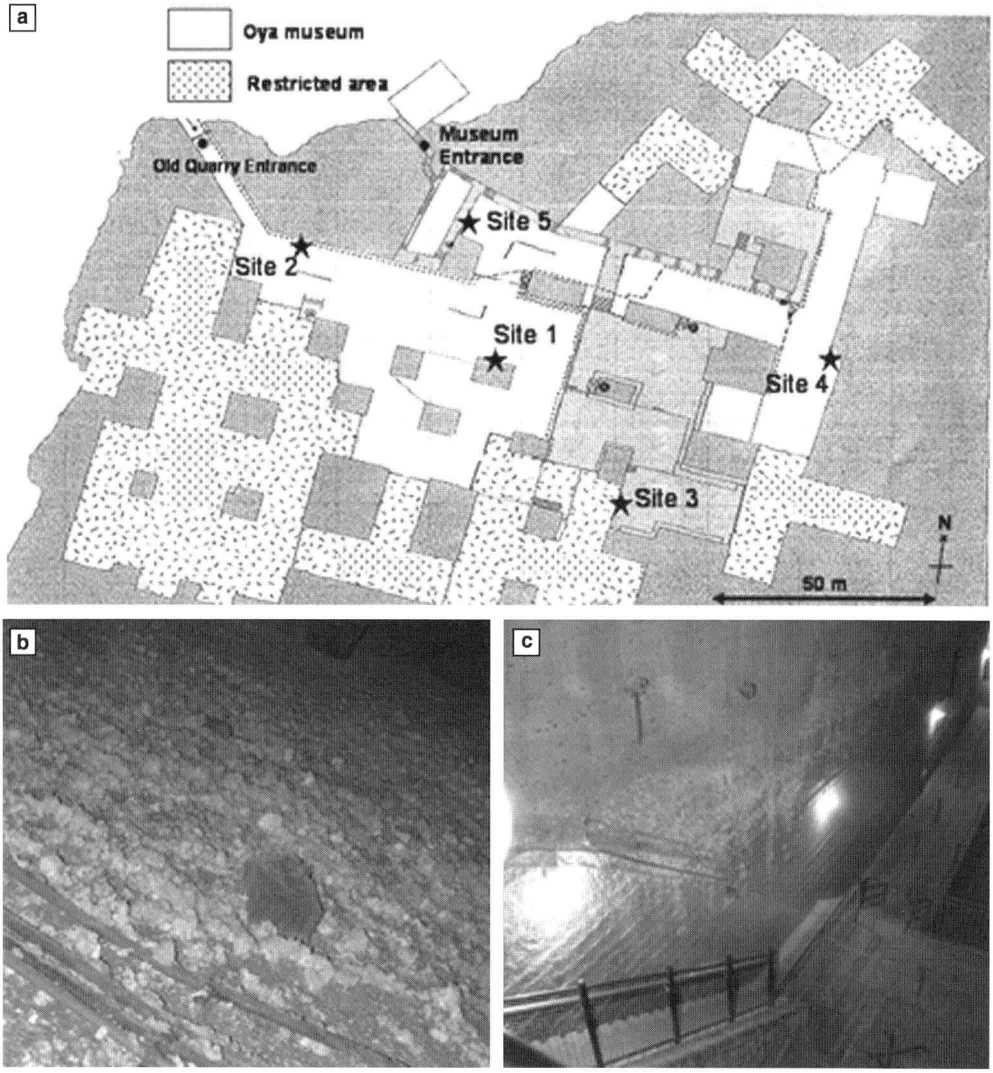

Fig. 2. Abandoned underground quarry: (**a**) plan map and sampling sites (☆), (**b**), (**c**) salt efflorescences on a wall of Oya tuff at site 1 (b) and site 2 (c).

volumes occupied by pores with diameters $10^{1.5}$–$10^{0.5}$ μm, $10^{0.5}$–$10^{-0.5}$ μm and $10^{-0.5}$–$10^{-1.5}$ μm, respectively.

Experimental procedures

Oya tuff is likely to be subjected to three kinds of weathering process: wetting/drying (slaking), freeze/thaw and salt crystallization. The experiments were designed to compare the three weathering processes on coarse, medium and fine types of Oya tuff. Since many previous studies have been performed by total immersion of cubic specimens, we partly submerged specimens of 5 cm length, 5 cm width and 15 cm height in solutions to a depth of 3 cm in order to investigate capillarity under certain environmental conditions.

The experimental designs are shown in Figure 5 and the experimental conditions are listed in Table 4. After measuring the rock properties in several specimens of each type of Oya tuff, one specimen was used for each weathering experiment. Each specimen was dried at 110 °C in a dry oven and weighed before we began the experiments. The salt crystallization experiment was set up in

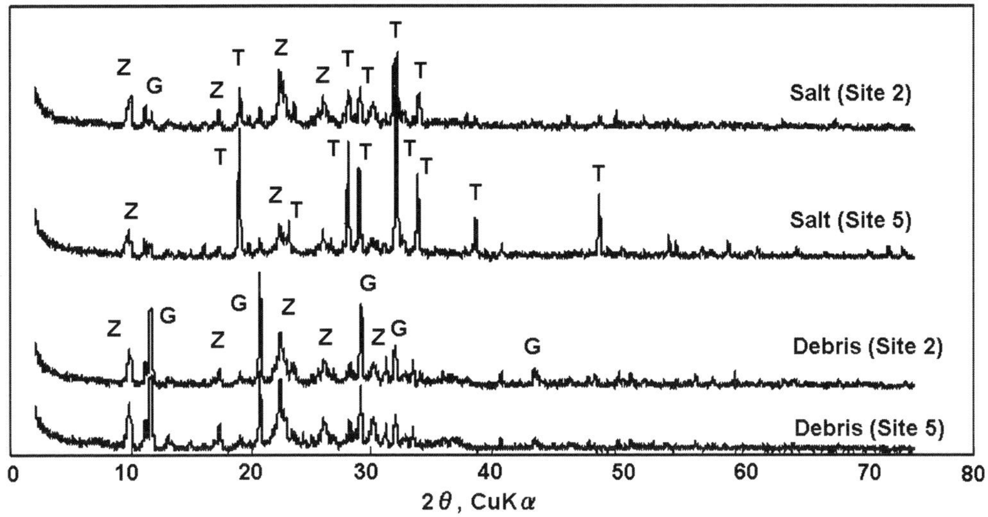

Fig. 3. Salt mineral and debris samples from sites 2 and 5 (Z, zeolite; G, gypsum; T, thenardite). The analysis was carried out by X-ray diffraction (Rigaku Co. RAD-X system) with the measurement conditions: voltage 40 kV; tube current 25 mA; scan speed 1°/m; 0.02° sampling; slit 1°–0.3 mm – 1°.

such a way that the bottom of the specimen was submerged in a saturated solution of sodium sulphate at 20 °C for 24 h; the specimen was then dried in an oven at 40 °C for 24 h. After the specimens had dried they were weighed and their visual condition photographed. The solution was then replenished to a depth of 3 cm to replace the evaporated solution. The amount of supplemental solution added at every cycle was weighed. This procedure was performed for 10 cycles. The wetting/drying (slaking) experiment was performed in the same way as the salt-weathering experiment, except that distilled water was used instead of saline solution. The environmental conditions were the same as for the salt-weathering experiment: 20 °C for 24 h of submersion and 40 °C for 24 h of drying. The freeze/thaw experiment was performed by using distilled water for submersion (thawing) at 20 °C for 24 h and then freezing at −15 °C in a freezer for 24 h. The other procedures were the same as in the previous experiments. At the end of the experiments, the total weight of debris of each type of rock in each experiment was measured and compared to the degree of weathering.

Results of experiments

The results demonstrated that salt weathering caused the most severe damage and produced many rock fragments (Fig. 6). The coarse-type Oya tuff showed the highest capillarity; the solution had reached the top of the specimen by the end of four cycles (Fig. 6a). After the same number of cycles in the medium type, the solution had moved only about 10 cm from the bottom. In the fine type the solution had moved only 5 cm from the bottom. The coarse and medium specimens were totally decomposed, whereas the fine type failed only at the bottom by the end of experiment.

In contrast, specimens were not visually decomposed in the wetting/drying (slaking) and freeze-thaw weathering experiments (Fig. 6b, c). A small amount of fine debris was produced by the freeze-thaw weathering. Although the clay patches flaked during the wetting/drying experiment, the specimens generally were undamaged.

The weight changes in dry specimens before and after the three experiments are shown in Table 5. The wetting/drying experiment produced 7.03 g of debris of the coarse type, 7.09 g of the medium type and 7.31 g of the fine type. In the freeze/thaw experiment, the debris weighed 14.03 g, 8.47 g and 16.7 g, respectively. Although the salt-weathering experiment appeared to cause the most severe damage, the debris without salts could not be weighed. The weight of the debris had to be calculated separately from that of the salt. The amount was calculated from the initial dry weight of the specimen ($w_{s=0}$), the initial weight of Na_2SO_4 saturated solution at 20 °C ($w_{Na_2So_4=0}$), the container weight (w_c), the weight of the supplemental solution ($w_{Na_2So_4=n}$) and the total weight of specimen, container and solution (w_{tn}) (Table 5). First, the initial ($n = 0$) total weight is calculated as:

$$w_{t=0} = w_{s=0} + w_{s=Na_2SO_4} + w_c \quad (1)$$

Table 1. *Bulk density determined by unit weight of dry condition and tensile strength of the three Oya tuffs*

	Specimen volume (10^{-6} m^3)	Weight of specimen (10^{-3} kg)	Bulk density (kg m^{-3})	Tensile strength (MPa)
Coarse type (dry condition)				
	35.68	50.15	1410	2.55
	34.73	49.49	1420	3.23
	35.17	48.51	1380	2.71
	32.95	44.72	1360	1.90
	35.51	45.62	1280	1.92
Average	37.00	47.70	1370	2.48
Coarse type (wet condition)				
	29.40	53.29	1813*	1.52
	34.78	62.7	1803*	1.53
	34.36	62.6	1822*	1.65
Average	34.81	59.53	1812*	1.56
Medium type (dry condition)				
	34.44	47.45	1380	2.30
	33.52	46.56	1390	2.55
	34.54	48.77	1410	2.52
	34.63	49.78	1440	2.71
	34.27	47.64	1390	2.47
Average	34.28	48.04	1400	2.51
Medium type (wet condition)				
	34.60	60.66	1753*	1.61
	33.40	59.07	1769*	1.24
	35.63	63.00	1768*	1.73
Average	34.54	60.91	1763*	1.53
Fine type (dry condition)				
	34.17	49.01	1430	2.99
	34.56	50.42	1460	3.08
	35.24	48.93	1390	2.84
	33.89	47.98	1420	2.29
	35.03	49.98	1430	3.23
Average	34.58	49.26	1420	2.89
Fine type (wet condition)				
	34.74	63.86	1838*	1.56
	34.40	62.75	1824*	1.71
	33.46	61.86	1849*	1.42
	33.95	61.84	1822*	1.33
Average	34.14	62.58	1833*	1.51

* = unit weight in wet conditions.

Table 2. *Mineralogical density and porosity of the three Oya tuffs; three specimens of each type were tested*

	Mineralogical density ρ_{true} (kg m^{-3})	Porosity n (vol. %)
Coarse-grained type	2500	45.2
Medium-grained type	2460	43.1
Fine-grained type	2500	43.2

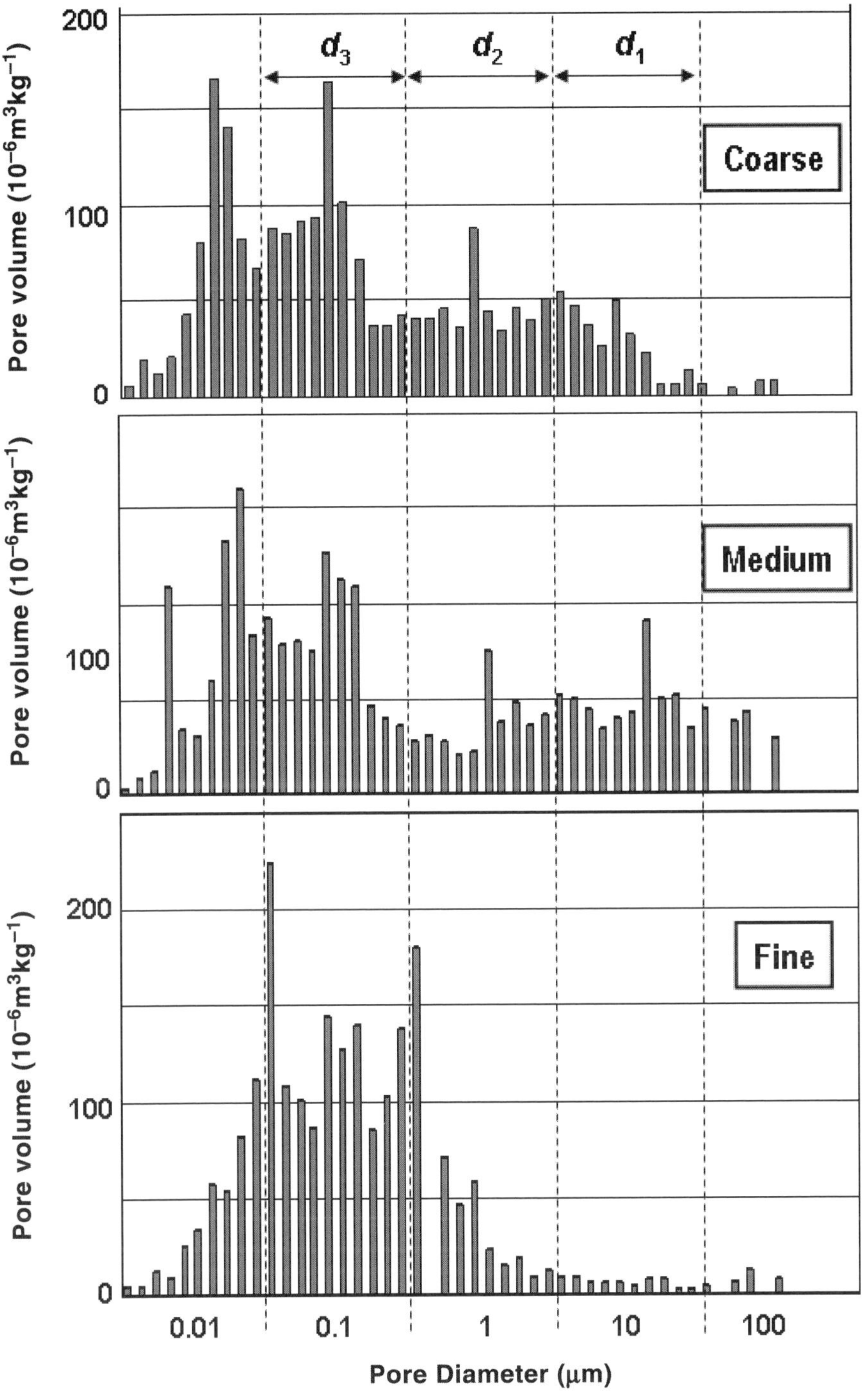

Fig. 4. Results of pore size distribution measurement.

Table 3. *Macropore, mesopore and micropore volumes of the three Oya tuffs*

	Volume of large-size pores, V_1 (10^{-6} m^3 kg^{-1})	Volume of medium-size pores, V_2 (10^{-6} m^3 kg^{-1})	Volume of small-size pores, V_3 (10^{-6} m^3 kg^{-1})	Connected pore volume, V_{total} (10^{-6} m^3 kg^{-1})
Coarse-grained type				
	29.2	46.6	82.1	223.1
	34.7	51.4	87.5	248.0
	17.7	48.5	73.9	213.8
Average	27.2	48.8	81.2	228.3
Medium-grained type				
	48.8	36.5	81.4	236.4
	31.0	38.5	72.5	208.5
	33.8	35.7	81.6	225.1
Average	37.9	36.9	78.5	223.3
Fine-grained type				
	6.1	43.8	128.4	218.4
	14.4	76.0	120.0	251.9
	4.6	39.9	129.4	216.4
Average	8.4	53.2	125.9	228.9

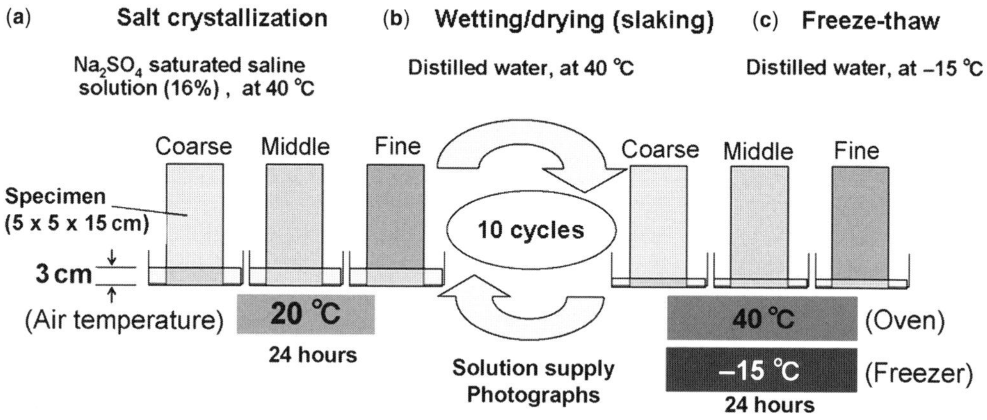

Fig. 5. Experimental design.

Table 4. *Conditions of the three weathering experiments*

	Salt crystallization	Wetting/drying	Freeze/thaw
Solution immersion condition			
Salt/solution	Na_2SO_4	DW	DW
Temperature (°C)	20	20	20
Duration (h)	24	24	24
Environment	room	room	room
Drying (crystallization) condition			
Temperature (°C)	40	40	−15
Duration (h)	24	24	24
Environment	Oven	oven	freezer

DW, distilled water.

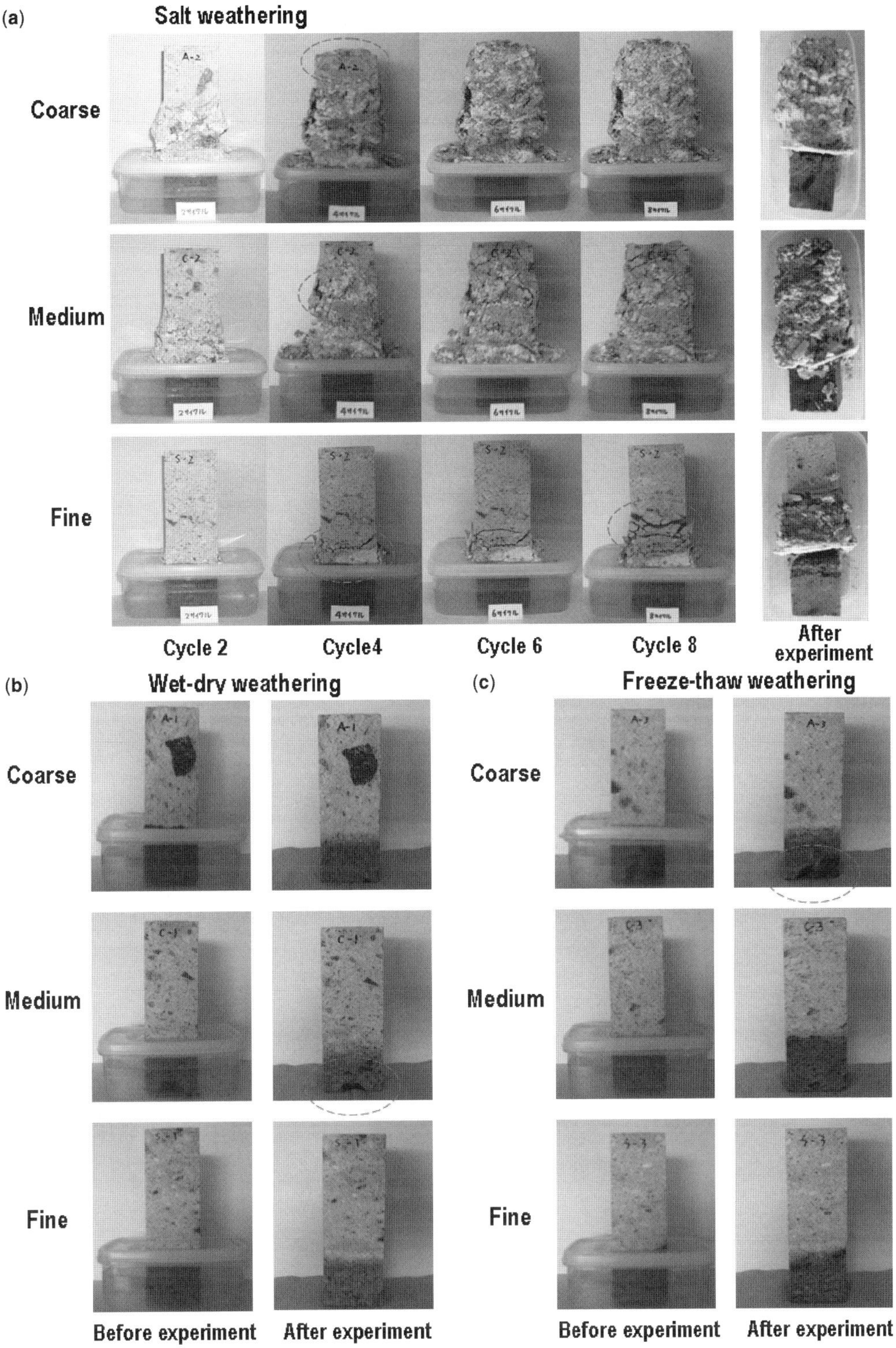

Fig. 6. Photographs of the three weathering experiments: (**a**) salt weathering, (**b**) wet/dry weathering, and (**c**) freeze-thaw weathering.

Table 5. *Results of weight measurements of specimens, solutions and debris*

	Salt crystallization			Wetting/drying			Freeze-thaw weathering		
	Coarse	Medium	Fine	Coarse	Medium	Fine	Coarse	Medium	Fine
Initial weight of dry specimen $w_{s=0}$ (g)	537.63	540.8	546.46	535.11	537.33	558.7	536.49	534.11	549.88
Container weight w_c (g)	63.75	67.66	67.98	68	67.71	68.07	68.84	68.12	67.98
Initial weight of Na_2SO_4 saturated solution at 20−C $w_{Na_2So_4=0}$ (g)	257.02	250.35	251.22	–	–	–	–	–	–
Initial weight of distilled water $w_{d.w.=0}$ (g)	–	–	–	233.78	226.36	232.22	220.68	221.4	215.69
Total Weight of specimen, container and solution $w_{t=n}$ after *n* cycles (g)									
$n = 0$ (initial weight)	858.40	858.81	865.66	810.29	796.89	843.33	824.01	820.65	832.33
$n = 1$	906.14	908.33	921.12	833.63	792.12	844.28	858.70	841.40	853.99
$n = 2$	934.77	916.66	932.63	839.39	822.92	863.40	855.69	836.65	850.19
$n = 3$	960.57	947.64	942.34	838.93	809.00	852.16	845.40	822.85	838.49
$n = 4$	1006.43	995.38	952.36	836.22	809.88	841.96	849.98	837.19	858.03
$n = 5$	1070.61	1041.71	952.74	826.71	807.17	836.23	844.52	829.56	850.59
$n = 6$	1135.50	1084.47	975.15	820.98	801.27	833.63	838.39	821.61	842.97
$n = 7$	1179.36	1134.63	981.83	829.22	816.94	841.29	864.82	853.03	871.11
$n = 8$	1199.66	1155.14	1003.11	830.04	814.44	844.92	858.28	844.30	863.87
$n = 9$	1229.45	1181.58	1000.35	827.97	816.87	846.25	852.75	837.40	857.34
$n = 10$ (final)	1244.35	1216.25	1005.69	841.64	826.29	857.10	862.80	851.57	871.83

Weight of supplement solution (Na_2SO_4/distilled water) solution after n cycles (g)									
$n = 1$	94.03	95.94	71.43	43.39	32.26	29.52	43.39	32.26	29.52
$n = 2$	95.61	80.18	34.03	0.00	0.00	0.00	0.00	0.00	0.00
$n = 3$	137.80	128.53	41.21	0.00	0.00	0.00	0.00	0.00	0.00
$n = 4$	165.22	140.70	32.73	13.90	28.30	31.09	13.90	28.30	31.09
$n = 5$	159.45	128.06	27.42	0.00	0.00	0.00	0.00	0.00	0.00
$n = 6$	130.50	116.72	44.84	0.00	0.00	0.00	0.00	0.00	0.00
$n = 7$	87.11	100.46	34.63	33.74	41.74	37.80	33.74	41.74	37.80
$n = 8$	59.34	85.06	43.70	0.00	0.00	0.00	0.00	0.00	0.00
$n = 9$	61.05	70.01	32.11	0.00	0.00	0.00	0.00	0.00	0.00
Weight of supplement Na_2SO_4 after n cycles (g)									
$n = 0$ (initial weight of Na_2SO_4)	41.12	40.06	40.20	–	–	–	–	–	–
$n = 1$	15.04	15.35	11.43	–	–	–	–	–	–
$n = 2$	15.30	12.83	5.44	–	–	–	–	–	–
$n = 3$	22.05	20.56	6.59	–	–	–	–	–	–
$n = 4$	26.44	22.51	5.24	–	–	–	–	–	–
$n = 5$	25.51	20.49	4.39	–	–	–	–	–	–
$n = 6$	20.88	18.68	7.17	–	–	–	–	–	–
$n = 7$	13.94	16.07	5.54	–	–	–	–	–	–
$n = 8$	9.49	13.61	6.99	–	–	–	–	–	–
$n = 9$	9.77	11.20	5.14	–	–	–	–	–	–
Cumulative Na_2SO_4 weight w_{ts} (g)	199.54	191.36	98.13	–	–	–	–	–	–
Final weight of dry specimen $w_{s=final}$ (g)	–	–	–	528.08	529.64	551.39	522.46	525.64	533.18
Debris weight (g)	529.99*	548.04*	260.1*	7.03	7.69	7.31	14.03	8.47	16.70

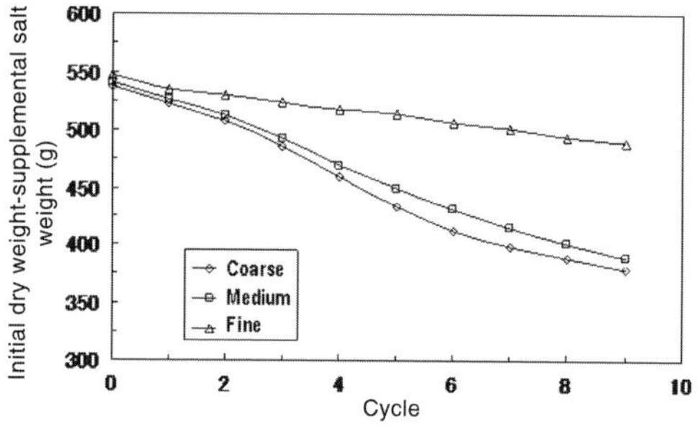

Fig. 7. Weight difference between initial dry weight of the specimen and supplemental salt.

As the Na_2SO_4 saturated solution at 20 °C has a concentration of 16%, the total weight of Na_2SO_4 (w_{ts}) is expressed:

$$w_{ts} = \left(w_{Na_2SO_4} + \sum_{n=1}^{9} w_{Na_2SO_4=n} \right) \cdot 0.16 \quad (2)$$

The difference between initial dry weight of the specimen and supplemental salt weight corresponds to the debris weight (w_d):

$$w_d = w_{t=10} - w_c - w_{ts} \quad (3)$$

Figure 7 depicts the results. The values gradually decrease with increasing cycle number. The grain size distributions of the debris, analysed by using sieves, are depicted in Figure 8. Coarse-type tuff produced the coarse debris, whereas fine types produced fine debris.

Discussion

Salt crystallization by sodium sulphate is a very effective approach in the deterioration of fabric of porous rocks such as the studied tuffs. The most important information sought from this kind of experiment is related to the determination of crystallization pressure. Two main equations for estimating crystallization pressure have been proposed. The first was proposed by Correns (1949) and is based on a thermodynamic concept, that is supersaturation:

$$P = \frac{RT}{V_c} \ln \frac{C}{C_s} \quad (4)$$

where P is crystallization pressure; R is a gas constant (8.31 J mol^{-1} K^{-1}), T is temperature, V_c is molar volume of the salt, C is the concentration of the saline solution; and C_s is saturated solubility. Correns' (1949) equation has been discussed in many studies (e.g. Scherer 1999; Steiger & Siegesmund 2007; Winkler & Singer 1972; Flatt *et al.* 2007). Flatt *et al.* (2007) provided elaborate comments on not only Correns' (1949) work but also his basic investigation (Corrrens & Steinborn 1939), written in Germany. According to Flatt *et al.* (2007), the crystallization pressure calculated by Correns' (1949) equation tends to be underestimated because he used potassium alum and did not consider the effects of ternary interaction parameters and low pH.

The other equation was proposed by Gauli *et al.* (1990) and is based on pore diameter and the surface tension of saline water:

$$p = \frac{4\sigma}{d} \quad (5)$$

where d is the pore diameter and σ is the surface tension between solid and liquid. This equation was derived by assuming that the stress due to crystallization is based on the chemical potential of a crystal growing from a solution (Everett 1961; Wellman & Wilson 1965; Fitzner & Snethlage 1982), and that the pore is cylindrical. Matsukura & Matsuoka (1996) used this equation to estimate crystallization pressure in the rock pores of coastal tafonis.

Sodium sulphate was the main salt used and it was necessary to estimate its crystallization pressure. However, it is difficult to quantify crystallization and/or hydration pressure in small pores. First, the surface tension of this saline water at a certain concentration and temperature is difficult to obtain. Second, the process of hydration of sodium sulphate during crystallization, that is thenardite–mirabillite transition, is still a mystery, although the transition has revealed the complete dissolution-subsequent recrystallization process rather than solid-state hydration revealed by the investigations using ESEM (e.g. Scherer 2000; Flatt 2002;

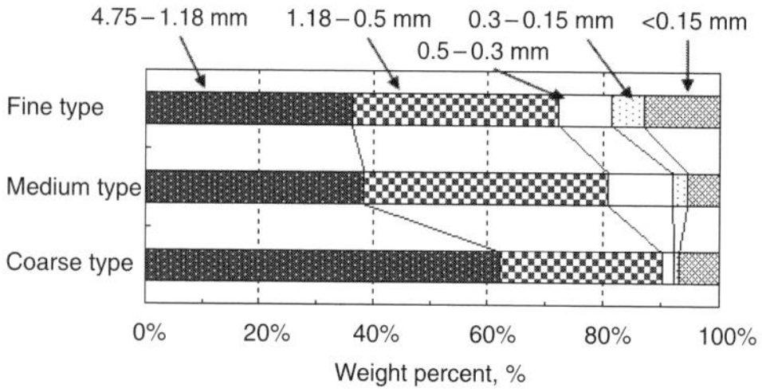

Fig. 8. Results of grain size analysis of debris produced by salt weathering.

Benavente *et al.* 2004). The damage attributed in the literature to hydration pressure of hydratable salts actually corresponds to the crystal growth pressure of hydrated phases (Rodriguez-Navarro *et al.* 2000).

Although this study did not examine methods of estimating the crystallization pressure of sodium sulphate, we did attempt to obtain a Weathering Susceptibility Index (WSI) (Matsukura & Matsuoka 1996). By using Correns' (1949) equation (4), the crystallization pressure of sodium sulphate can be estimated (although the values are underestimated). According to Shimada *et al.* (2002) and Oguchi *et al.* (2006), the crystallization pressure in the pores P_c is expressed by the equation:

$$P_c = P\sum_{i=1}^{3}\left(\frac{d_{max}}{d_i}V_i\gamma_d\right) \quad (6)$$

where d_{max} is the pore diameter of the largest pore ranges (median pore diameter is about 10 μm), d_i is the pore diameter of the i-th pore range classified as large (d_1-size pores with a diameter from $10^{1.5}$ μm to $10^{0.5}$ μm) medium (d_2-size pores with a diameter from $10^{0.5}$ μm to $10^{-0.5}$ μm) and small (d_3-size pores with a diameter from $10^{-0.5}$ μm to $10^{-1.5}$ μm). V_i is the pore volume of these pore ranges and γ_d is the unit weight of rock under dry conditions. Remaining pores are not considered to cause excessive damage to the rock, since water does not percolate into pores with a diameter less than 0.3 μm (Oguchi & Matsukura 1999). The product of V_i and γ_d corresponds to the pore volume in a unit volume of a specimen. Using equation (6), the crystallization pressure was calculated for each range of pores (Table 6), assuming that there is no time delay in pressure generation as a function of pore size. The parameter P_c/P describes the validity of the salt crystallization pressure in pores.

It is known that tensile strength S_t resists crystallization pressure P_c (e.g. Wellman & Wilson 1968). The ratio of P_c to S_t was obtained as a WSI (Matsukura & Matsuoka 1996):

$$\mathrm{WSI} = \frac{P_c}{S_t} \quad (7)$$

Table 6. *Na_2SO_4 crystallization pressure P_c and WSI of three types of Oya tuff*

	P_c	WSI
Coarse type	660.2	2.09
Medium type	382.4	0.064
Fine type	386.6	1.43

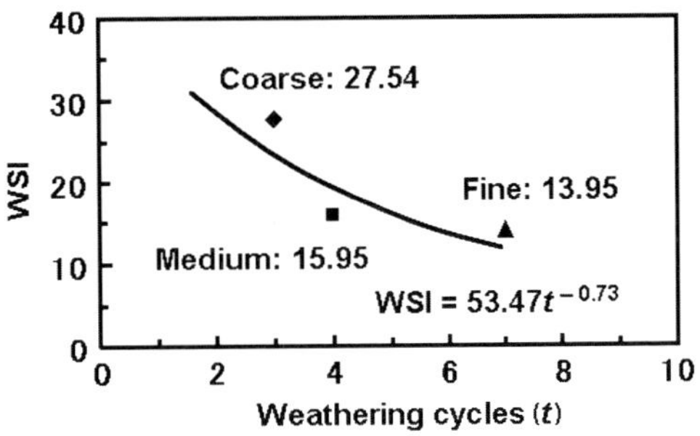

Fig. 9. WSIs of the three types of Oya tuff.

The WSI values were plotted against the number of cycles when the first deterioration occurred (Fig. 9). The coarse-type Oya tuff had the largest WSI value and the damage to this rock appeared earlier than that of the other two types. The fine type had the lowest WSI value and its deterioration occurred latest. The WSI values are therefore proportional to the number of weathering cycles necessary for rock damage.

Conclusions

The most damaging process in the Oya tuff distributing area was salt weathering, although the tuff is also weathered by various other processes including slaking and freeze-thaw. In salt weathering, the main salt was thenardite and the subdominant salt was gypsum. Large amounts of thenardite were observed on the walls of the underground tuff mine. The three kinds of weathering experiment revealed that salt weathering with sodium sulphate was the most destructive to Oya tuff. The coarse-type Oya tuff sustained the most severe damage, whereas the fine type received the least. There was a large amount of debris in the coarse type, but less in the fine type. The grain size analysis revealed that the coarse-type Oya tuff produced coarse debris, whereas the fine type produced fine debris. The WSI was also calculated and reconciled with the results of the experiment. The phenomena of weathering of Oya tuff were explained by three weathering experiments on three kinds of tuff. The WSI may be useful as a practical indicator of rock weathering.

We are grateful to Dr Takahashi and Dr Sato of the Research Center for Deep Geological Environments for their help in collecting the data on pore size distribution. We also thank Dr Hatta of the Japan International Research Center for Agricultural Sciences for access to the X-ray diffractometer used. We thank Professor Matsukura of the University of Tsukuba for valuable discussions. This study was supported financially by the Science Research Fund of the JSPS (No. 18680054) to C. Oguchi.

References

AOKI, H., SASAKI, T., OGUCHI, C. T. & MATSUKURA, Y. 2005. Stability analysis on a rock collapse over an abandoned mine cave in Miocene tuff. *Transactions, Japanese Geomorphological Union*, **26**, 423–437 (in Japanese with English abstract).

ANGELI, M., BIGAS, J. P., BENAVENTE, D., MENÉNDEZ, B., HÉBERT, R. & DAVID, C. 2007. Salt crystallization in pores: quantification and estimation of damage. *Environmental Geology*, **52**, 205–513.

BENAVENTE, D., GARCÍA DEL CURA, M. A., FORT, R. & ORDÓNEZ, S. 2004. Role of pore structure in salt crystallization in unsaturated porous stone. *Journal of Crystal Growth*, **260**, 532–544.

CORRENS, C. W. 1949. Growth and dissolution of crystals under linear pressure. *Discussions of the Faraday Society*, **5**, 267–271.

CORRENS, C. W. & STEINBORN, W. 1939. Experimente zur Messung und Erklärung der sogennanten Kristallisationskraft. *Zeitschrift für Kristallographie*, **101**, 117–133.

DOEHNE, E. 2002. Salt weathering: a selective review. *In*: SIEGESMUND, S., WEISS, T. & VOLLBRECHT, A. (eds) *Natural Stones, Weathering Phenomena, Conservation Strategies and Case Studies*. Geological Society of London, Special Publication, **205**, 43–56.

EVERETT, D. H. 1961. The thermodynamics of frost damage to porous solids. *Transactions Faraday Society*, **57**, 1541–1551.

FITZNER, B. & SNETHLAGE, R. 1982. Über zusammen hange zwischen saltzkristallisationsdruck und porenradienverteilung. *GP Newsletter* **3**, 13–24.

FLATT, R. J. 2002. Salt damage in porous materials: how high supersaturations are generated. *Journal of Crystal Growth*, **242**, 435–454.

FLATT, R. J., STEIGER, M. & SCHEERE, G. W. 2007. A commented translation of the paper by C. W. CORRENS and W. STEINBORN on crystallization pressure. *Environmental Geology*, **52**, 187–203.

GAULI, K. L., CHOUWDHURY, N. P., KULSHERESHTHA, N. P. & PUNNURU, A. R. 1990. Geologic features and durability of limestones at the Sphinx. *Environmental Geology Water Science*, **16**, 57–62.

IMOGAWA, A., NAKATA, M. & HONMA, H. 2000. The mode of occurrence of mirabilite and thenardite from Oya-ishi and its weathering process of tuff. *Bulletin of Tokyo Gakugei University, Section 4, Mathematics & Natural Science*, **52**, 31–36.

MATSUKURA, Y. & MATSUOKA, N. 1996. The effect of rock properties on rates of tafoni growth in coastal environments. *Zeitschrift fur Geomorphologie. Supplement Band*, **106**, 57–72.

NAKAMURA, Y., MATSUI, S. & SUZUKI, A. 1981. Geology of the Oya Area, Utsunomiya City. *Bulletin of Faculty of Education, Utsunomiya University*, **31**, 105–116 (in Japanese with English abstract).

NICHOLSON, D. T. & NICHOLSON, F. H. 2000. Physical deterioration of sedimentary rocks subjected to experimental freezing and thawing. *Earth Surface Processes and Landforms*, **25**, 1295–1307.

OGUCHI, C. T. & MTASUKURA, Y. 1999. Microstructural influence on strength reduction of porous rhyolite during weathering. *Zeitschrift für Geomorphologie. N.F., Supplement Band*, **119**, 91–103.

OGUCHI, C. T., MATSUKURA, Y., SHIMADA, H. & KUCHITSU, N. 2006. Application of weathering susceptibility index to salt damage on a brick monument. *In*: FORT, A. B., GOMEZ, H. & VAZQUEZ, C. (eds) *Heritage, Weathering and Conservation*. Taylor & Francis Group, London, 217–227.

RODRIGUEZ-NAVARRO, C. & DOEHNE, E. 1999. Salt weathering: influence of evaporation rate, supersaturation and crystallization pattern. *Earth Surface Processes & Landforms*, **24**, 191–209.

RODRIGUEZ-NAVARRO, C., DOEHNE, E. & SEVASTIAN, E. 2000. How does sodium sulfate crystallize? Implications for the decay and testing of building materials. *Cement & Concrete Research*, **30**, 1527–1534.

RUEDRICH, J. & SIEGESMUND, S. 2007. Salt and ice crystallisation in porous sandstones. *Environmental Geology*, **52**, 225–249.

SCHERER, W. G. 1999. Crystallization in pores. *Cement and Concrete Research*, **29**, 1347–1358.

SCHERER, W. G. 2000. Stress from crystallization of salt in pores. *In*: FASSINA, V. (ed.) *Proceedings of the 9th International Congress on Deterioration and Conservation of Stone*. **1**, Elsevier, Amsterdam, 187–194.

SHIMADA, H., OGUCHI, C. T. & MATSUKURA, Y. 2002. A laboratory experiment on salt weathering of bricks. *Bulletin of Terrestrial Environmental Research Center, University of Tsukuba*, **3**, 59–65 (in Japanese).

STEIGER, M. & SIEGESMUND, S. 2007. Special issue on salt decay. *Environmental Geology*, **52**, 185–186.

THOMACHOT, C., MATSUOKA, N., KUCHITSU, N. & MORII, M. 2006. Dilation of bricks submitted to frost action: field data and laboratory experiments. *In*: FORT, A. B., GOMEZ, H. & VAZQUEZ, C. (eds) *Heritage, Weathering and Conservation*. Taylor & Francis Group, London, 507–512.

THOMACHOT, C. & JEANNETTE, D. 2002. Evolution of the petrophysical properties of two types of Alsation sandstone subjected to simulated freeze-thaw conditions. *In*: SIEGESMUND, S., WEISS, T. & VOLLBRECHT, A. (eds) *Natural Stone, Weathering Phenomena, Conservation Strategies and Case Studies*. Geological Society, London, Special Publications, **205**, 19–32.

WELLMAN, H. W. & WILSON, A. T. 1968. Salt weathering or fretting. *In*: FAIRBRIDGE, R. W. (ed.) *The Encyclopedia of Geomorphology*. Stroundsburg, Pennsylvania, 968–970.

WINKLER, E. M. & SINGER, P. C. 1972. Crystallization pressure of salts in stone and concrete. *Geological Society of America Bulletin*, **83**, 3509–3514.

ZEHNDER, K. & ARNOLD, A. 1989. Crystal growth in salt efflorescence. *Journal of Crystal Growth*, **97**, 513–521.

Fire damage of trachyte: investigations of the Teplá monastery building stones

STEPHANIE GILLHUBER[1], GERHARD LEHRBERGER[1]* & JÜRGEN GÖSKE[2]

[1]*Technische Universität München, Lehrstuhl für Ingenieurgeologie, Arcisstraße 21, D-80333 München, Germany*

[2]*Zentrum für Werkstoffanalytik Lauf GmbH, Hardtstrasse 39b, D-91207 Lauf, Germany*

**Corresponding author (e-mail: lehrberger@tum.de)*

Abstract: The building stones of Teplá monastery (founded 1193) experienced a huge fire in 1677. Trachyte as a major rock type responded by changing colour from light beige to pink-red. Laboratory tests, during which the fresh unaltered stone from the original quarry was heated, proved the same reddened surface. Three varieties of local trachyte were examined: a greyish, naturally very fresh type (TA), a yellowish, slightly weathered and therefore iron hydroxide-bearing type (TB) and a trachyte typical with black manganese oxide dendrites and patches (TC). The changes of the physical properties and composition of these rocks were examined by ultrasonic velocity, thin section analysis, SEM observation and XRD tests. The experimental studies showed that Teplá trachyte is generally fire resistant up to 1000 °C.

The changes in physical properties and colour (namely reddening) which typically occur on dimension stones after fire have recently been described for many monuments (Goudie *et al.* 1992; Allison & Goudie 1994; Chakrabarti *et al.* 1996; Hajpál & Török 2004; Zier & Weise 2005; Hajpál 2006; Obojes *et al.* 2006; Sippel *et al.* 2007). The first systematic studies on natural stone altered by fire are associated with war events in Vienna during first half of 20th century (Kieslinger 1932, 1949). The reduction of strength and alteration of visual appearance through rock fabric and mineralogical changes are typical features of stones affected by fire. Further studies mainly on fire-related damage of building stone during the Second World War were compiled by Winkler (1997).

The alteration of stone appearance and properties attributed to fire can be observed on numerous monuments that remained untouched during the war events of the 20th century. The observed changes can, however, be associated with other alteration processes such as chemical changes (e.g. oxidation of Fe^{2+} to Fe^{3+}). To be certain of, which process was responsible for the observed changes, we simulated expected decay phenomena in the laboratory. In this study, we tried to experimentally prove the process of reddening of trachytic rocks by oxidative conditions during fire. The rocks were taken from an important central European monument – the Teplá monastery in southwest Bohemia (Czech Republic) – and also from the nearby quarry where the stone was originally exploited. This monument provides an excellent opportunity to study the use of local stone resources for architectural and sculptural purposes, and also long-term decay sequences of these stones. The series of experiments and analyses of heated trachytes from the quarry and samples from the monument proved that the rocks were affected by fire.

Teplá monastery and its building stone

History and archive studies

The Teplá monastery (located in the western part of the Czech Republic close to Mariánské Lázně spa) is one of the most important historical monuments in Central Europe (Fig. 1). The oldest buildings can be dated to the year of its foundation by the Bohemian nobleman Hroznata in 1193. The 800 year history of the monastery was continuously documented without interruption in the monastery archives. The most important facts of the building history are summarized in Table 1.

Concerning the major aims of current study, the archive study proved that the monastery was affected by several fires mainly during the 17th century. The major buildings of the monastery were damaged by fire in the year 1611. After that, the prelature and the convent had to be rebuilt. The monastery was again set on fire by troops lead by Captain Mansfeld at the beginning of the Thirty Years War in 1621. The economy buildings were destroyed during the blaze; the church, prelature and convent were spared. The most destructive fire occurred in the year 1659. The church, prelature and convent, as well as the brewery and the mill, were in flames. The following fire in the year 1677 ignited the hospital, the visitor buildings, the prelature and the outside walls of the church and the cloister.

From: PŘIKRYL, R. & TÖRÖK, Á. (eds) *Natural Stone Resources for Historical Monuments.* Geological Society, London, Special Publications, **333**, 73–79.
DOI: 10.1144/SP333.7 0305-8719/10/$15.00

Fig. 1. Panoramic view of the impressive building complex of the Teplá monastery in 1908; drawing and painting of Weeser & Krell Kunstanstalt, Linz.

Table 1. *Important facts in the building history of the 800 year old Teplá monastery*

Year	Important event
1193	Foundation of the monastery by the Bohemian nobleman Hroznata
2nd half of the 15th century	Bloom of the monastery, construction of the library
1611	Destruction of major parts of the monastery buildings through a disastrous conflagration
From 1612 onwards	Reconstruction of the buildings by Antonio de Consei
1621	Thirty Year's War; destruction of the economy buildings of the monastery
1639	Construction of the hospital by Antonio de Consei
1659	Devastating fire which affected the whole monastery including the church, prelature and the convent
1660–1667	Reconstruction under Domenico Pinchetti, construction of a new library, rebuilding of the prelature
1677	Big conflagration which affected the economy and convent buildings and the outside walls of the monastery
After 1677	Reconstruction of the wing between the church and the prelature
1680	Construction of a new storage
1686	Renovation of the abbey, construction of the brewery
1690–1704	First construction stage of the convent under Christoph Dientzenhofer and the foreman Wolfgang Braunbock
1709–1724	Second construction stage, construction of the prelature
c. 1750	Baroque rearrangement of the church
1852–1852	Renovation efforts at the convent and at the steeple
1890–1893	Construction of the mill, the brewery and other economy houses, implementation of the electrical illumination
1892	Finishing of the new building of the hospital, construction of a post office and a telegraph office in the monastery
1893	Refurbishments at the church and the convent; replacement of the church portal from the Renaissance period to a Neoromanesque portal
c. 1900	The pull down of Wenzel's chapel on the northern wall of the church, safety measurements on the northern wall with the construction of abutment
1902–1905	Construction of a Neobaroque library and museum by Josef Schaffer
1918	Formation of the Czechoslovakian Republic, deprivation of the monasteries property
1950	Socialization of the monasteries properties
1990	Restitution of the monastery to the Premonstratensian order

Local building stone

The most important local source of building stone used is an igneous rock of trachytic composition. It has been exploited in a hill-slope quarry located *c*. 3 km east of the Teplá monastery on the former grounds of the monastery.

The subvolcanic intrusion of alkali trachyte created a dome-shaped deposit 12.5 million years ago. The rock itself is composed of alkali feldspar and plagioclase with insets of sanidine phenocrysts up to a size of 12 mm. Biotite, Mn- and Fe-oxides form minor phases. Titanite, apatite, magnetite, goethite and/or hematite can be found as accessories. The groundmass is very fine-grained.

The colour of the rock varies from light grey to yellowish-beige. Brownish striae and black manganese-oxide and hydroxide dendrites are typical features of the Teplá trachyte.

Experimental material has been sampled from the still active original quarry. Samples of each variety of the Teplá trachyte were obtained: a greyish (TA), a yellowish (TB) and a type with black manganese oxides (TC).

On-site study of weathering phenomena

The weathering phenomena on the façade of the church, constructed mainly of Teplá trachyte, were mapped systematically during a project supported by the DBU (German Federal Environmental Foundation) (Lehrberger & Gillhuber 2007). During this mapping campaign, trachyte blocks with an intense red colour were found on the southern façade of the church. The colour changes were accompanied with Centimetre thick scaling locally accompanied by spalling and cavities. The scales have already partially fallen off and have been lost. Samples were taken from the façade and studied using scanning electron microscopy (SEM) to prove the assumption of fire decay.

The surface zone of the trachyte shows intense heating features such as molten feldspar crystals (Fig. 2, left). In comparison to unaltered trachyte from the quarry, the twin lamellas of the feldspars were clearly more rounded and not so sharp edged. This phenomenon is accompanied with sintered crusts and melted globules with sinter necks (Fig. 2, right) which allow differentiation from the similar fly ash pellets. The fact that globules can only be found on the southern part of the façade and are absent from other blocks supports the evidence of their origin by fire.

Laboratory tests

Heating experiment (fire simulation test)

For each of the sampled stone varieties, two test series with nine cubes (4 × 4 × 4 cm each) were prepared. Six specimens (two of each stone variety) were heated for three hours in a laboratory oven at temperature stages of 200 °C, 300 °C, 400 °C, 500 °C, 600 °C, 700 °C, 800 °C, 900 °C and/or 1000 °C. The oven was preheated to the desired temperature stage before inserting the specimen. The initial temperature difference between the rock and the oven temperature thereby simulated the heating shock which occurs when a fire develops suddenly. After the heating procedure, the samples cooled down at room temperature and were visually examined.

The first colour changes from grey to yellowish-beige took place at a temperature of 400 °C for the trachyte type TA. At 700 °C, its colour changed from yellowish-beige to pink until an intensive red colour was reached at 1000 °C. The trachyte variety TB, originally of yellow colour due to

Fig. 2. SEM-image: sample from the southern façade of the church with melt globules on the inner side of a scale (left side) and sinter crusts (right side). They result from the melting of the silicates in the intense heat of a large fire. Picture width: 0.02 mm (left) and 0.01 mm (right).

weathering, developed a reddish pigmentation at 200 °C. A colouring of the matrix to yellowish-beige followed at 400 °C. An increase in the red colouring could be observed at a temperature range of 600–1000 °C. Type TC also changed colour from yellowish-beige at 400 °C to intensive red at 1000 °C. With increasing temperature, the cubes seemed to become lighter and, when hit with a hammer, sound hollow. At a temperature of 1000 °C, the sound is similar to that of a brick.

Ultrasonic velocity measurement

The change in physical properties was determined by investigating the changes in P-wave velocity. The measurement was conducted by using Geotron Electronics UKS 12 system which was combined with a compressed air stand to reach a regular contact pressure of 3–4 bar (DIN EN 14579). Fleischer (2002) showed that lower contact pressures should be avoided to reduce the coupling defect and falsification. The coupling medium was avoided due to the perfect flat surface of the specimen. The planar probes UPG 250 (transmitter) and UPE (receiver) were used with a frequency of 250 kHz. The P-wave velocity was calculated using the formula:

$$v_P = \frac{l}{t}$$

where v_P is longitudinal wave velocity in km s^{-1}, l is distance through the rock sample in mm and t is the delay which is necessary for the impulse to recline the distance through the sample in μs.

A decrease in the velocity with increasing temperature could be observed for all trachyte types. This results from the increasing porosity caused by the dehydration and dehydroxilation reactions of the rock-forming minerals in the trachyte. The most striking change in P-wave velocity was recorded for the TC trachyte variety from *c.* 2700 m s^{-1} to 900 m s^{-1} (Fig. 3).

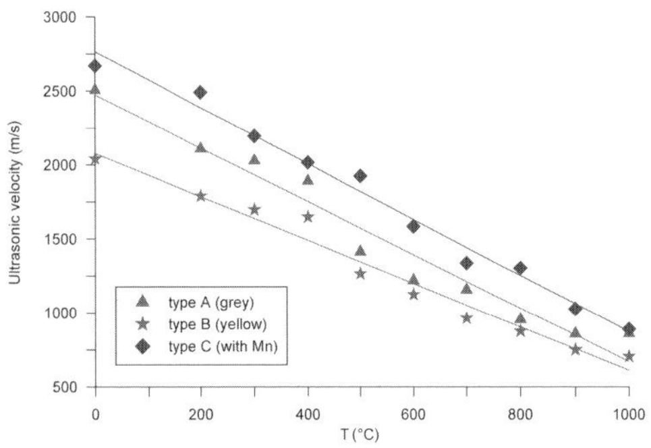

Fig. 3. Relation between the ultrasonic velocity and temperature during the firing experiments.

Microscopic analyses

Thin sections of the former grey TA cubes (burned at temperature stages of 200 °C, 500 °C, 700 °C and 1000 °C) were bedded in blue resin under vacuum for the purpose of microscopic study. Additional thin sections were prepared from the types TB and TC, both fired at a temperature of 1000 °C.

The thin sections show that the cleavages of the feldspar phenocrysts were widened with increasing temperature. This widening is clearly observable in the TA stone variety in the temperature range 700–1000 °C. The widening of intergranular space between the sanidine phenocrysts and the groundmass causes differential expansion of crystals during heating (Winkler 1997).

At 500 °C, the minute microcracks within a groundmass occur very rarely. At 700 °C they are rare but they occur quite often at 1000 °C. The thermal expansion of the minerals also caused a distinct disaggregation of the matrix, striking at a heating stage of 1000 °C. A disaggregation could not be observed at a temperature lower than 500 °C. These results correspond well with experimental data provided by Gomez-Heras *et al.* (2008).

Another phenomenon is the transformation of Fe- and Mn-oxides and oxihydroxides. In thin sections of the samples heated up to 500 °C, the Fe- and Mn-oxides and oxihydroxides constantly appear in dark-brown colours or are opaque. When heated above *c.* 700 °C, the minerals become translucent and show yellowish-brown colours. After being heated to 1000 °C, the yellowish-brown Fe-oxides disappeared in all three trachyte varieties and intense reddish hematite formed by the dehydration of the iron-hydroxides (limonite and goethite).

Phase changes

The XRD analyses performed on samples of the grey variety (TA) show that tridymite is formed at temperatures of about 1000 °C. In contrast to the densely packed structure of quartz, the crystal structure of tridymite is very open and its density (2.26 g cm^{-3}) is considerably lower than that of quartz (2.65 g cm^{-3}). Although the Teplá trachyte contains less than 5% quartz, the transformation from low to high quartz and the subsequent formation of tridymite, which is accompanied by a distinct volume increase of *c.* 10%, is regarded to be partly responsible for the disaggregation of the trachyte (Winkler 1997). The relatively low content of quartz in the Teplá trachyte is responsible for the fact that the damage to the building stones in the cathedral's walls is fairly limited. In general, the stone must be considered to be stable at moderate temperatures and can be even used as a heat storage stone in wood-burning stoves.

Discussion

The main influence of fire on natural stone is characterized by a rapid and intense heating, when flames burn directly against wall stones or staircases (Kieslinger 1932). The extent of fire damage is dependent on the thermal conductivity of the rock, the heating expansion coefficient of the single minerals and the rock as a whole (Winkler 1997), the heating absorption and the intrinsic factors such as porosity, grain size and boundaries, grain/grain or matrix/grain contacts (Hajpál 2006; Dionísio 2007). Allison & Goudie (1994) showed with their experiments that a heterogeneous material is less susceptible to thermal stress than a homogeneous material. The type and the duration of the fire impact are also important for the extent of the damage. Hajpál (2002) distinguishes two kinds of fires:

- local fires with low heat development that cause only superficial soot and decay; and
- intensive and expanded fires with huge heat development that change the physical-chemical quality of the rock fabric.

An important factor is also the short-term and one-sided influence of the fire on the rock surface (Kieslinger 1949). Temperatures of several hundred degrees to much more than 1000 °C can arise during such conflagrations (U. Obojes 2008, pers. comm.). The maximum temperature is usually only reached for a short period during these events (Schwarz 1986). The development of such high temperatures also occurs through the stack effect during conflagrations. The firestorm evolves from the rising hot air which results in the down movement of fresh air which arouses the fire once more. Temperatures of around 2000 °C can be reached during such firestorms.

These test results lead to the conclusion that the fire which caused the reddened trachyte blocks at the church wall must have been a large fire with the development of high temperatures. The melting and sintering processes which – as proven with the SEM – doubtlessly took place in the samples, also point towards a large disastrous fire: the melting points of feldspars occurred at temperatures of *c.* 1100 °C. This is in accordance with for example Brearley (1986), who observed the formation of melt globules in similar rocks from 1000 °C upwards.

The development of scales is caused by an increase in thermal stress during a fire. The one-sided heating of the surfaces results in stress which can exceed the tensile strength of the rock. The uniaxial strength of the trachyte determines the thickness of the scale. During a fire, tensions can build up within the heated rock which lead to intense spalling; this has been known and exploited by man in mining since the antiquity as the method of fire setting (Winkler 1997). This can be caused by the expansion of the minerals during heating. The expansion effect can be suddenly disturbed through shock cooling (e.g. from fire extinguishing work). A scaling is generated by the temperature difference of the highly heated surface of the rock and the cooler inner core. The development of steam pressure within the heated pore water could be a differential stress. If any of these tensions exceed the tensile strength of the rock, superficial scaling and flaking occurs. This process releases the built-up stresses. Also important for the development of scales is the distribution of stress *after the built-in of single constructional elements* (Kieslinger 1949).

All these scale-developing processes occur faster than the colouring of the rock. The transmitting of the heat will be interrupted by the existing air crack between the scale and the rock surface which acts as an isolation gap. This is also the reason why the red colouring could not be found deeper in the rock. Kieslinger (1949) stated that the red colouring cannot be found deeper than *c.* 2 cm in the rock. However, in some cases it can also be found to a depth of 3–4 m. The burned cubes from the heating test with a width of 4 cm confirmed the results of Kieslinger: they showed a regular colouring. Mineral changes are responsible for the colouration, mainly the oxidation process of Fe^{2+} to Fe^{3+} through the transformation of the existing goethite and limonite to hematite (Winkler 1997).

The results are confirmed through the heating test where the heated cubes at a temperature stage of 1000 °C are still in a solid state. A decay of the fabric and hints for mineral transformations can only be seen in a decreasing of the ultrasonic velocity and the hollow sound.

Kieslinger (1949), Winkler (1997) and Gomez-Heras *et al.* (2008) describe the first colour changes at 200–300 °C due to the excursive deficit of water loss of limonite. Changes in the colourations of the trachyte samples in the heating test can first be seen from yellow-beige at a temperature of around 400 °C. This is caused through the starting transformation of goethite and limonite to hematite which begins at a temperature of *c.* 250–300 °C (Chakrabarti *et al.* 1996; Sippel *et al.* 2007). The intensive red colouring of the trachyte begins at a temperature of *c.* 600 °C and is caused by the ongoing transformation to hematite. This is considerably higher than could be expected according to Winkler (1997). At a temperature of 900 °C, goethite, the last reaction product of the iron containing oxyhydroxides, disappears completely (Hajpál & Török 2004; Hajpál 2006).

The ultrasonic velocity in trachyte decreases by the firing by one-third, according to the increasing temperature in Figure 3. Such a result was only reached for different limestones in the experiments of Hajpál (2006) and is connected to the thermal expansion of the minerals and the generated microcracks in dense rock. Furthermore, the fabric can be damaged by the steam pressure of the pore water. An increase in fabric decay can also be confirmed by the decrease of the bulk density as well as by the increase of the utilizable pore volume (Fig. 4).

A semi-quantitative determination of the fabric damage by microcracking can be carried out with ultrasonic velocity measurement and microscopy of thin sections. The microcracks followed mostly grain boundaries of feldspar crystals at a temperature above 600 °C. Hajpál & Török (2004), Hajpál (2006) and Allison & Goudie (1994) described how microcracks could be observed at a temperature of *c*. 500 °C. The development of cracks is responsible for the reduction of ultrasonic velocities observed in the experiments. Cracks are caused by the expansion and shrinkage processes during the heating and cooling down phases, widely known from the literature (e.g. Winkler 1997). An apparent decrease of the accessible pore volume can be observed for all three types of Teplá Trachyte at 900–1000 °C, explained by the initial melting of the rock which leads to the closing of pores and microcracks.

The firing experiments with trachyte samples improve the knowledge of the reaction of feldspar-rich rocks at different temperatures, since there have been almost no data published previously. The results are in good accordance with other results from sandstones (Obojes *et al.* 2006; Gomez-Heras *et al.* 2008). In particular, the changes in porosity and the decrease of ultrasonic velocities provide a basis for the interpretation of damage observed on historical buildings.

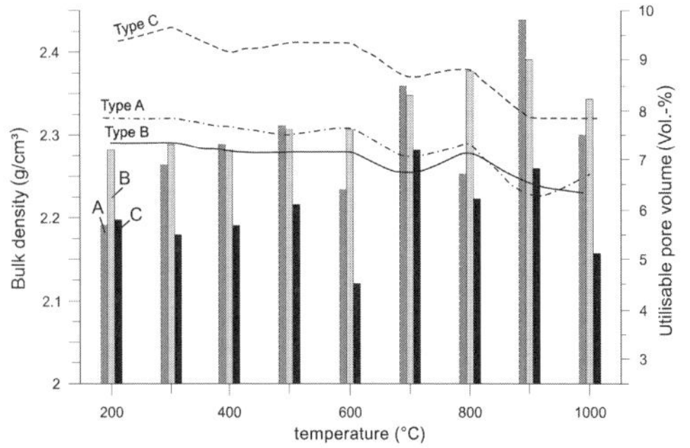

Fig. 4. Negative correlation of the bulk density (lines) and the accessible pore volume (bars) in relation to firing temperature.

Conclusions

Study of the literature and laboratory and field tests prove that the reddened trachyte blocks from the southern façade of the Teplá monastery church are the result of a fire during the 17th century. The fire presumably occurred in 1677, when it was reconcled that a big fire affected the outside walls of the monastery. Damage of the outer church walls is described in the archives and the decay of the building stone surfaces can be still seen today.

The fire changed the yellow-beige and grey trachyte blocks to red by a transformation of the goethite and limonite to hematite. Laboratory tests showed that red colouring of the trachyte samples was achieved at a temperature above 700 °C. It can be assumed from the bright sound when touching the samples that transformations of clay minerals had occurred.

The heating of the dimension stones caused a change in the physical properties. The best tools to examine the grade of the fabric damage are the ultrasonic velocity, thin sections, SEM and XRD tests. The decay studies at the monastery and the laboratory test showed that Teplá trachyte is fire resistant up to 1000 °C. Initial decay features could be observed in the form of colour changes, formation of microcracks, decreasing ultrasonic velocity and scaling or spalling.

We gratefully acknowledge the Deutsche Bundesstiftung Umwelt (DBU, German Federal Environmental Foundation; DBU-Project AZ: 20725) and Max Bögl Bauunternehmung GmbH & Co. KG for their financial support. We also thank Dr Erhard Westiner from the CBM (Centrum Baustoffe und Materialprüfung) of the Technische Universität München for support with the firing experiments.

References

ALLISON, R. J. & GOUDIE, A. S. 1994. The effects of fire on rock weathering: an experimental study. *In*: ROBINSON, D. A. & WILLIAMS, R. B. G. (eds) *Rock Weathering and Landform Evolution*. Chichester, New York, 41–56.

BREARLEY, A. J. 1986. An electron optical study of muscovite breakdown in pelitic xenoliths during pyrometamorphism. *Mineralogical Magazine*, **50**, 385–397.

CHAKRABARTI, B., YATES, T. & LEWRY, A. 1996. Effect of fire damage on natural stonework in buildings. *Construction and Building Materials*, **7**, 539–544.

DIN EN 14579 2005. *Prüfverfahren für Naturstein – Bestimmung der Geschwindigkeit der Schallausbreitung*. Berlin.

DIONÍSIO, A. 2007. Stone decay induced by fire on historic buildings: the case of the cloister of Lisbon Cathedral (Portugal). *In*: PŘIKRYL, R. & SMITH, B. J. (eds) *Building Stone Decay: From Diagnosis to Conservation*. Geological Society, London, Special Publications, **271**, 87–98.

FLEISCHER, G. 2002. *Beurteilung von Ultraschallmessungen an Natursteinobjekten in der Denkmalpflege*. PhD Thesis, Technical University Wien.

GOMEZ-HERAS, M., FORT, R., MORCILLO, M., MOLPECERES, C. & OCAÑA, J. L. 2008. Laser heating: a minimally invasive technique for studying fire-generated heating in building stone. *Materiales de Construcción*, **58**, 203–217.

GOUDIE, A. S., ALLISON, R. J. & MCLAREN, S. J. 1992. The relations between modulus of elasticity and temperature in the context of the experimental simulation of rock weathering by fire. *Earth Surface Processes and Landforms*, **17**, 605–615.

HAJPÁL, M. 2002. Changes in sandstones of historical monuments exposed to fire or high temperature. *Fire Technology*, **38**, 373–382.

HAJPÁL, M. 2006. The behavior of natural building stones by heat effect. *In*: KOURKOULIS, S. K. (ed.) *Fracture and Failure of Natural Building Stones*. Springer, Dordrecht, 439–445.

HAJPÁL, M. & TÖRÖK, Á. 2004. Mineralogical and colour changes of quartz sandstones by heat. *Environmental Geology*, **46**, 311–322.

LEHRBERGER, G. & GILLHUBER, S. 2007. Teplá-Trachyt: Herkunft, Verwendung, Verwitterung und Konservierung in der Klosteranlage von Teplá in Westböhmen. *Münchner Geologische Hefte*, **22**: 1–264.

KIESLINGER, A. 1932. *Zerstörungen an Steinbauten, ihre Ursachen und ihre Abwehr*. Verlag F. Deutike, Leipzig/Wien.

KIESLINGER, A. 1949. *Die Steine von St. Stephan*. Herold, Wien.

OBOJES, U., TROPPER, P., MIRWALD, P. W. & SAXER, A. 2006. The effects of fire and heat on natural building stones: first results from the Gröden Sandstone. *In*: FORT, R., ALVAREZ DE BUERGO, M., GOMEZ-HERAZ, M. & VAZQUES-CALVO, C. (eds) *Heritage, Weathering and Conservation*. Taylor and Francis/Balkema, Leiden, 521–524.

SCHWARZ, U. 1986. *Bestandsaufnahme der Natur-werksteine und ihres Verwitterungszustandes in der Innenstadt Münchens*. PhD Thesis, Ludwig-Maximilians-Universität München, Munich.

SIPPEL, J., SIEGESMUND, S., WEISS, T., NITSCH, K.-H. & KORZEN, M. 2007. Decay of natural stones caused by fire damage. *In*: PŘIKRYL, R. & SMITH, B. J. (eds) *Building Stone Decay: From Diagnosis to Conservation*. Geological Society, London, Special Publications, **271**, 139–151.

WINKLER, E. 1997. *Stone in Architecture*. 3rd edn. Springer, Berlin.

ZIER, H.-W. & WEISE, G. 2005. Brandschäden an Natursteinen - dargestellt am Beispiel des Kirchenbrandes in Riethnordhausen. *WTA-Journal for Technology and Applications in Building Maintenance and Monument*, **1**, 35–63.

Serpentinites from Cabo Ortegal (Galicia, Spain): a search for correct use as ornamental stones

DOLORES PEREIRA[1]*, MERCEDES PEINADO[1], MARIANO YENES[1], SERAFIN MONTERRUBIO[1], JOSE NESPEREIRA[2] & JOSE ANTONIO BLANCO[1]

[1]*Department of Geology, Universidad de Salamanca, 37008 Salamanca, Spain*

[2]*Investigación y Control de Calidad, S.A., Parque Tecnológico de Boecillo R-102, Valladolid, Spain*

**Corresponding author (e-mail: mdp@usal.es)*

Abstract: Different varieties of serpentinites present at Cabo Ortegal were studied. For many years, the 'Verde Pirineos' type has been quarried and sold commercially, but its physical characteristics do not fulfil the requirements for its use as an ornamental stone. 'Piedra de Doelo' is the local name for a serpentinite that has been quarried and used for many centuries in a large number of historical buildings distributed throughout the area. The preservation status of the rock is very poor and the stone is severely affected by weathering. A third variety, similar to the 'ophicalcite' described in the literature, is currently under investigation with a view to studying the possibility of the resumption of serpentinite quarrying. Although serpentinites are commercially known as 'green marbles', Galician serpentinites do not fulfil the mineral requirements to be described as such. Study of the characteristics of serpentinites, including their mineralogy, may offer a clue to the correct use of Galician serpentinites.

The mineralogy of ultramafic rocks is severely transformed by metamorphic reactions (Moody 1976; O'Hanley 1996), changing pre-existing anhydrous minerals such as olivine, pyroxene and other Mg-rich silicates and carbonates into assemblages such as calcite/dolomite-tremolite, calcite/dolomite-diopside-quartz, calcite/dolomite-tremolite-tal, calcite/dolomite-olivine-diopside-serpentine-brucite-magnetite, etc. (Pereira *et al.* 2007). Once transformed into serpentinites, these rocks exhibit beautiful textures that make them very valuable as ornamental dimension stones. However, once in place, serpentinites may be severely affected by weathering; this is an acute problem when they are used in exterior environments. Studies of the serpentinites from Cabo Ortegal, NW Spain (Fig. 1, see Pereira *et al.* 2007) have shown that these rocks do not fulfil the requirements for use as a dimension stone. The present work is based on historical examples in Galicia that reflect how this rock has been affected by time and how this problem could be avoided.

Mineralogy

Serpentinization can obliterate the primary mineralogy of rocks, and only a few remnants of this mineralogy can be identified in the rocks studied here (Figs 2 & 3a, b). Different serpentine phases can occur, depending on the conditions of transformation (i.e. pressure, temperature and fluid origin). Some rocks show carbonate transformation: that is, the serpentinite has been converted to a talc-carbonate paragenesis, and no evidence of the previous mineralogy is visible (Pereira *et al.* 2007). These latter serpentinites are known commercially as 'green marbles'. However, if some of the mineral precursors are still present in the rock, weathering may affect the rock selectively. Olivine is therefore the first mineral to disappear. It is impossible to find remains of this mineral in most cases; although pyroxene may be very well preserved, only a few fractures affect the mineral phases. Serpentinization takes advantages of these fractures, which allow fluids to circulate and alter the rock. In fact, the first symptoms of serpentinization is seen in these fractures (Pereira *et al.* 2007).

Serpentinites can also be affected by shearing during different episodes, producing veins filled with calcite. These veins act in different ways in response to weathering because they are weaker than the host rock. The rock can break along these veins once it has been installed as a facing stone on a building, resulting in degradation of the whole slab.

In this study we followed the sequence of alteration of the ultramafic rocks; the results are depicted in Figures 3 and 4. Different degrees of carbonate replacement can be observed. The process is not

From: PŘIKRYL, R. & TÖRÖK, Á. (eds) *Natural Stone Resources for Historical Monuments*. Geological Society, London, Special Publications, **333**, 81–85.
DOI: 10.1144/SP333.8 0305-8719/10/$15.00

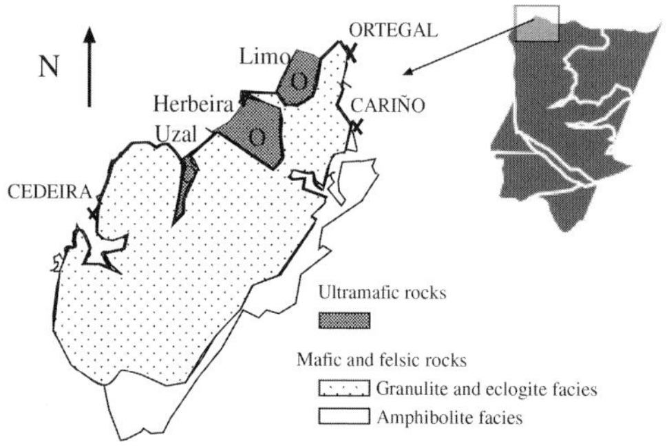

Fig. 1. Location of Cabo Ortegal complex and the three different massifs studied in this work (from Pereira *et al.* 2007).

always complete: carbonates are found mainly in veins, although the carbonation may also affect the whole rock. The carbonation of serpentine based on the olivine texture is a common feature.

Behaviour of serpentinites once emplaced

Serpentinites exhibit a broad spectrum of colours (typically ranging from light to dark green to almost black) and rock fabric because they are formed by the alteration of rock types of diverse bulk-rock compositions and structures.

The colour of serpentinite also varies with the extent of hydration of the protolith and with the extent of deformation. Once in place as a dimension stone, serpentinites can weather in different ways according to their composition (Fig. 5). If the rock is cut by veins and shears, weathering may begin to occur through these structures. For example, two slabs with apparently the same characteristics can behave differently when subjected to changes in humidity and temperature. The weathering will be more evident in the slab affected by veins and shears, possibly leading to complete destruction of the slab.

Fig. 2. Photomicrographs showing primary minerals in ultramafic rock from Cabo Ortegal. Shears are cutting the pyroxene, which act as conduits for serpentinization of the mineral. Crossed polars. Scale: bar = 250 μm.

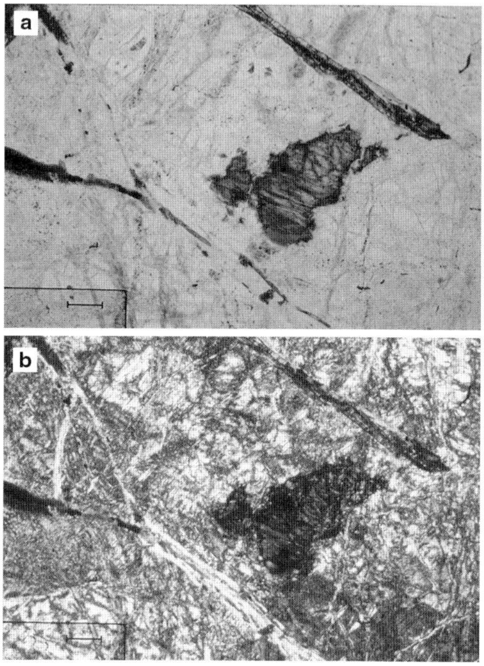

Fig. 3. Photomicrographs showing remnants of primary mineralogy in serpentinite from Moeche (Galicia). The main mineralogical composition of the rock is serpentine, produced from olivine. Some orthopyroxene pseudomorphs can be distinguished as a trace: (**a**) plane polarized light; and (**b**) crossed polars. Scale bar = 250 μm.

To properly characterize serpentinites for use as dimension stones, their study (and the dissemination of the results) should include aspects of geochemistry, mineralogy and mechanical properties. All of these properties vary depending on the rock in question. The compositions of serpentinites from different locations also differ [e.g. *Verde Macael* and *Verde Pirineos*, from Spain; *Verde Alpi* and *Verde Prato*, from Italy (Cimmino *et al.* 2004; Marino *et al.* 2004); *Rajasthan Green*, from India]. Knowledge of the behaviour of serpentinites is important not only for the conservation of historical buildings (Malesani *et al.* 2003), but also in the prevention of stone decay since this dimension stone is now widely used for tiling in certain buildings (Pereira *et al.* 2007).

Varieties of Galician serpentinites

Three different varieties of serpentinites can be distinguished in Galicia, one of them already described

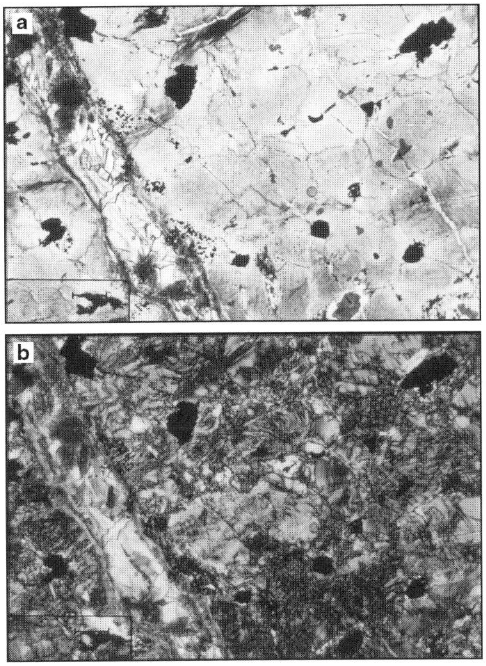

Fig. 4. Photomicrographs showing serpentinite cut by veins filled by carbonate. Carbonate veining has occurred in different, overlapped, episodes: (**a**) plane polarized light and (**b**) crossed polars. Scale bar = 640 μm.

by Pereira *et al.* (2007). This latter type is composed of serpentine (lizardite), a few remains of pyroxene and an abundance of shears filled by both serpentine and carbonates. It is known commercially as *Verde Pirineos*. The technical properties of *Verde Pirineos* do not fulfil the requirements for its use as an ornamental stone (Table 1). The properties are referred to the US standards (ASTM 2002) bacuse no European specification for serpentinites exists (Pereira *et al.* 2007). *Verde Pirineos* has a very high absorption value compared to that required for both interior and exterior use. The compressive and flexural strengths are very low, which would prevent these rocks from being used in construction. This is

Fig. 5. Different weathering behaviour of serpentinite used as decorative cladding on a building face. The left-hand panel is highly weathered while the right-hand panel preserves the initial characteristics. The picture was taken from an outside wall of a clinic in Salamanca (Spain).

probably the reason why most quarries ceased production several years ago (Fig. 6).

The second variety is the rock locally known as 'Piedra de Doelo' (sometimes 'Toelo'). Piedra de Doelo was described by Macpherson in 1881 (Tomkeieff 1983) as a 'local name for a manganese-rich listvenite, composed of giobertite (Mg-Fe-carbonate), talc, chlorite and magnetite' (Fettes & Desmons 2007). We found outcrops where the rock has been completely altered to secondary minerals, mainly talc, which makes the stone very easy to work (Fig. 7). This stone has traditionally been used for building purposes in Galicia (Fig. 8a–c), but has been strongly affected by weathering.

The main ornamental features of several historical buildings, dating from the 16th to the 19th century, are made of this rock. The chapel at San Xixo de Osos has a coat of arms showing two bears beside an evergreen oak. This element is on the altar of the chapel and is characteristic of Baroque art. Two anthropoid sarcophagi made of the same serpentinite are located outside the chapel. The San Xurxo church is neoclassic,

Table 1. *Physical properties of the serpentinites 'Verde Pirineos' and 'Ophicalcite' from Moeche (Galicia) and the requirements for the use of serpentinite as dimension stone (ASTM 2002)*

Physical properties	ASTM requirements	Verde pirineos	'Ophicalcite'
Water absorption (wt. %) Ext./Int.	0.20 max/0.60 max	0.93	0.70
Density (kg m^{-3})	2560 (min)	2700	2600
Compressive strength (MPa)	69 (min)	34.46	35.1
Flexural strength (MPa)	6.9 (min)	5.92	n.a.

n.a.: not available.

Fig. 6. Abandoned quarry of 'Verde Pirineos' in Moeche (Galicia).

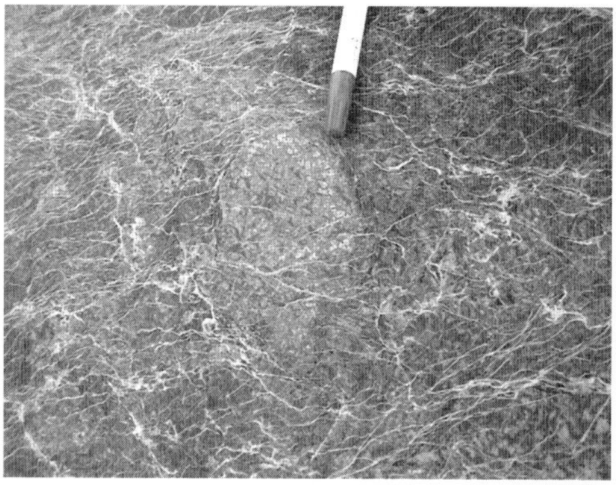

Fig. 7. Piedra de Doelo (a typical variety of serpentinite from Galicia), showing the characteristic colour and rock fabric resulting from its mineralogical composition (secondary minerals, mainly talc).

but the renovation of the façade dates from the 19th century. The façade of the church at Labacengos dates from the 16th century. However, in all three buildings many details have been lost due to weathering of the serpentinite.

The third variety found is very similar to that described in the literature as ophicalcite (Cimmino *et al.* 2004; Fig. 9). It is a brecciated rock that appears to have greater consistency than the above two varieties, and will be addressed in future research. If it can be shown to fulfil the physical requirements, it could lead to the resumption of serpentinite quarrying in Galicia. Nevertheless, preliminary results have determined very low values for its compressive strength (Table 1). This third variety shows high mechanic anisotropy, determined by a highly penetrative foliation that affects most rocks found at the quarry. Mechanical strength mainly depends on the orientation of the anisotropy

Fig. 8. Examples for the use of Piedra de Doelo in Galicia: (**a**) detail of the entrance of San Juan de Moeche Church; (**b**) the city council of San Ramón (close to Moeche); and (**c**) a fountain in San Ramón.

Fig. 9. Third variety of serpentinite found in Cabo Ortegal. Its physical aspect resembles the Ophicalcite described in literature. This rock is currently under investigation to determine its physical and mineralogical properties.

surfaces with respect to the direction of the application of force. To obtain the compressive strength value (ASTM C-170), the cylindrical samples with a height/diameter ratio of 2:1 were employed. The main foliation of the rock was oriented perpendicular to the maximum axial force. This orientation was chosen because theory dictates it should be the one showing the highest strength values. The results obtained (Table 1) show that this variety does not fulfil the ASTM requirements for serpentinites, and its compressive strength values of 35.1 MPa are much lower than the required 69 MPa. The values obtained are similar to those obtained for '*Verde Pirineos*' (Pereira *et al.* 2007, Table 1). Work is in progress to determine whether the mineralogy is also similar.

Conclusions

Serpentinites from Cabo Ortegal show poor performance when used as facing material on buildings because they tend to crumble and disintegrate when exposed to the weather. This can be explained by their mineralogical composition. Both the *Verde Pirineos* type and the 'Piedra de Doelo' type (a local name for a serpentinite transformed into talc) are severely affected by the humid climate of Galicia. Although the quarries which provided stone for historical buildings have been identified, the poor quality of the rock does not support the possibility of restoring damaged buildings with this material. Indeed, the renovation work carried out on one of the buildings (San Xurxo chapel) at the beginning of the 19th century now shows a high degree of decay.

A third variety of serpentinite has been identified at Cabo Ortegal, with a similar physical aspect to the ophicalcites described in the literature. Work in progress on the mineralogy and technical properties of this rock should provide information regarding the possibility of resuming serpentinite quarrying in Galicia. However, preliminary results have shown that this rock does not fulfil the geo-mechanical requirements for its use as an ornamental stone. A detailed study of the mineralogy should show whether there is a common mineralogical composition for all three varieties and whether the mineralogical transformation follows the same evolution and weathering behaviour in all of them.

This work was funded by the Spanish government through project BTE CGL2006-05128/BTE. N. Skinner helped with English grammar. Comments from R. Přikryl, E. Hyslop and R. Sandrone further improved the manuscript.

References

ASTM 2002. *Standard Specification for Serpentine Dimension Stone.* C 1526-02. ASTM International, West Conshohocken, PA.

CIMMINO, F., FACCINI, F. & ROBBIANO, A. 2004. Stones and coloured marbles of Liguria in historical monuments. *Periodico di Mineralogia*, **73**, 71–84.

FETTES, D. & DESMONS, J. 2007. *Metamorphic Rocks. A Classification and Glossary of Terms.* Cambridge University Press, Cambridge.

MACPHERSON, J. 1881. Apuntes petrográficos de Galicia. *Anales de la Sociedad Española de Historia Natural*, **10**, 49–87.

MALESANI, P., PECCHIONI, E., CANTISANI, E. & FRATINI, F. 2003. Geolithology and provenance of materials of some historical buildings and monuments in the centre of Florence (Italy). *Episodes*, **26**, 250–255.

MARINO, L., CORTI, M., COLI, M., TANINI, C. & NENCI, C. 2004. The 'Verde di Prato' stones of cathedral and baptistery of Florence (abstract). *Proceedings of 32nd IGC Florence*, T16.03.

MOODY, J. B. 1976. Serpentinization: a review. *Lithos*, **9**, 125–138.

O'HANLEY, D. 1996. *Serpentinites.* Oxford University Press, New York.

PEREIRA, M. D., BLANCO, J. A., YENES, M. & PEINADO, M. 2007. Characterization of serpentinites to define their appropriate use as building stones. *In*: PŘIKRYL, R. & SMITH, B. J. (eds) *Building Stone Decay: From Diagnosis to Conservation.* Geological Society, London, Special Publications, **271**, 55–62.

TOMKEIEFF, S. I. 1983. *Dictionary of Petrology.* Wiley, Chichester.

A legacy of mistreatment: conceptualizing the decay of medieval sandstones in NE Ireland

STEPHEN MCCABE*, BERNARD J. SMITH & PATRICIA A. WARKE

School of Geography, Queen's University, Belfast, BT7 1NN, Northern Ireland, UK

**Corresponding author (e-mail: stephen.mccabe@qub.ac.uk)*

Abstract: Sandstone is commonly used as a building material in medieval monuments throughout NE Ireland. This paper explores the reasons why, and the ways in which, medieval sandstone monuments decay in the temperate Atlantic maritime environment of the north Antrim coast, using Bonamargy Friary, Ballycastle, as a case study. Monumental stone decay is placed in the context of inheritance and sensitivity to change, that is, the ability or inability of a sandstone to absorb change as a result of the past stress events it has experienced. A consideration of the combined impact of background environmental factors (such as salt accumulation, temperature cycles, frost, chemical alteration, soiling of the surface, changes in surface morphology and biological colonization) and 'exceptional' factors (such as lime rendering, fire, climate change, abandonment and conservation intervention) is used to formulate alternative decay pathways of the sandstones identified at the Friary. Discussion focuses on the value of identifying conceptual event sequences such as the cumulative impact of past events, individual and combined, to produce recognizable decay features seen in the present day. The possible impact of future climate change on the decay of medieval sandstone monuments is discussed.

Sandstone has been a common building material in NE Ireland from medieval times to the present day. It is commonly used in medieval castles and ecclesiastical monuments, especially as a dressing stone. Studying and understanding the reasons why and the ways in which these stones decay is vital to the sustainable conservation of this cultural heritage, given that the decay of this historically and culturally significant stone is tantamount to the irrevocable loss of cultural heritage.

Inheritance is a key concept in stone decay studies. All that a stone experiences from the quarry to the present day will impact on its performance. This inheritance can reduce stone strength and compromise the ability of the stone to resolve stresses experienced in its lifetime (Warke 1996). Inheritance is especially important when considering ancient buildings; medieval buildings, for example, will experience many exceptional high-magnitude/low-frequency events as well as background low-magnitude/high-frequency stress factors that will determine how the stone decays through the centuries and how it performs in the present day (McCabe *et al.* 2007*a*). Present day manifestations of decay can therefore be conditioned by past stress events. Because of this it is likely that medieval sandstone will follow complex decay pathways based on the disturbing forces that have acted upon it over time and the response of the stone to those cumulative stresses.

This paper makes preliminary observations on cumulative stresses ('exceptional' and background factors) and their influence on the decay pathways of medieval sandstone both indirectly, through conditioning one or more of the other stress factors, and directly. By considering the interaction of the disturbing and resisting forces experienced by the stone since the construction of the friary, the decay seen on the stone in the present day can begin to be accounted for. Stone blocks can be viewed as environmental systems (Smith 1996) where complex feedback loops, some positive and some negative, act together to determine the state of the stone. If negative feedback is dominant, the stone will maintain a steady state and will exhibit characteristic form. However, if positive feedback predominates then destructive change is promoted and the stone exhibits a more transitory form (White *et al.* 1992). The paper is concerned with both the gradual accumulation of stresses and with extreme 'exceptional' events, but special emphasis is placed on the complex history of the stones (McCabe *et al.* 2007*a*; Smith & Přikryl 2007) where stress accumulation and strength decrease (increase in susceptibility to stresses) is neither smooth nor continuous. In this context, 'exceptional' factors can be defined as events that have the potential to cause an abrupt step change in the equilibrium of the stone system. This is the first of a series of papers that will seek to explore the decay of medieval sandstone monuments in the context of their sensitivity to change, that is, their ability or inability to absorb change in the environment.

From: PŘIKRYL, R. & TÖRÖK, Á. (eds) *Natural Stone Resources for Historical Monuments.* Geological Society, London, Special Publications, **333**, 87–100.
DOI: 10.1144/SP333.9 0305-8719/10/$15.00

Sandstones used at Bonamargy

Bonamargy Friary (Fig. 1) is a medieval ecclesiastical monument near the town of Ballycastle on the north coast of County Antrim, Northern Ireland. Present day climate is temperate maritime and relatively mild due the proximity to the Atlantic coast. Average July temperatures are 16 °C (maximum) and 9 °C (minimum), with 72 mm of rainfall. Average January temperatures are 6 °C (maximum) and 1 °C (minimum), with 114 mm rainfall. Average annual rainfall is 1076 mm. Climatic data were taken from the 1961–1990 base period. Three sandstone types have been identified in the structure of the Friary (Fair Head A, B and C). Each of the stones was quarried from Fair Head, a large and varied Carboniferous succession approximately 2 km NE of the Friary (Fig. 2).

Fig. 1. East gable wall of Bonamargy Friary.

The characteristics of the sandstones used at Bonamargy can be seen in Table 1. Fair Head A is light-brownish-grey in colour (Munsell 2.5Y 6/2). It is composed of coarse quartz grains and is very poorly sorted, with large pore spaces. Fair Head A is often severely weathered and grains disaggregate very easily. Fractured quartz grains can be seen under the microscope. Iron cementing is also visible under the microscope and severe iron staining is seen on the surface of many blocks of Fair Head A at the site. Fair Head B is yellow in colour (Munsell 2.5Y 7/6). The stone is regularly used as a dressing stone around windows and doorways in Bonamargy and at medieval sites along the north Antrim coast. It comprises mainly subrounded, moderately sorted, quartz grains in a carbonate cement. Iron spots are often visible on the surface of the stone. Fair Head C is brown/dark brown in colour (Munsell 7.5YR 4/4). This stone is much finer grained than the former two (quartz grains of less than 0.25 mm) and is very well sorted with grains very tightly packed.

History and inheritance

It is believed that Bonamargy was constructed in 1500 and originally used by the 3rd order Franciscans. Figure 3 shows an event timeline for Bonamargy. In January of 1584, a battle was fought between the English MacQuillans who had possession of the Friary and the Scottish MacDonalds. During the attack on the friary, the MacDonalds 'sett the roofe of the churche, being thatched, on fyer' (Bigger 1898, 15). Bonamargy was subsequently abandoned and unoccupied between

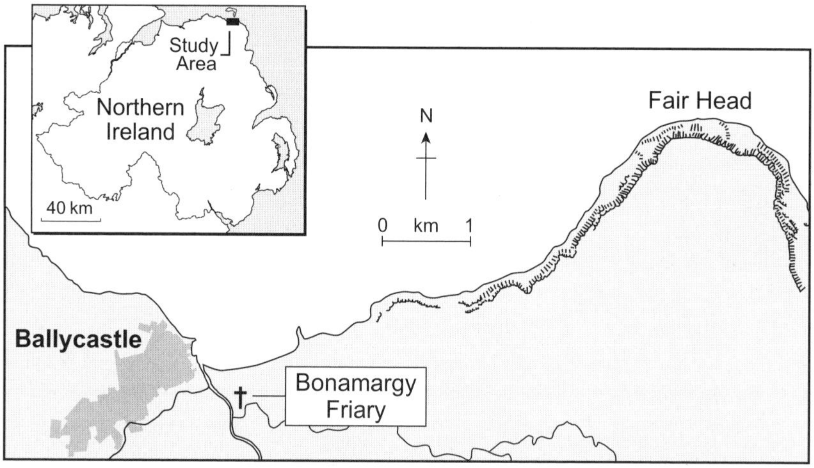

Fig. 2. Map of the study area.

Table 1. *Sandstones at Bonamargy Friary and their characteristics*

Name	Age	Colour	Mineralogy	Grain size	Water absorption (wt. %)
Fair Head A	Carboniferous	Light brownish grey (Munsell 2.5Y 6/2)	Quartz, iron cement, koalinite	0.5–2 mm; coarse grained, poorly sorted	9.3
Fair Head B	Carboniferous	Yellow (Munsell 2.5Y 7/6)	Quartz, iron cement, kaolinite	0.2–0.5 mm; fine grained, well sorted	8.0
Fair Head C	Carboniferous	Brown/dark brown (Munsell 7.5 Y 4/4)	Quartz	0. 2 mm; fine grained, very well sorted	4.8

1584 and 1621. In 1621, Randal MacDonald, the then Earl of Antrim, made the Friary habitable again. Bonamargy was given to the 1st order Franciscans in 1637, but by 1639 it was abandoned completely by the monks. There is some evidence that it was used as a local parish church in the 17th century (Bell & McNeill 2002).

Inheritance is important when thinking about the decay of stone over a long period of time. Background environmental stress factors and exceptional stress factors (generally related to human impact) are likely to have had a profound influence on the decay of the sandstone at Bonamargy Friary.

Background factors – their role in inheritance

Background factors and their possible long-term impacts on the decay of sandstone at Bonamargy Friary are outlined briefly in Table 2. These factors promote a gradual decline in the strength of the stone, and in the ability of the stone resolve stresses. As strength/stress thresholds are approached by the fatigue of the stone, the likelihood of stone surface failure and retreat is increased due to the occurrence of high magnitude 'exceptional' stress events, which have the capacity to dramatically breach thresholds, or the convergence of decreasing stone strength and increasing background stress levels (McCabe *et al.* 2007*a*).

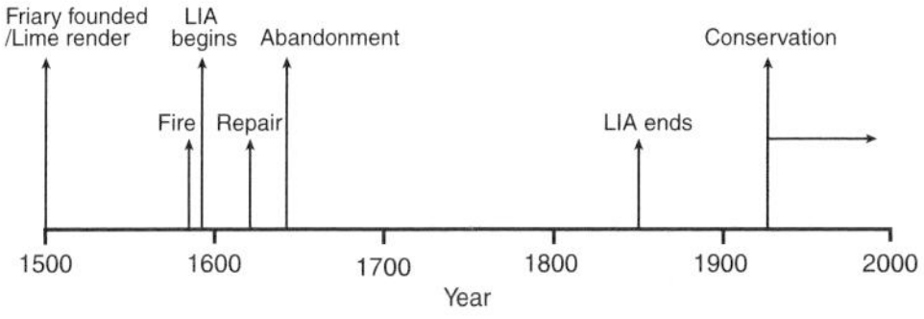

Fig. 3. Bonamargy Friary event timeline.

Salt accumulation and deposition

One of the most important factors contributing to the decay of the sandstone at Bonamargy is the presence of salts. There has been evidence of salt weathering reported in nearby coastal outcrops on Fair Head, especially by gypsum. The failure to identify halite may not reflect its absence, but rather highlight its transitory nature in a wet, maritime environment. This means it may not be readily identified by *ad hoc* surface sampling (McGreevy 1984).

Salt weathering is particularly effective in an environment that is rich in both salts and moisture. Moisture is essential because it provides a means of transport for salt throughout a porous material. The salt can then be deposited on stone surfaces in cracks and pores, with the potential to cause considerable damage. It follows that salt attack is often seen in coastal regions, where salt and moisture are abundant (Goudie & Viles 1997). Decay features frequently linked to the physical mechanisms of salt decay are common at Bonamargy. For example, granular disaggregation, surface flaking and scaling and alveolar (honeycomb) weathering are widely associated with the action of salts; crystallization and expansion and contraction of salts due to hydration and dehydration are common. It is likely that complex daily temperature and moisture cycles encourage expansion and contraction of salt crystals several times a day. This behaviour can cause grains to be pushed apart. It ultimately leads to granular disaggregation and flaking when salts are concentrated at the surface, or contour scaling when salts are present at depth (Smith & McGreevy 1988; Warke & Smith 2000) (Fig. 4).

Thermally assisted microfracturing

Although the role of thermal fatigue in microfracturing is contentious, it has been suggested that

Table 2. *Factors influencing decay at Bonamargy Friary (adapted from Warke 1996)*

Background factors	Significance
Salt accumulation and deposition	Interstitial deposition of salt contributes to stone decay through the mechanisms of intergranular crystallization, hydration/dehydration and thermal expansion/contraction.
Thermally assisted microfracturing	Possible differential thermal expansion of surface mineral grains and interstitial salt deposits in response to long- and short-term temperature fluctuations may eventually lead to microfracture development.
Frost-induced microfracturing	Repeated freezing of moisture in pore spaces and microfractures may eventually lead to shattering of stone and loss of material.
Chemical alteration	Dissolution of stone fabric alters surface pore dimensions and may facilitate the subsequent ingress of salt and moisture.
Sulphate crusts	Soiling of stone surface by the development of sulphate crusts can change albedo, increasing absorption of solar radiation and hence surface/subsurface temperature conditions. These crusts also act as salt reservoirs.
Surface induration/iron precipitation	Iron migration and precipitation can contribute to a decrease in substrate strength as material is leached out. Dark crusts can also affect albedo, impacting on the surface/subsurface temperature gradient.
Changes in surface morphology	Prolonged exposure to weathering processes leads to increased stone surface roughening allowing accumulation of moisture, salts and general particulate material which facilitate surface weathering processes.
Biological colonization	Can cause physical and chemical decay, often efficacious in synergy with other decay processes and mechanisms.
Post-emplacement 'exceptional' factors	Significance
Lime rendering (1500)	Although initially stabilizing the surface of the stone, lime rendering is likely to cause calcium loading of the sandstone, providing an internal source and supply of calcium.
Fire (1584)	Historical fire has long-term implications on stone strength and ability to resolve stresses.
The Little Ice Age (*c*. 1590–1850)	Increased frequency and intensity of extreme frost events.
Conservation/intervention (20th century)	Inappropriate interventions can accelerate decay. For example, the application of modern mortar on the chapel at Bonamargy has led to box-work.

differential thermal expansion of surface mineral grains and interstitial salt deposits in response to long- and short-term temperature fluctuations may eventually lead to microfracture development

Fig. 4. Flaking associated with salts.

(Warke 1996). Thermally induced breakdown of rocks is often associated with aggressive desert conditions and researchers have previously concentrated on the role of extreme diurnal temperature variability in weathering (Smith 1994; Smith *et al.* 2005; Hall 1999). Rock breakdown, however, is rarely the product of a single weathering mechanism. Rocks in the desert environment have a cumulative history of stresses that create weaknesses to be exploited by thermally induced expansion and contraction at the surface (Jenkins & Smith 1990). In the less aggressive Atlantic maritime environment of NE Ireland, the role of temperature in stone decay is less well understood. Even less extreme temperature cycling, however, is likely to encourage decay when juxtaposed with other processes and mechanisms of decay (Turkington & Paradise 2005). A stress caused by temperature changes at the rock surface 'should enhance fatigue failure and the operation of a number of mechanical weathering mechanisms' (Turkington & Paradise 2005).

Frost-induced microfracturing

Freeze-thaw weathering is a well-documented process of stone decay (Hall 1999). Repeated freezing of moisture in pore spaces and microfractures may eventually lead to the shattering of stone and loss of material; this will be most effective when acting synergistically with other processes and mechanisms. The importance of frost weathering in the decay of monuments is clearly related to environment: 'in some cold environments ... the role of thermal stress may predominate', (Hall 1999). While freeze-thaw may not be a dominant process in the current decay of Bonamargy, extreme events can occur in a comparatively mild maritime climate. In the harsher environment of the Little Ice Age (LIA, discussed further below) it is likely that freeze-thaw weathering could have exerted a more significant influence on the deterioration of the monument through increased frequency and intensity of freeze-thaw cycles as well as possible prolonged episodes with temperatures below the freezing point of water.

Chemical alteration

The dissolution of silica can be accomplished by alkali salt solutions, and is often common in coastal and saline environments (Goudie & Viles 1997). This can be significant if concentrated at grain contacts and can result in the loss or weakening of silica cements.

More evident in the stones of Bonamargy Friary, especially Fair Head A, is the selective mobilization and subsequent surface and near-surface precipitation of iron (and possibly manganese). Iron migration and its consequent impact on surface induration and subsurface weakening is an issue often neglected by stone decay researchers. When iron cement migrates from the substrate and precipitates in the near-surface zone, the interior of the stone can become significantly weakened. The surface becomes stabilized and hardened due to the formation of an iron crust (Fig. 5), but once this is breached rapid surface retreat may ensue (McAlister *et al.* 2003).

Soiling of the stone surface

Surface soiling has the potential to change albedo and hence increase absorption of solar radiation, influencing surface/subsurface temperature gradients and increasing stress in the material. Soiling can be caused by gypsum crusts (as well as the migration of minerals from the substrate to the surface), as well as biological growths. Gypsum crusts are frequently black in colour because of co-precipitation of particulate pollutants, and commonly comprise a mix of organic material (such as plant remains), inorganic particles (including partly burnt or un-burnt coal) and organic growths on the crusts. Gypsum acts as a binder to hold the crust together (Whalley *et al.* 1992). In this context it may be worth noting that there is a history of coal mining at Fair Head and this may very well have been used locally as a fuel source in addition to the more ubiquitous peat. These crusts can act as stores of salts that have the potential to be mobilized and cause significant damage to the stone.

Fig. 5. Iron precipitation on the stone surface.

Changes in surface morphology

Surface roughening allows the accumulation of moisture, salts and general particulate material which can facilitate surface weathering processes. Roughening of the surface increases the specific surface area exposed and hence susceptible to weathering. When surface crusts are breached, hollows (alveoli) can develop in the surface of the stone (Fig. 6) through processes not yet fully understood (Turkington & Paradise 2005). In this scenario, self-reinforcing decay can be triggered – it has been suggested that stone decay occurring in caverns may be encouraged by the particular microclimatic conditions influenced by the morphology of the cavern (Turkington *et al.* 2002).

Biological colonization

Biological action is a well-cited and widespread cause and form of decay. Physical and chemical decay can be caused by bacterial communities, algae and lichen, as well as higher plants (Chen *et al.* 2000). Biological decay is often efficacious in synergy with other decay processes and mechanisms, for example, the action of lichen and salt

Fig. 6. Alveolar weathering and iron staining.

associated with blistering and detachment of grains at Bonamargy Friary (McCabe *et al.* 2007*b*) (Fig. 7).

Exceptional factors: their role in inheritance

Pre-emplacement factors

Although not an 'exceptional' factor *per se*, the history of the stone prior to its use in the building may nonetheless exert a significant influence upon subsequent patterns of decay. This study is not seeking (ultimately it may not be possible) to isolate the pre-emplacement history of individual boulders and stone blocks. However, the factors that may influence later decay could include quarrying techniques, how long quarried stone was left to 'cure' before use and the methods employed to 'dress' the stone for specific uses. Added to these factors, there is the possibility that, in a post-glacial landscape, some of the material used may have been 'field stones' with complex and possibly multiple histories of natural excavation, transport, deposition and prolonged burial and/or exposure. For this study it is assumed that the pre-emplacement stress histories are broadly similar; possible differences will be explored in later papers.

Fig. 7. Blistering associated with the action of lichen and salts.

Lime render

When the Friary was founded, the walls were built and rendered with a lime mortar. This is still visible on some portions of the south wall of the church. It is possible that this use of lime mortar, while protecting the stone in the first instance, will have inadvertently loaded the stone with calcium. During the construction and rendering process, the sandstone blocks would have absorbed moisture from the render by capillary action. This process creates an internal source of calcium that allows the possible formation of calcareous salts in a non-calcareous stone through, for example, reaction with atmospheric sulphur. Lime used as a mortar between sandstone blocks has also been shown to significantly alter the physical characteristics of a stone, reducing the permeability of sandstone blocks and effectively sealing the mortared surfaces (McCabe *et al.* 2006; Smith *et al.* 2001).

Fire

Fire has long been recognized as an important agent of change in the natural environment (Dorn 2003). However, the effects of fire are not well understood and often unquantified, based on anecdotal evidence (Allison & Bristow 1999). The roof of the church at Bonamargy was burned down in 1584. The impact that fire can have on stone decay has been somewhat neglected by researchers, but fire is likely to be an important factor in the decay of many medieval structures. Physically, fire has the potential to subject building stones to extreme and sudden temperature changes. This is related to the generation of internal stresses in the material below stone surfaces exposed to fire, for example, the differential thermal expansion of surface grains. Chemically, extreme heat can cause changes in the mineral matrix or cement of a sandstone, leading to loss of material by disaggregation (Allison & Goudie 1994). According to Chakrabarti *et al.* (1996), heating sandstone to a temperature in excess of 573 °C will usually cause internal fracturing of quartz grains, weakening the structure of the stone and causing the surface to become friable.

It is likely that the impact of fire will not have been uniform across the monument. Some stone may never have acquired a 'memory', whereas other sandstone blocks may have lost the 'memory' of the fire through loss of affected material. Equally, other blocks may still harbour embedded stresses and weaknesses that have yet to be exploited by, for example, lower magnitude stresses associated with current environmental exposure (McCabe *et al.* 2007*c*). After simulating fire on sandstone in the laboratory, Allison & Goudie (1994) noted that there was a general weakening of grain boundaries, especially around the edges of the sample blocks. This suggests that the 'memory' of fire may not be held throughout the entire block. Further research is needed to determine how deeply the influence of fire penetrates a sandstone block. Fire should be considered not only as an agent of decay, but also as a pre-cursor to decay: the possible long-term impacts of historical fire need to be more fully understood.

Abandonment

We can be sure that the Franciscan monks would have maintained the Friary during periods of habitation and it is likely that their occupancy slowed decay. However, periods of abandonment (1584–1621, 1639–present) could have ultimately accelerated decay of the monument through associated biological colonizations and exposure to background environmental processes associated with exposure to enhanced environmental cycling for example, daily temperature and moisture regimes (causing heating and cooling and wetting and drying) and accumulation of marine salts in the stone.

Climate change – the Little Ice Age

External influences on stone decay systems never remain static and in addition to, for example, diurnal and seasonal environmental cycles, it is possible that local climate will vary over the lifetime of a building (Smith 1996). Indeed, in the case of medieval buildings and monuments, it is known that climate has varied significantly over the last 600 years. Different climatic regimes will cause different levels of stress to the stonework. In particular, the Little Ice Age (*c.* 1590–1850) could have had a profound impact on the possible decay pathways experienced by the stone types at Bonamargy Friary. Recent research has shown that the frequency and intensity of extreme frost events increased in central England during the Little Ice Age (Nesje & Olaf Dahl 2003) and, although there are no data relating to the north of Ireland, it is likely that temperatures were significantly lower and that freeze-thaw was a more important factor than under the current environmental regime. A high frequency of freeze-thaw events can exploit microfractures within the stone, applying pressure to vulnerable areas and thus propagating weaknesses in the substrate (Nicholson & Nicholson 2000). This exploitation of pre-existing weaknesses can lead to shattering of the stone and loss of surface material (Warke 1996).

Conservation intervention

Interventions by conservators can either slow decay or, all-to-often (after initially appearing to stabilize stone), accelerate decay. A visit to the site reveals evidence of what appears to be inappropriate intervention that has impacted on sandstone performance. One example of this is the use of modern mortar. Deployment of hard cement-based mortar on the softer sandstone walling has resulted in the creation of a box-work effect that is, the sandstone surface has retreated dramatically in a matter of decades, leaving the mortar to protrude from the façade (Fig. 8). The 'rigid' mortar has the effect of confining the sandstone blocks leading to relative compressive loading upon block expansion in response to weathering (swelling of blocks in response to salt weathering is described by Smith *et al.* 2008). The chapel at Bonamargy is dominated by this feature, causing a negative aesthetic impact as well as triggering stone surface retreat by back weathering through granular disaggregation.

Another example of inappropriate intervention is the use of iron grills by those with responsibility for care of the monument to block doorways and windows in the Friary. In 1932 Schaffer noted that the embedding of iron rods within stonework could be regarded as an error in craftsmanship (Schaffer 1932). Factors influencing decay at Bonamargy are summarized in Table 1.

Complex decay pathways

An important strategy for unravelling the complex decay of the medieval sandstones at Bonamargy

Fig. 8. Box-work caused by rigid mortar on sandstone walling.

Friary is to consider the possible decay pathways that the stones might have experienced. Conceptual models of decay pathways can be deciphered by considering environmental conditions, decay processes and stone response that is, disturbing and resisting forces in the stone decay system (Brunsden & Thornes 1979; Smith 1996). Both the disturbing and resisting forces acting on any object vary in time and space (Thomas 2001).

To enable this analysis at Bonamargy, it is assumed that each block in a façade can be largely viewed as a separate system, or perhaps subsystems with varying degrees of connectivity. This approach is facilitated by the observation that in lime-mortared structures, especially those in which internal drainage is allowed through the network of mortar joints, mortar between blocks effectively limits moisture and salt movement (Turkington & Smith 2004; Smith *et al.* 2001). The 'memory' of each block will vary – 'no two systems receive exactly the same number, sequence, frequency, duration and magnitude of events' (Brunsden 2001) – and each block will have a different threshold of change for each process acting on it. These small variations in 'memory' can encourage divergent behaviour in the different stone types and blocks (McCabe *et al.* 2007*a*; Brunsden 2001). Because of this there are many possible responses of the stone to any stress impulse; 'this is especially true if the factor causing change affects more than one process at different rates and with varying reaction and relaxation times' (Brunsden & Thornes 1979).

The reaction time of a system is the time that passes between the event inducing change and the beginning of the adjustment of the system. Relaxation time is the time taken by a system to adjust to a change in input and achieve a new equilibrium. When change is brought about it can be gradual or episodic and rapid according to the ratio of disturbing and resisting forces unique to a particular stone decay system. This underlines the inherent complexity of the stone decay system and the possible decay pathways of different stone types and different stone blocks.

Conceptual modelling of decay pathways, although subjective, can produce a profound understanding of stone decay with a careful weighing-up of the relevant factors (Smith 1996).

Stress factors

The effects of 'exceptional' factors that have applied stress to Bonamargy Friary over time are hypothesized in Figure 9. These factors are climate change, fire, lime rendering and inappropriate conservation. The importance of these events has been considered when dealing with the concept of inheritance (Table 2). In contrast, the effects of

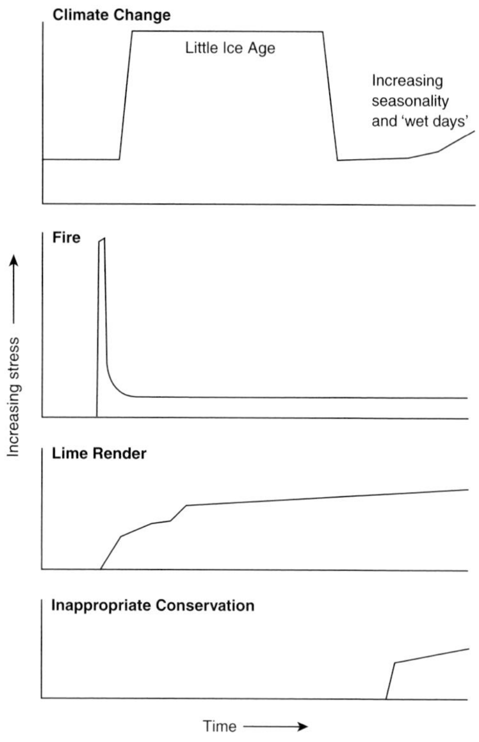

Fig. 9. Schematic diagram showing stress caused by 'exceptional' factors over time.

background stress factors, including daily temperature and moisture regimes (heating and cooling, wetting and drying and hydration and dehydration could be experienced more than once in a day) along with 'random catastrophic' events (e.g. severe frosts), are depicted in Figure 10. The dotted lines in the projections indicate the potential impact of 'exceptional' stress factors on the background stresses. In this context, each factor can influence stone decay directly or indirectly (through one or more of the other factors). For example, the Little Ice Age is likely to have increased the intensity and frequency of frost events and overall climate change impacts on daily temperature and moisture regimes, effectively increasing the frequency of 'random catastrophic events'. If they form, calcium salts have a greater potential to cause damage after the lime render has been removed from the sandstone walling and the stone has been loaded with calcium. Stress caused by biological growths has possibly been reduced by 'exceptional' factors such as the 1584 fire and the Little Ice Age. Confinement of blocks that on expansion leads to compressive loading of the stone will have been exaggerated by the presence

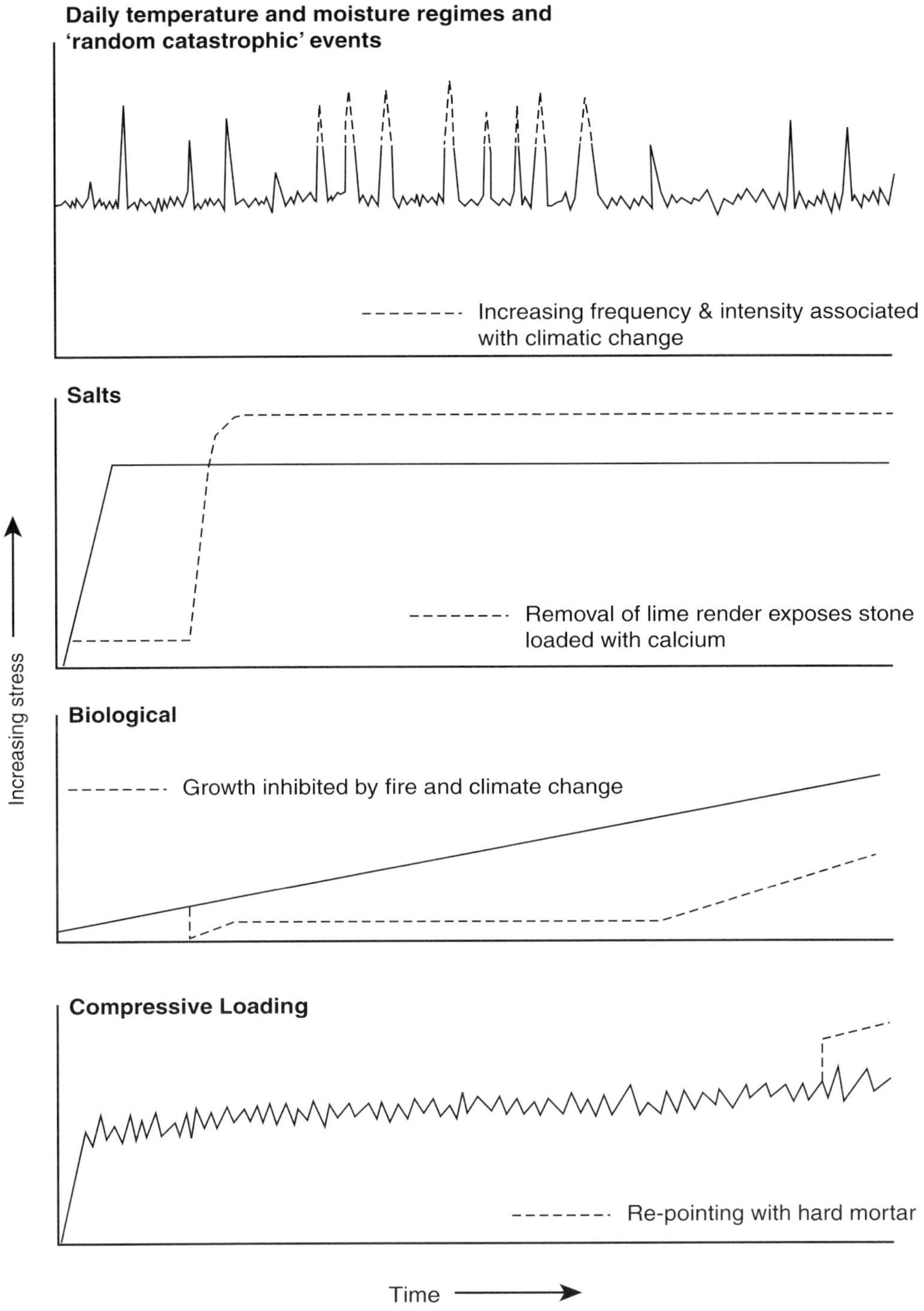

Fig. 10. Schematic diagram showing impact of 'exceptional' factors on stress caused by background factors.

of 'rigid' cement-based mortar on parts of the Friary walling.

Strength factors

Resistance to change comes from the properties and characteristics of the stone types. The properties of Fair Head A, B and C have been described above (summarized in Table 1). Fair Head C appears to have the highest resistance to, or ability to absorb, change; the surface of the stone is stable throughout Bonamargy. It is likely that Fair Head C will spend long periods of time exhibiting characteristic form (Brunsden & Thornes 1979). Fair Head A and B have a much greater propensity for change and exhibit transient behaviour. The stone types experience the same macro-environment, so the transitional forms in evidence in Fair Head A and B must be due to weaknesses or instabilities within the stone (Brunsden & Thornes 1979).

Stone response to stress

By combining projections of background and 'exceptional' stresses and changes in stone strength related to stone properties, possible decay pathways related to stone type can be identified. Figure 11 illustrates the possible responses of Fair Head A, B and C in terms of material loss from the surface over time; after the detachment of lime render, surface loss occurs. Fair Head A and B appear to exhibit significant material loss in the form of granular disaggregation, flaking and scaling. Fair Head A and B have iron in their cement. They often show iron spots or staining on the surface of the stone, which can result in iron crusts (described previously). The surface hardening and subsurface weakening can lead to rapid retreat of blocks once the crust detaches or is breached. The detachment and re-development of surface crusts is seen on Fair Head A and B, illustrated in Figure 11, and is akin to the detachment and re-growth of secondary and tertiary gypsum crusts seen on the Matthias Church, Budapest (Smith *et al.* 2003). Material loss is minimal from Fair Head C, with only some granular disaggregation present.

Figure 12 shows the response of the stone in terms of surface strength. The surface of fresh stone initially increases in strength. This is followed by a gradual decrease in strength until a threshold is reached and surface material is lost (Fig. 11). As thresholds are approached, the likelihood of failure is increased due to the superimposition either of exceptional stress factors or a gradual convergence of decreasing stone strength and increasing background stress levels. After each event when detachment of the surface layer takes place, the stone can behave in several ways (numbers relate to Fig. 12).

1. The surface layer detaches, exposing 'fresh' stone and thus 'resetting the clock to zero'. The process of surface hardening begins again.
2. If the detached crust was composed of iron leached from the substrate, then the substrate will have been significantly weakened and the stone strength will decrease sharply.
3. As surface hardening begins again, stability of the stone will not be attained if the frequency of high magnitude events, sufficient to cause material loss, is greater than the relaxation time of the system (the time taken for the stone to adjust to the change and achieve equilibrium).

Because of their iron content, Fair Head A and B are more likely to follow path 2 in Figure 12. It is likely that path 3 would have been common during the LIA, when high magnitude frost events were more common than in the present day climate. The diagram illustrates how the likelihood of stone failure is increased when paths 2 and 3 are followed: stone strength decreases so that even small stress events may cause failure. These paths encourage a positive feedback loop, promoting destructive change (Chorley & Kennedy 1971).

It may be possible that we are only seeing a period of relative quiescence in the response of Fair Head C to its environment: 'if a harsh regime precedes a gentle process domain, then all the possible work may have been done and no change is possible until there is a further change in the controls' (Brunsden 2001). The LIA may have accomplished so much 'work' that the present day environment rarely produces an event large enough to cause any change in form. In this situation only events of very high magnitude will be effective in causing stone failure and change is likely to be very episodic.

Conclusions and the way forward

This study emphasizes the importance of considering the cumulative impact of stresses on stone response and establishes hypotheses to be tested by future detailed study and analysis (especially the spatial mapping of stone response). A conceptual framework in which to view the decay of sandstone monuments is proposed, and draws heavily on previous geomorphological research.

'Exceptional' events impact not only the sandstones but also the other background stress factors. Understanding how these disturbing factors affect each other and how they, in turn, impact on the stone is essential to understanding and conceptualizing the complex decay pathways of the different

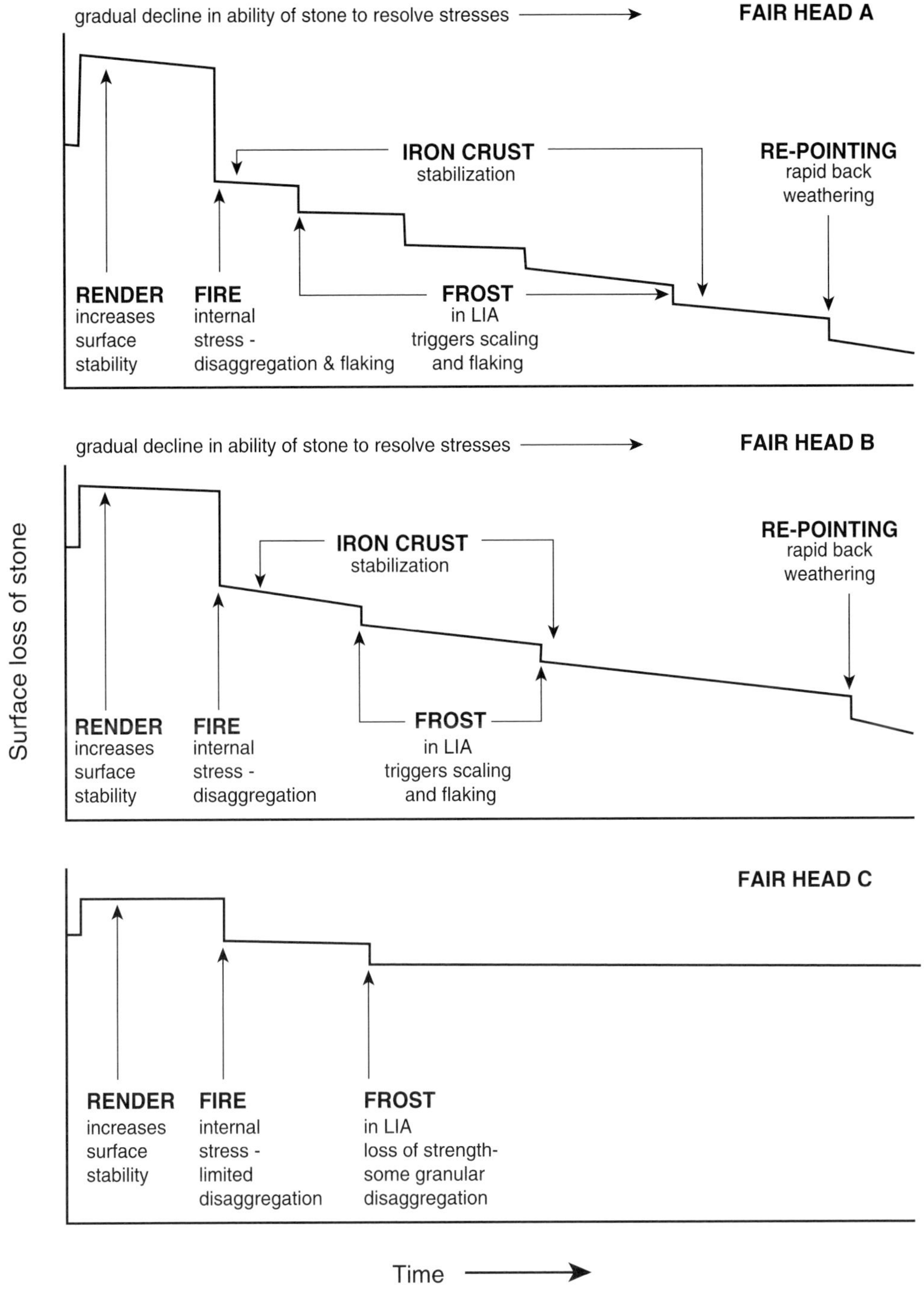

Fig. 11. Schematic diagram showing stone response: material loss.

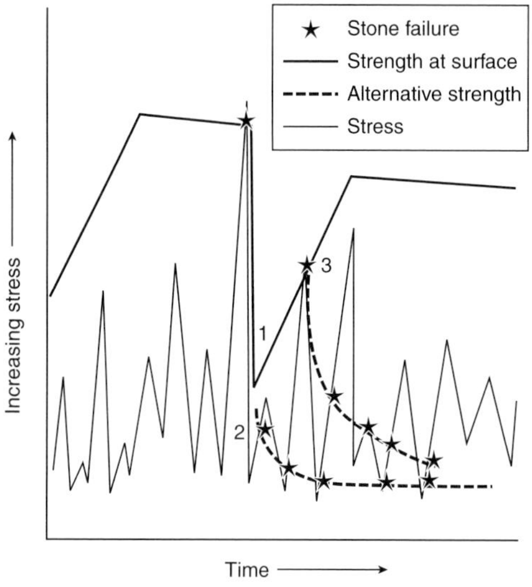

Fig. 12. Schematic diagram showing stone response: strength at the surface.

stone types at Bonamargy Friary. This enables us to take steps towards identifying an event sequence at Bonamargy; 'an event sequence is the combination of events at any frequency, magnitude, and duration, which occurs at a place and which achieves a recognizable effect as a sequence ... the system is regarded as having a memory of effective events' (Brunsden 2001).

Stone response has often been linked to an individual stress event or factor; identifying event sequences, the cumulative impact of stress events over time on the stone decay system (that reflects the reality of exposing stone to a complex range of stresses experienced within a building), is a challenge for future research. To this challenge is added the consideration both of future extreme events and changes in environmental conditions, especially climate change. Climate change scenarios for the northwest UK show a trend towards wetter, warmer and longer winters. It is likely that this will have an impact on the ways in which sandstone monuments decay in the northeast of Ireland. With warmer and wetter conditions, sandstones will be subject to an increased 'time-of-wetness' (Smith *et al.* 2004).

This could mean deeper moisture penetration into blocks for longer periods and thus a greater potential for salts to migrate through entire blocks, encouraging the rapid salt-fuelled retreat of block surfaces. If blocks become completely saturated there is the potential for salts to be transported to depth by ionic diffusion.

Prolonged wetting is also likely to promote surface and subsurface biological growth. Algal growths can already been seen on the sandstone at Bonamargy Friary – these are moist and spongy in the winter and dry to crusts in the summer months. 'Greening' of monuments may become more widespread (Fig. 13), causing aesthetic problems as well as having the effect of retaining any moisture that has penetrated into the stone (Smith *et al.* 2003). This moisture retention may create a positive feedback loop in which the surface and subsurface of sandstone blocks are kept damp for long periods of time, with implications for other processes including salt migration and chemical alteration. The impact of future climate change and 'time-of-wetness' is to be explored further in an EPSRC-funded project ('Climate change, "greening" of masonry and implications for built heritage').

Fig. 13. 'Greening' of stonework at Bonamargy during the winter.

Considering the stress history of Bonamargy, (the background and 'exceptional' factors that have acted to change the stone over the centuries) and determining conceptual decay pathways for the response of different sandstone types in the context of their sensitivity to change has proved to be a useful approach in understanding and accounting for the decay seen on the stone in the present day. This paper did not set out to establish a particular decay pathway applicable to all buildings. In reality, all buildings are different and have different exposure regimes and stress histories. However, the ideas discussed provide a conceptual framework by which complex decay histories can be rationalized. In essence, it encourages conservators to ask the relevant questions and moves beyond uniformitarian views of stone decay expressed in terms of rates of decay.

This research was funded by EPSRC grant EP/G01051X/1 and the Department of Employment and Learning NI (DEL). We are grateful to the Environment and Heritage Service (EHS) for access to monuments. Thanks are also due to Gill Alexander (School of Geography, QUB) for diagrams and advice. Climate data was supplied by the British Atmospheric Data Centre (BADC).

References

ALLISON, R. J. & GOUDIE, A. S. 1994. The effects of fire on rock weathering – an experimental study. *In*: ROBINSON, D. A. & WILLIAMS, R. B. G. (eds) *Rock Weathering and Landform Evolution*. John Wiley & Sons Ltd, Chichester, 41–56.

ALLISON, R. J. & BRISTOW, G. E. 1999. The effects of fire on rock weathering: some further considerations of laboratory experimental simulation. *Earth Surface Processes and Landforms*, **24**, 707–713.

BELL, J. L. & MCNEILL, T. 2002. Bonamargy Friary, County Antrim. *Ulster Journal of Archaeology*, **61**, 98–116.

BIGGER, J. F. 1898. Friary of Bun-na-margie. *Ulster Journal of Archaeology*, Special Volume, Belfast.

BRUNSDEN, D. 2001. A critical assessment of the sensitivity concept in geomorphology. *Catena*, **42**, 99–123.

BRUNSDEN, D. & THORNES, J. B. 1979. Landscape sensitivity and change. *Transactions of the Institute of British Geographers*, **4**, 463–484.

CHAKRABATI, B., YATES, T. & LEWRY, A. 1996. Effect of fire damage on natural stonework in buildings. *Construction and Building Materials*, **10**, 539–544.

CHEN, J., BLUME, H. & BEYER, L. 2000. Weathering of rocks by lichen colonisation – a review. *Catena*, **39**, 121–146.

CHORLEY, J. & KENNEDY, B. A. 1971. *Physical Geography: A Systems Approach*. Prentice Hall International Inc., London.

DORN, R. I. 2003. Boulder weathering and erosion associated with wildfire, Sierra Ancha Mountains, Arizona. *Geomorphology*, **55**, 155–171.

GOUDIE, A. S. & VILES, H. A. 1997. *Salt Weathering Hazards*. John Wiley and Sons, Chichester.

HALL, K. 1999. The role of thermal stress fatigue in the breakdown of rock in cold regions. *Gemorphology*, **31**, 47–63.

JENKINS, K. A. & SMITH, B. J. 1990. Daytime rock surface temperature variability and its implications for mechanical rock weathering: Tenerife, Canary Islands. *Catena*, **17**, 449–459.

MCALISTER, J. J., SMITH, B. J. & CURRAN, J. A. 2003. The use of sequential extraction to examine iron and trace metal mobilisation and the case-hardening of building sandstone: a preliminary investigation. *Microchemical Journal*, **74**, 5–18.

MCCABE, S., SMITH, B. J. & WARKE, P. A. 2006. Calcium loading of building sandstones by lime rendering: implications for decay. *In*: FORT, R., ALVAREZ DE BUERGO, M., GOMEZ-HERAS, M. & VAZQUEZ-CALVO, C. (eds) *Heritage, Weathering and Conservation*. Taylor & Francis Group, London, 177–182.

MCCABE, S., SMITH, B. J. & WARKE, P. A. 2007*a*. Preliminary observations on the impact of complex stress histories on the response of sandstone to salt weathering: laboratory simulations of process combinations. *Environmental Geology*, **52**, 269–276.

MCCABE, S., SMITH, B. J. & WARKE, P. A. 2007*b*. An holistic approach to the assessment of stone decay: Bonamargy Friary, Northern Ireland. *In*: PŘIKRYL, R. & SMITH, B. J. (eds) *Building Stone Decay: From Diagnosis to Conservation*. Geological Society, London, Special Publications, **271**, 77–86.

MCCABE, S., SMITH, B. J. & WARKE, P. A. 2007*c*. Sandstone response to salt weathering following simulated fire damage: a comparison of the effects of furnace heating and fire. *Earth Surface Processes and Landforms*, **32**, 1874–1883.

MCGREEVY, J. P. 1984. A preliminary scanning electron microscope study of honeycomb weathering of sandstone in a coastal environment. *Earth Surface Processes and Landforms*, **10**, 509–518.

NESJE, A. & OLAF DAHL, S. 2003. The 'Little Ice Age' – only temperature? *The Holocene*, **13**, 139–145.

NICHOLSON, D. T. & NICHOLSON, F. H. 2000. Physical deterioration of sedimentary rocks subjected to experimental freeze-thaw weathering. *Earth Surface Processes and Landforms*, **25**, 1295–1307.

SCHAFFER, R. J. 1932. *The Weathering of Natural Building Stones*. Donhead facsimilie, 2004.

SMITH, B. J. 1994. Weathering processes and forms. *In*: ABRAHAMS, A. D. & PARSONS, A. J. (eds) *Geomorphology of Desert Environments*. Chapman & Hall, London, 39–63.

SMITH, B. J. 1996. Scale problems in the interpretation of urban stone decay. *In*: SMITH, B. J. & WARKE, P. A. (eds) *Processes of Urban Stone Decay*. Donhead, Shaftesbury, 3–18.

SMITH, B. J. & MCGREEVY, J. P. 1988. Contour scaling of a sandstone by salt weathering under simulated hot desert environments. *Earth Surface Processes and Landforms*, **13**, 697–706.

SMITH, B. J., TURKINGTON, A. V. & CURRAN, J. M. 2001. Calcium loading of quartz sandstones during construction: implications for future decay. *Earth Surface Processes and Landforms*, **26**, 877–883.

SMITH, B. J., TÖRÖK, A., MCALISTER, J. J. & MEGARRY, Y. 2003. Observations on the factors influencing stability of building stones following contour scaling: a case study of oolitic limestones from Budapest, Hungary. *Building and Environment*, **38**, 1173–1183.

SMITH, B. J., WARKE, P. A. & CURRAN, J. M. 2004. Implications of climate change and increased 'time-of-wetness' for the soiling and decay of sandstone structures in Belfast, Northern Ireland. *In*: PŘIKRYL, R. (ed.) *Dimension Stone*. Taylor & Francis Group, London, 9–14.

SMITH, B. J., WARKE, P. A., MCGREEVY, J. P. & KANE, H. L. 2005. Salt-weathering simulations under hot desert conditions: agents of enlightenment or perpetuators of preconceptions? *Geomorphology*, **67**, 211–227.

SMITH, B. J. & PŘIKRYL, R. 2007. Diagnosing decay: the value of medical analogy in understanding the weathering of building stones. *In*: PŘIKRYL, R. & SMITH, B. J. (eds) *Building Stone Decay: From Diagnosis to Conservation*. Geological Society, London, Special Publications, **271**, 1–8.

SMITH, B. J., GOMEZ-HERAS, M., MENEELY, J., MCCABE, S. & VILES, H. A. 2008. High resolution monitoring of surface morphological change of building limestone in response to simulated salt weathering. *In*: LUKASZEWICZ, J. W. & NIEMCEWIZC, P. (eds) *11th International Congress on the Deterioration and Conservation of Stone*. Nicolaus Copernicus University Press, Torun, 275–282.

THOMAS, M. F. 2001. Landscape sensitivity in time and space – an introduction. *Catena*, **42**, 83–98.

TURKINGTON, A. V., SMITH, B. J. & BASHEER, P. A. M. 2002. The effect of block retreat on subsurface temperature and moisture conditions in sandstone. *In*: PŘIKRYL, R. & VILES, H. A. (eds) *Understanding and Managing Stone Decay*. The Karolinum Press, Prague, 113–126.

TURKINGTON, A. V. & SMITH, B. J. 2004. Interpreting spatial complexity of decay features on a sandstone wall: St. Matthew's Church, Belfast. *In*: SMITH, B. J. & TURKINGTON, A. V. (eds) *Stone Decay: Its Causes and Controls*. Donhead, Shaftesbury, 149–166.

TURKINGTON, A. V. & PARADISE, T. R. 2005. Sandstone weathering: a century of research and innovation. *Geomorphology*, **67**, 229–253.

WARKE, P. A. 1996. Inheritance effects in building stone decay. *In*: SMITH, B. J. & WARKE, P. A. (eds) *Processes of Urban Stone Decay*. Donhead, Shaftesbury, 32–43.

WARKE, P. A. & SMITH, B. J. 2000. Observations of three-dimensional salt distribution in building sandstone. *Earth Surface Processes and Landforms*, **25**, 1317–1332.

WHALLEY, B., SMITH, B. & MAGEE, R. 1992. Effects of particulate air pollutants on materials: investigation of surface crust formation. *In*: WEBSTER, R. G. M. (ed.) *Stone Cleaning and the Nature, Soiling and Decay Mechanisms of Stone*. Donhead, Shaftesbury, 227–234.

WHITE, I. D., MOTTERSHEAD, D. N. & HARRISON, S. J. 1992. *Environmental Systems: An Introductory Text*. London, Chapman & Hall.

Oxford stone revisited: causes and consequences of diversity in building limestone used in the historic centre of Oxford, England

MIGUEL GOMEZ-HERAS[1]*, BERNARD J. SMITH[1] & HEATHER A. VILES[2]

[1]*School of Geography, Archaeology and Palaeoecology, Queen's University Belfast BT7 1NN, UK*

[2]*Oxford University Centre for the Environment, South Parks Road, Oxford OX1 3QY, UK*

**Corresponding author (e-mail: m.gomez@qub.ac.uk)*

Abstract: Stone decay is the result of the interaction of stone with its environment. It is therefore important to understand why certain materials, sometimes not the most suitable, were used to shape the built heritage of specific areas. The historical evolution of these areas conditioned many of the combinations of materials we see today, which in some cases can interact to accelerate decay. These combinations were driven by availability during construction, architectural fashion or the simultaneous utilization of materials that are aesthetically similar but differ significantly in their physical and chemical properties. A microcosm of the complex decisions that determine stone selection and subsequent interactions is provided by the City of Oxford, which is an excellent example of how such historic evolution can work with material characteristics to accelerate decay.

There is a wide literature on how explanations of urban stone decay require an understanding of the interactions between the stone – conditioned by a range of physical, petrological and chemical properties – and the often complex weathering environments in which it is placed (e.g. McCabe *et al.* 2007; Smith *et al.* 2005, 2008 for recent overviews on the subject). The environments combine to cause decay through their control of a range of physical and chemical decay processes. From an architectural standpoint, however, it is evident that stone type (and often quite subtle variations within what is perceived as one stone type) exerts a major control on the long-term aesthetic and durability of buildings. Consideration of the physical environment, which can be more or less aggressive to buildings, must therefore be viewed in conjunction with stone and the major controls that it exerts in its own right, not just on its immediate visual impact, but also on long-term change.

In addition, one has to consider the 'historical environment' whereby stone provides the personality of both individual buildings and the vernacular architecture by which towns, cities and regions are identified within the popular psyche. Stone therefore provides personality to heritage and leads to common associations between stone type and traditional architecture. Within the British Isles, one of the prime examples of this relationship is found in the 'honey-coloured' limestone buildings of the Cotswold Hills. This association has been a major driver of tourism associated with the 'sense of place' that it imbues. Within regions such as the Cotswolds, however, the apparent uniformity of the vernacular stone masks considerable variation in stone characteristics other than colour. This results in a diversity of performance and requires, in theory at least, careful choice of stone by architects to meet, for example, specific structural needs. The choice of stone is therefore a crucial decision for any architect, builder or building owner. However, decisions as to which stone is used are rarely, in practice, an objective exercise based upon the matching of desired characteristics with the properties of the most appropriate stone. Instead this choice is mediated through a range of cultural, economic and political constraints that are, in turn, superimposed upon the personal preferences of architect, builders and owners, often based on previous experience of use.

When it comes to analysing the choice of stone used in historic buildings, additional factors such as consideration of the historically available sources of material, which are strongly conditioned by the geological setting of the surrounding area, are often ignored or neglected. The physical availability of stone, and everything that relates to it, is then factored through the evolution of architectural criteria for selection of the stones (e.g. changes in durability testing and limits) and changes in transport technologies (e.g. from horse-drawn carts to canal barge to steam train) that made different stone types progressively more widely available. As a result of this overlaying of factors that have influenced choice of stone, the built environment often shows a variability of materials that can be placed together

From: PŘIKRYL, R. & TÖRÖK, Á. (eds) *Natural Stone Resources for Historical Monuments*. Geological Society, London, Special Publications, **333**, 101–110.
DOI: 10.1144/SP333.10 0305-8719/10/$15.00

in a seemingly non-structured fashion. The importance of understanding these apparently illogical combinations is significant because the superimposition and interdigitation of physically and chemically incompatible stone types can accelerate and, in some instances, trigger completely 'unanticipated' forms of stone decay.

Only by understanding these factors is it possible to begin to understand the often complex mixture of stone types found within historical buildings. Once this understanding is achieved, it may then be possible to fully appreciate the past and present environmental controls on stone decay and begin the process of identifying appropriate conservation interventions.

To address these issues, this paper explores a number of the key factors that have conditioned the complex arrangements of materials presently found in the collegiate buildings of Oxford. Despite a perceived lack of change, there have been many different types of stone used over the centuries for both construction and replacement. It is also evident that these complex combinations are one of the key factors that have contributed to the accelerated decay of many buildings.

Geological setting and stone types

Oxford lies in the centre of the Jurassic formations that crop out SW–NE across England from the southern coast of Devon and Dorset through to Yorkshire. Geomorphologically, the Mesozoic in southern England forms a series of clay vales and SW–NE trending limestone ridges from the Midland Trias plains in the northwest to the Cretaceous chalk of the Chiltern in the southeast. The succession, and therefore the associated limestone ridges, gently dips towards the SE and forms successive cuestas with gentle dip-slopes to the SE and steeper scarps to the NW. Oxford itself lies in the Oxford clay vale and is flanked by the Cretaceous outcrops of the Chiltern chalk to the southeast and the Jurassic Cotswold limestones to the north and west.

These limestone formations are the most immediate and obvious source for building stone in the area. The building stones used in Oxford have been generally labelled 'oolitic limestone'. This broad categorization does not, however, take into account the great variety of different lithofacies associated with a shallow epicontinental sea deposition regime with episodes of subaerial, deltaic and estuarine deposition.

Building limestones used in Oxford belong mainly to the Middle Jurassic Inferior Oolite group (Aalenian–Bajocian) and Great Oolite group (Bathonian) and the Late Jurassic Corallian Group (Oxfordian) and Portland and Purbeck formations (Portlandian), which are largely sourced from three regions (Fig. 1). The first of these (a) and historically the most important is Oxfordshire

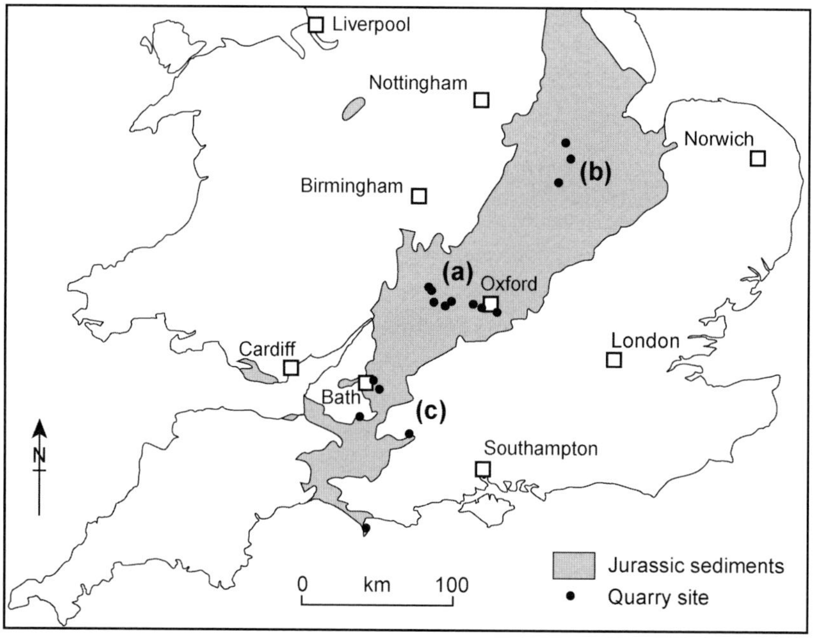

Fig. 1. Map of south Britain showing some of the main quarries which have supplied stone to Oxford.

and Gloucester (including the immediate surroundings of Oxford and the Cotswold hills), where stones from different formations of the Inferior and Great Oolite and the Corallian Group have supplied stone to Oxford for more than 800 years. This area (Fig. 2) includes two main sources for stone: the first corresponds to locally extracted stone from the outskirts of the original city, which have now been overtaken by urban growth. These quarries exploited material from the Corallian group including Headington, one of the most iconic building stones in Oxford. The second area close to Oxford corresponds to the Cotswold Hills, in a band that runs mainly from Oxford through the Windrush and Evenlode valleys to a maximum distance of 50 km. This area has been the most constant supplier of stone to Oxford right up until the present day. The quarries located here extract materials from several units of the Inferior and Great Oolite, and show the greatest lithological variability. Taynton (Middle Bathonian) is the most widely used of these Cotswold stones, even beyond the confines of Oxford in, for example, Windsor Castle.

The second region (b), to the northeast of Oxford, comprises the counties Rutland and North Northamptonshire. Within this region, Clipsham, which is from the Inferior Oolite, is the best-known stone. Finally, in the South of England from Bath to Portland (c), stones mainly originate from the Bath Oolite Member (Great Oolite) and the Portland formation.

The stratigraphy of the deposits is quite complicated, particularly in the case of the Inferior and Great Oolite groups, with rapid variations in lithology and thickness which are related to changing

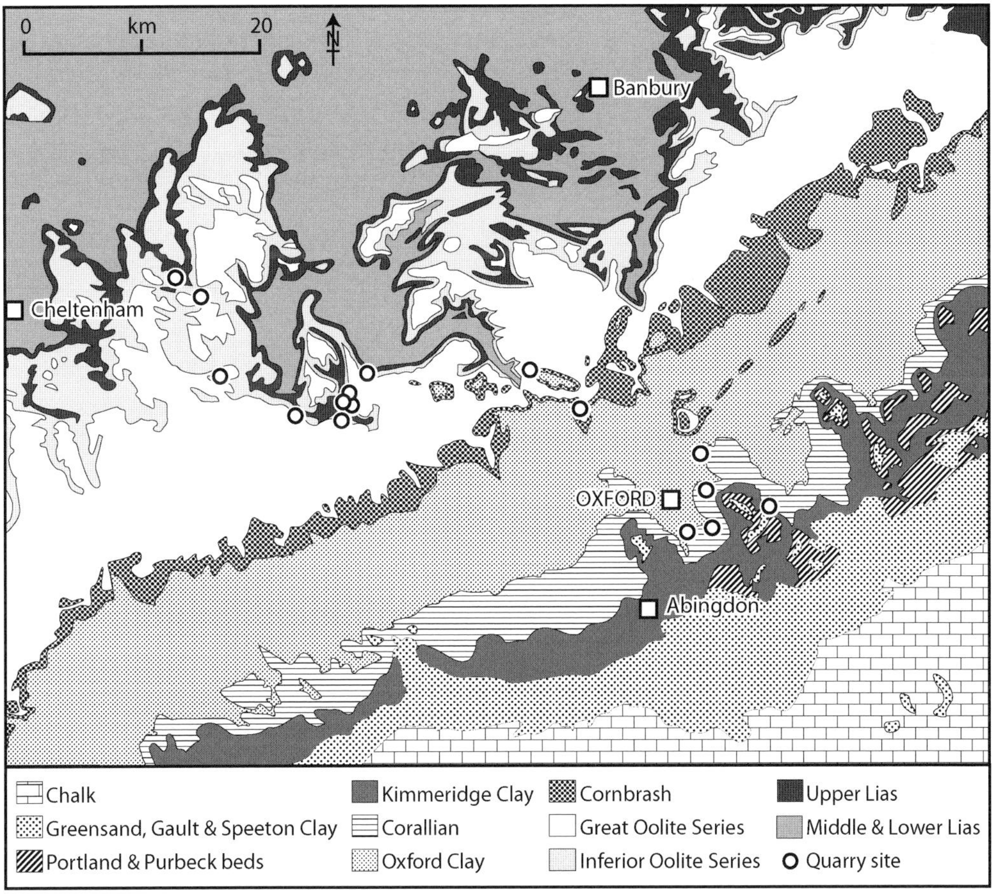

Fig. 2. Detailed geological map of the Oxford area showing the mid-Jurassic (from bottom to top: Inferior Oolite, Great Oolite, Cornbrash and Oxford Clay) and late Jurassic (from bottom to top: Corallian, Kimmeridge Clay and Portland & Purbeck beds) sediments and the location of the main quarries from which Oxford building limestones were sourced (cartographic base after Geological Survey of Great Britain 1957).

environments of deposition across the region (Sumbler 1996). This has an influence on the type of quarries found in this area, as they are mostly of small to medium size and exploit beds of a few metres thickness (Fig. 3) which in many cases cannot guarantee a continuous supply for many years. There is also the question of the petrological variability within the quarries. These limestones are prone to heterogeneities and frequently there is the presence of abundant terrigenous content and/or intercalated clay beds, which lead to considerable structural and mineralogical variability even in individual stone blocks within buildings.

These regional and local lithological variations gave rise to a wealth of local names related to the great number of quarries that provided stones to Oxford. The multiplicity of quarries and names complicates stone identification as some of the different names refer to the same lithostratigraphic units accessed at different locations. Conversely, stone of the same name can encompass a range of petrologies. Indeed, it has been known for the

Fig. 3. Typical exposure of quarry bank in Cotswold limestone illustrating typically thinly bedded and fragmented structure within a narrow seam. Height of exposure *c*. 4 m.

name of one stone (e.g. Milton) to be changed to another (e.g. Taynton) purely as a marketing ploy.

The evolution of use

Any study of 'Oxford Stone' has to acknowledge the encyclopaedic study by Arkell (1947). This study draws heavily on this source in reconling the change in stone use over time.

Figure 4, largely obtained from an analysis of Arkell's data on the use of different stone types, shows a timeline for the main stone types employed in Oxford. The first documented sources of stone used in Oxford were the quarries situated towards the east of the city that extracted material from coral reefs of the Mid-Oxfordian Upper Corallian Group (Coral Rag). These materials were used to build rubble walls and can be seen, for example, in the tower of St Michaels at the North Gate in Cornmarket Street, which is the oldest architectural feature remaining in Oxford and dates from 1040.

Associated with the coral reefs in growth position are bedded detrital deposits in the gaps between them. These form the so-called Wheatley limestone. During the 13th and 14th centuries, this was the main building stone used. Although there is some evidence for the use of Wheatley as ashlar work (Arkell 1947), this stone is relatively hard. Freestone for dressings and sculptural details was obtained from other sources such as Taynton and other quarries near Burford, a little farther away in the Cotswold area.

Cotswold stones were the preferred freestones during most of the 14th and 15th centuries. Taynton is the most renowned name within the Cotswold area, and even material from other quarries was sometimes sold as Taynton, for example Clypeus Grit and White Limestone (Arkell 1947).

Taynton quarries are first mentioned in 1086, although they are thought to have been used from Roman times (Arkell 1947). Some buildings from the 13th century in Oxford, such as Christ Church Cathedral, are thought to have been built with Taynton stone. However, the first documented use of this stone in Oxford refers to the Merton Rolls (1310) (Arkell 1947).

Headington, which is also part of the Corallian Group, is highy variable between beds and the so-called Headington hard stone took the place of Wheatley in the 15th century. In the late 16th and 17th century, Headington Freestone began to be increasingly used for large-scale projects.

By the end of 18th century, substitutes for Headington freestone were sought. This marked the arrival of stone from new supply areas outside of the Oxford district. Bath stone, the best of which was found near Bath at Box, appeared in Oxford around 1820. This was by far the most widely used stone during the Gothic Revival period and throughout the 19th century, with its heyday from 1850 to 1870.

Portland stone, although very popular in other parts of the country, was not widely used in Oxford; it was used only in very specific buildings such as the Ashmolean Museum.

Towards the end of the Gothic Revival, architects returned to using local materials and Taynton and Milton stone were again used more widely during the last quarter of the 19th century.

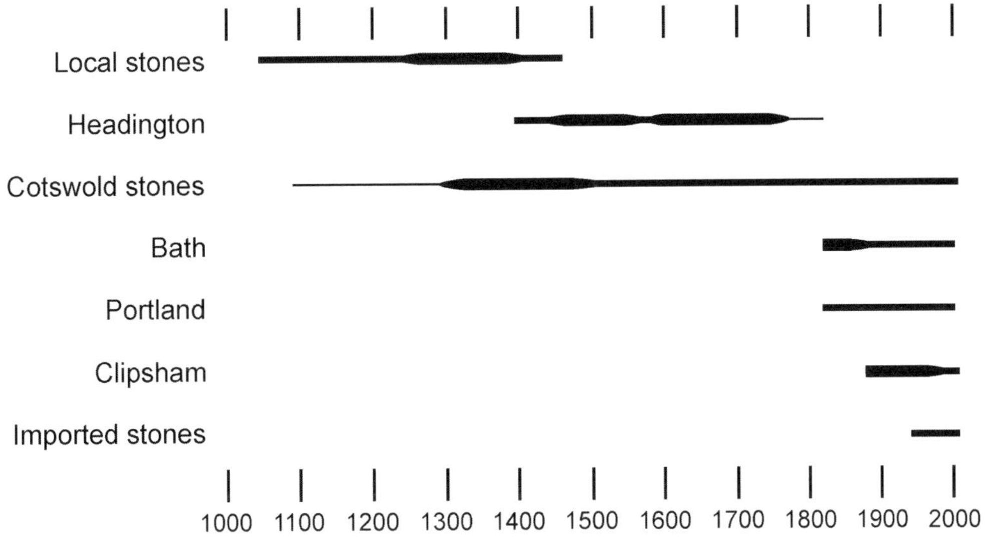

Fig. 4. Timelines for the use of different stone types in Oxford.

Clipsham (Upper Inferior Oolite) was introduced by Sir Thomas Jackson in 1876–78 for constructing the examination schools and since then it has been used extensively. Its use was especially prominent for the refacing of many buildings during the extensive campaigns of restoration of the middle-to-late 20th century.

In response to a more globalized market for stone, many other new stone types (including stones from more distant parts of the country or overseas) appear in the buildings of Oxford from one middle of the 20th century.

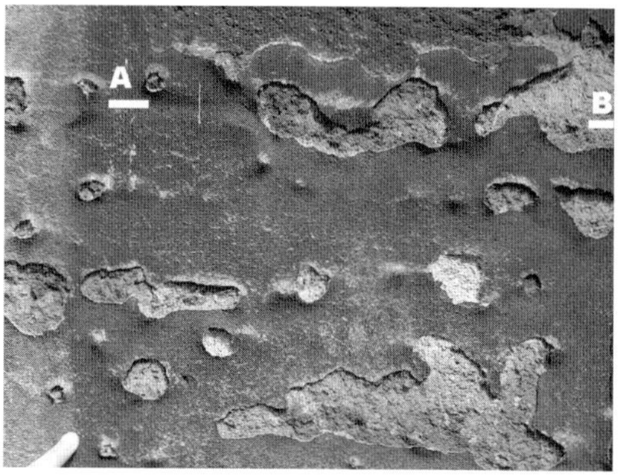

Fig. 5. Stone blocks separated by very fine joints, with very little intervening mortar, exhibiting patterns of surface spalling that cross to adjacent blocks almost as if the joint did not exist (e.g. A–B on photograph).

The causes of change

The evolution of stone types used in Oxford is a response to several factors, the convergence of which resulted in the complex arrangements of materials observed in many of its buildings.

Architectural style and the architect's choice

There have been a number of times when the ornamental needs of particular styles have dictated the type of stone to be used. For example, as the prestige of Oxford grew and the university became increasingly important (University College was created in 1249), larger building projects began to appear and the Coral Rag was regarded as unsuitable. This marks the change from rubble to ashlar construction. Quarries which could supply increasing quantities of suitable material, leading to the advent of Cotswold stones such as Taynton, were sought.

During the late 16th century and 17th centuries, Renaissance projects required freestone for entire walls. Taynton could not meet these demands, so Headington freestone began to be used on a large scale. Interestingly, architects using Headington freestone viewed this as the discovery of a new source of stone, either neglecting to remember or failing to realize that it was in fact the rediscovery of an old source of stone that had previously been abandoned.

In many respects, this marked the introduction of stone types that did not perform very well, especially under conditions of increasing atmospheric pollution, most notably from the myriad of coal fires typical of the Oxford colleges (Viles 1993), and the beginning of extensive decay that often required subsequent replacement of complete façades.

Another fine example, in this case dominated by the requirements of construction technology rather than preference for a particular stone type, is the Gothic Revival style. This was the beginning of the use of Bath stone; the large size and ease of working large blocks of stone led to 'the indulgence of fashionable fancies in masonry, such as extremely fine joints' (Arkell 1947). These fine joints were not mortared and the masonry would be fortified by iron clamps which, as can be seen in the present day (Fig. 5), led to problems of 'contagion' of decay across the joints. This process was assisted where surface crusts had also developed across the joints.

The industrial revolution: rise of Victorianism and changing transport technologies

When several stone types have been available in an area, the knowledge that comes from the experience of their use typically exerts a major influence on which stones were used continuously and which were discarded. During the Victorian period, possibly related to the rise of the professional class, architects began to have a stronger say on material choice. They increasingly sought to employ scientific criteria for stone selection and had no difficulty in breaking from traditional stones and traditional practices.

This rise of the 'technocrat' is not only found in England and associated with Victorianism, but in other European countries that implemented the 'technocratic decrees'. For example the Marquis de Cubas (1886), the Spanish architect in charge of many important buildings during the second half of the 19th century in Spain, said in a note regarding selection of the material for the Cathedral of Madrid that he wanted a thorough selection of stone. This was in contrast to what he thought was the main tendency in previous centuries, when selection of the nearest and softest stone was more important than the longest-lasting stone. He criticized architects who, in the past, chose material hastily with little regard to its durability.

The possibility of selecting more suitable materials (or indeed more fashionable, in some cases) from a wider selection was only possible through the appearance of new ways of transport which allowed stone to be brought from areas that were previously inaccessible or uneconomic. There has been always a relation between the means of transport and the accessibility of material. In earliest times, even when quarries were near building works, the transport of the stone could still be one of the biggest contributions to its overall cost (Knoop & Jones 1938).

This has been amply demonstrated in other countries through the close correlation between the appearance of new transport technologies and the use of new stone types (Gomez-Heras & Fort González 2004). This was not always an improvement as, coupled with the availability of 'better materials', it was sometimes forgotten that the use of vernacular types of stone had in many cases resulted from their proven history of good performance in a given environment. In contrast, the selection of new stones was at times more related to the fashion of being able to use a new available material rather than changing to a better stone.

In the case of Oxford there were two developments that contributed to the opening of a wider network of stone supply: the Oxford Canal in early 1786 and the railway from 1844. The Duke's Cut was completed in 1789 to link the new canal with the River Thames and in 1790 the Oxford Canal connected the city with Coventry. In the 1840s, the Great Western Railway and London and North Western Railway linked Oxford with London.

In observing the timeline of Figure 4, it is noticeable that the establishment of these new methods of transport coincided with the appearance in the city of stone types from different non-local locations, for example Clipsham, Bath or Portland.

Private ownership: the impact of the university and colleges

The collegiate structure of Oxford University has often led to a lack of unified decisions on what stone to use for construction. Each of the colleges had their own decision-making structures and individual policies. In addition, they are living institutions that often planned construction and repairs more from a practical point of view rather than from that of conserving the structures. Many of the oldest medieval colleges, for example Queen's College, do not preserve any remaining examples of the oldest constructions as over the centuries they were destroyed to make room for bigger and 'better' buildings. This lack of co-ordination is in marked contrast to many other Heritage Cities in which a centralized administration and/or inexhaustible supply of the same stone type has resulted in much greater uniformity and continuity of stone use; for example, the similarly historical university city of Salamanca (Spain) and the Villamayor Sandstone.

In many of the colleges, the Provost oversees decisions regarding new constructions and then liaises with the college architect and the contractors. For repairs to college stonework, the college architect might take charge and make informed decisions about what stone is best to use and which methods to utilize. Some figure in charge (usually the Bursar or Clerk of Works) would then liaise with the external architects and contractors. In the past, however, it is highly likely that overall decisions were made more centrally by college fellows who together made up the governing body of the college.

Each governing body or architect would have had his or her own views on the best stone materials and repair strategies, which contributed to the patchwork of methods used around Oxford today. Occasionally, colleges have had their own in-house masonry staff. In this case, the work would be done almost entirely by college staff.

For some of the richest colleges the extent of required building works necessitated owning a quarry to guarantee stone supply. There are numerous records of this ownership of quarries by colleges. For example, it is recorded that in 1525 half an acre of a quarry in Headington was bought by Christ Church College (Arkell 1947) and Headington quarries were partly, for some of the time, in university ownership (Oakeshott 1975). We therefore find records that describe, for example, how the rusticated lower stage of the Sheldonian Theatre was made of Headington freestone from 'our own quarry' at Shotover (Arkell 1947).

The conservation programmes of the 20th century

In addition to the factors that encouraged changes in stone type over the centuries, evolution of historical conservation theory and practice also led to changes in stone use. The Oxford Historic Fund was established from 1957 to 1974 to carry out extensive restoration of most of the Oxford buildings. This was necessitated by the cumulative impact of atmospheric pollution, the use of inappropriate materials (Headington freestone among others) and, on occasions, damage resulting from deficient craftsmanship which had resulted in inadequate conservation of the university buildings, affecting the institutional image of Oxford (Oakeshott 1975).

In carrying out these and subsequent restorations throughout the 20th century, a wide range of materials was again used, reflecting the ongoing individuality of decision making within a largely decentralized collegiate structure. In addition to the formerly described stone types, these works introduced the use of stones from beyond Britain (e.g. French Savonnier).

The strategies undertaken ranged from the complete replacement of large sections of buildings and/or complete façades to the replacement of individual blocks or patches of decayed stonework. One consequence of the latter piecemeal approach is that it placed new and old stone in direct contact. Even when stone for the same quarry was used, the different stress histories of adjacent blocks frequently resulted in divergent patterns and rates of decay (Fig. 6). However, where stones with distinctly different structural characteristics were placed next to each other, this opened the possibility of decay acceleration. The most obvious example of this is where original soft or weathered stone is completely encased within a rigid framework of new stone. This combination of relatively soft stone within a rigid framework could place a constraint on any expansion and contraction of the original stone (in response, e.g. to environmental cycles of wetting/drying and heating/cooling). This could be one explanation of why such blocks

Fig. 6. Different examples from Oxford buildings showing an acceleration of decay of original stone blocks as a result of combining them with replacement stone.

are seen to retreat preferentially to create a so-called 'boxwork' effect on the façade. Such decay is inevitably accentuated if, during replacement, hard, cement-based mortars are used instead of traditional lime mortars.

The present-day consequences of a complex history

Oxford built heritage is characterized by the preponderance of the University which, in turn, is characterized by a deeply collegiate structure with numerous decision-making bodies. These buildings were constructed with a deep sense of functionality as well as the desire for a conspicuous display of wealth and status, both of these characteristics necessarily being maintained during centuries as these buildings were, and are, living structures.

In addition to this, these buildings have attained a heritage status. While having to keep their functionality, they are meant to be preserved as a witness to the past and as a cultural resource; the latter has also become an economic asset, especially through tourism.

The result of this complex history of changing materials and continuous renewal of the collegiate buildings over nearly eight centuries of the history of Oxford University reveals a very complex arrangement of stone types often coexisting in the same buildings (Fig. 7). Rebuilding, renewals and different generations of stone replacement are superimposed over each other.

A whole façade built with a homogeneous material arguably tends to decay more homogeneously, in some instances almost organically, and therefore tends to attain a dynamic equilibrium. When stones with different properties and/or different stages of decay coexist within a façade, this equilibrium may be broken and some blocks may therefore perform more poorly than if they coexisted with more compatible materials.

The incompatibility of their properties has broken the equilibrium of the façades and, in numerous cases, has accelerated the decay of individual blocks with sometimes catastrophic results. This has caused a situation that is very difficult to manage because of the continuous disequilibrium induced in the buildings by replacing individual blocks that, in turn, trigger decay in adjacent blocks and initiate a downward spiral of decay that can spread across a façade.

Conclusions

Oxford provides a fine example of the complexities of stone decay in response to changes in the availability of stone and the historical evolution of the criteria applied to stone selection and building management. It also demonstrates how ill-conceived stone selection, especially with choice of replacement stone, may eventually favour accelerated decay through the generation of an unwanted heterogeneity within facades.

Fig. 7. The Radcliffe Camera: an iconic Oxford structure illustrating the combination of a variety of stone types in one building: (**a**) Headington Hard, (**b**) Headington Freestone, (**c**) Taynton with Clipsham patches and (**d**) Portland.

At a simplistic level, this heterogeneity is marked by an apparent stone homogeneity that presents an image of a city built of a single stone type. Oxford stone has often been labelled together under the same denomination of 'oolitic limestone' based on superficial appearance. This simplification has hindered the understanding of the often major differences between the different materials that shape the city nowadays. This aggregation of the geology hides, however, the internal complexity and variability of the wide range of 'oolitic limestones'. Within individual stone blocks it is internal heterogeneity that invariably triggers decay. At another scale, the effectively chaotic combination of stone within façades – largely a product of complex socio-political and economic histories – triggers heterogeneity that can adversely affect the aesthetic and historical integrity of structures.

A major factor in the complex evolution of stone use in Oxford is its geological setting which on the one hand provides a variety of limestones within comparatively easy reach, while frequently providing insufficient quantity to ensure a long-term supply. In addition to this, the evolution of architectural fashions and the pressure to build, based not only on the functionality of the buildings but also as a symbol of status of the different colleges, conditioned a constant change of criteria for the selection of the stone. Many stakeholders have been involved (architects, masons and fellows) in this as well as a continuous search for new materials to keep building and repair, update and maintain previous buildings.

The appearance of new methods of transport widened the scope of selection of material even further and included new sources of stone. The resulting coexistence of new materials with vernacular types of stone did not always result in stability.

The superimposition of different stone types and different generations of repairs increased the heterogeneity in these buildings, and this heterogeneity resulted in a breaking of the natural trend towards dynamic equilibrium that more homogeneous sets of materials tend to acquire. One possible consequence of this heterogeneity could be when, within a façade, individual blocks retreat more rapidly than surrounding stonework. This is a pattern of decay that is very common across many Oxford buildings where widespread patchwork replacement of stone has taken place (e.g. Fig. 6).

This complex history highlights the need to consider the importance of addressing the holistic management of historic structures and a clearer reconciliation of the ongoing conflicts between the function and historic preservation of living structures. In this dynamic, it is sometimes necessary to acknowledge the primacy of maintaining function through modernization over the preservation of historical integrity and the need to incorporate social and political factors into decision making including, for example, consideration of health and safety requirements. In cases such as Oxford, this has been difficult because of the many stakeholders involved in decisions concerning construction and conservation. The situation has also been exacerbated by a tendency to apply piecemeal solutions to decay problems, both in terms of individual buildings and individual blocks within buildings.

This research was supported by a grant from the UK Engineering and Physical Sciences Research Council (EP/D008603/1). The authors are indebted to G. Alexander and M. Pringle for preparing the figures.

References

ARKELL, W. J. 1947. *Oxford Stone*. Faber & Faber, London.

CUBAS, F. MARQUES DE. 1886. *Boletín eclesiástico del obispado de Madrid-Alcalá*, **39**, 688–696.

GEOLOGICAL SURVEY OF GREAT BRITAIN 1957. *Geological Map of Great Britain*, Sheet 2, 2nd edn. England and Wales.

GOMEZ-HERAS, M. & FORT GONZÁLEZ, R. 2004. Location of quarries of non-traditional stony materials in the architecture of Madrid: the Crypt of the Cathedral of Santa Maria la Real de la Almudena. *Materiales de Construccion*, **54**, 33–49.

KNOOP, D. & JONES, G. P. 1938. The English medieval quarry. *The Economic History Review*, **9**, 17–37.

MCCABE, S., SMITH, B. J. & WARKE, P. A. 2007. Preliminary observations on the impact of complex stress histories on sandstone response to salt weathering: laboratory simulations of process combinations. *Environmental Geology*, **52**, 269–276.

OAKESHOTT, W. F. ed. 1975. *Oxford Stone Restored*. University Press, Oxford.

SMITH, B. J., WARKE, P. A., MCGREEVY, J. P. & KANE, H. L. 2005. Salt-weathering simulations under hot desert conditions: agents of enlightenment or perpetuators of preconceptions? *Geomorphology*, **67**, 211–227.

SMITH, B. J., GOMEZ-HERAS, M. & MCCABE, S. 2008. Understanding the decay of stone-built cultural heritage. *Progress in Physical Geography*, **32**, 439–461.

SUMBLER, M. G. 1996. *British Regional Geology: London and the Thames Valley*. 4th edn, HMSO, London.

VILES, H. A. 1993. Observations and explanations of stone decay in Oxford, UK. *In*: THIEL, M. J. (ed.) *Conservation of Stone and Other Materials*. E & FN Spon, London, 115–120.

Evaluation of the compatibility of building limestones from salt crystallization experiments

KÉVIN BECK & MUZAHIM AL-MUKHTAR*

Centre de Recherche sur la Matière Divisée, Université d'Orléans - CNRS-CRMD, 1B rue de la Férollerie 45071 Orléans Cedex 2, France

**Corresponding author (e-mail: muzahim@cnrs-orleans.fr)*

Abstract: Salt absorption in porous building stones contributes greatly to the degradation of monuments. Two French porous limestones with similar main characteristics (total porosity, densities and mechanical resistance) were studied: white Tuffeau and Sebastopol stone. Accelerated ageing (weathering) tests were carried out by applying immersion-drying cycles with water containing sodium sulphate or sodium chloride. Samples of the two stones were tested separately and then in sets containing both rock varieties. The results facilitated interpretation of the observed and measured responses of these two limestones to the cycling salt crystallization. The durability of studied stones was evaluated by determining the normalized weight changes during the applied cycles. The Sebastopol stone amplified the amount of salt stored in the Tuffeau with increasing number of cycles performed, inducing its more rapid degradation. Water retention and water transfer in the pore space were found to be two main factors controlling the rate and the type of stone decay due to the salt crystallization.

Limestones used in monuments gradually deteriorate with time due to several physical, chemical and biological complex processes. It is well known that water plays a key role in the phenomena of stone deterioration. Indeed, water can act directly with freezing–thawing or dissolution of some minerals such as calcite. Water can also act indirectly by the transport of soluble salts which are major agents of the deterioration of building stones. Sodium sulphate and sodium chloride are the most frequently found salts and the most damaging to historical monuments (Goudie & Parker 1998; Rodriguez-Navarro & Doehne 1999; Benavente *et al.* 2001; Angeli *et al.* 2006). Salts can originate from various sources: air pollution, soil, sea spray, inappropriate chemical treatment or interaction between building materials such as mortars which may contain salts (Price 1995).

The stone degradation is closely related to its geomechanical properties and weathering agent. The petrophysical properties that characterize the durability of a stone are mainly the tensile strength and the pore space (porosity and pore size distribution). Nevertheless, part of the degradation of monuments built with limestones originates from an incompatible association between the original construction stone and the stone of replacement in restoration works. The replacement of the damaged stones is a crucial issue in the restoration of historical monuments. In fact, the choice of compatible stones constitutes a major difficulty in France because some quarries of original stones are now exhausted or unidentified. The aim is to find a new stone similar to and compatible with the old stones if it is impossible to use exactly the same. The aim of this work is to study the compatibility between stones gathered during ageing tests. The behaviour of each stone was previously studied during the same ageing test.

The durability of the stones is studied through the measurement of the stone resistance to salt crystallization by means of accelerated ageing tests. Stone which retains its original size, shape and appearance over an extensive period of time during these tests is considered to have a high durability. These tests are carried out by applying immersion-drying cycles using water polluted by salts (sodium sulphate Na_2SO_4 and sodium chloride NaCl) in accordance with the EN 12370 European standard test. Two French building limestones with a similar high total porosity (around 45 vol.%) are tested in this study: the white Tuffeau and Sebastopol stone. The compatibility of the two stones is analysed with the observation of reaction to the ageing test for two types of samples: isolated stones and stones linked together (Tuffeau and Sebastopol stone). A comparison of the behaviour of the two stones under the same accelerated weathering test allows the compatibility these stones to be determined.

Materials and methods

Materials

The studied limestones were characterized using different complementary techniques. Mineralogical

From: PŘIKRYL, R. & TÖRÖK, Á. (eds) *Natural Stone Resources for Historical Monuments.* Geological Society, London, Special Publications, **333**, 111–118.
DOI: 10.1144/SP333.11 0305-8719/10/$15.00

Table 1. *Main physical properties of two tested stones*

	Tuffeau stone	Sebastopol stone
Mineralogical composition	Calcite ≈50% Quartz ≈10% Opal CT ≈30% Clays and micas ≈10%	Calcite ≈80% Quartz ≈20%
Skeletal density (g cm^{-3})	2.55	2.71
Bulk dry density (g cm^{-3})	1.31	1.58
Porosity (vol.%)	48	42
Median pore access diameter (μm)	2.7	17.9
Tensile strength at saturated state (MPa)	0.45	0.59
Compressive strength at saturated state (MPa)	4.5	5.8

composition was determined by X-ray diffraction, thermogravimetric analyses, infra-red spectrometry and scanning electron microscopy. The characterization of pore space was carried out by mercury intrusion porosimetry. Tensile strength was calculated from the results of the Brazilian splitting method with an Instron 4485 press. The main physical characteristics of stones are summarized in Table 1. The two limestones studied (Tuffeau and Sebastopol stone) were selected because they have several close characteristics: total porosity of about 45 vol.% and a compressive strength and a tensile strength at the saturated state of the order 5 MPa and 0.5 MPa, respectively. Mechanical properties (compressive and tensile strengths) were measured in the saturated state since mechanical strength is much weaker than dry stone strength (West 1994; Beck *et al.* 2007).

White Tuffeau is commonly used in most monuments built along the Loire valley such as the famous Loire Castles (Beck *et al.* 2003) and the Sebastopol stone is used near Paris. The major minerals of these limestones are calcite (i.e. $CaCO_3$) and quartz (i.e. SiO_2). Moreover, in the white Tuffeau, predominantly clayey minerals and opal cristobalite/tridymite are found. The size and shape of grains in white Tuffeau vary significantly and consist of large grains of clay (glauconite), quartz, sparitic calcite and mica arranged with several types of fine grains such as clay (smectite), micritic calcite, small fossil shells and spherules of CT-opal (Beck *et al.* 2003). The Sebastopol stone comprises calcite and quartz grains that are coarser with a more uniform size (Beck & Al-Mukhtar 2005). The white Tuffeau is a relatively light building material. It has an apparent density of 1.31 g cm^{-3} and a total porosity of 48 vol.% under total dry conditions (Table 1). Results of mercury porosimetry demonstrated the multiple porosity nature of white Tuffeau stone with a very wide pore-size distribution (from 6 nm to 20 μm) with a median pore access diameter of 2.7 μm. The pore-size distribution of Sebastopol stone is mainly unimodal (from 1–40 μm) with a median pore access diameter of 17.9 μm and predominant pore access diameter of about 20 μm (Beck & Al-Mukhtar 2005).

Hydraulic properties of these stones are also very different. Indeed, water transfer properties are related to the porosity and to the porous network characteristics constituted by pore sizes, pore shapes and connectivity of the porous structure. Both stones used in this study have relatively high permeabilities and high capillary kinetics. Nevertheless, capillary coefficients of Sebastopol stone are about twice as large as those of Tuffeau and the permeability coefficient is about 15 times greater. This confirms the influence of pore size distribution on the main hydraulic properties (Beck & Al-Mukhtar 2005). Moreover, water absorption is higher in the case of the Tuffeau stone and this limestone absorbs very large amounts of moisture in an environment of high humidity. Indeed, the degree of saturation of Tuffeau is around 20% in a wet atmosphere (98% of relative humidity) while that of Sebastopol stone is about 2% in the same conditions (Table 2). The Tuffeau stone exhibits a very high water retention capacity. Such behaviour can be attributed to the presence of clays and the fine pore space with nanometric pores (Table 1).

Experimental method

Two types of salts were used for the aging experiment: sodium chloride which crystallize to form the halite (NaCl) and sodium sulphate which can crystallize to form the mirabilite $Na_2SO_4 \cdot 10H_2O$ at high humidity or the thenardite Na_2SO_4 at low humidity (Flatt 2002). The experimental material consisted of two different arrangements of specimens: (1) single isolated cubes having dimensions $5 \times 5 \times 5$ cm (three specimens for each rock type), and (2) a specimen prepared by joining together (with a high quality hemp cord) one cube of Tuffeau and one cube of Sebastopole stones.

Table 3. *Normalized weights at the end of the ageing test (weight loss corresponds to normalized weight <1)*

	NaCl			Na_2SO_4		
	Tuf	Seb	Seb-Tuf	Tuf	Seb	Seb-Tuf
Normalized weight at the drying stage of the 15th cycle	1.160	1.083	1.140	–	0.688	0.486
Normalized weight at the dry state (1 month at 105 °C)	1.121	1.082	1.112 Tuf 1.086 Seb 1.131	–	0.688	0.478 Tuf 0.208 Seb 0.785

visible rupture. Degradation is gradual with a homogeneous matter (particles) loss on the surface (Fig. 4). This causes a progressive weight loss (Fig. 2) which in a dry state' is approximately 30%. The total porosity and the tensile strengths, which evaluate cohesion between the particles, are almost the same for the two stones (approximately 45 vol.% and 0.5 MPa respectively). Even if the tensile strength is somewhat higher for the Sebastopol stone, the difference in the deterioration process is mainly due to the large difference in the water transfer properties such as capillarity, permeability and water retention. These parameters are controlled by the structure of pore space, the size and shape of pores and connectivity of the porous network (Beck & Al-Mukhtar 2005). The Tuffeau stone contains about 10% of clay minerals and has an important fraction of very small pores which can retain water at low relative humidity. This could generate high salt crystallization pressure in the pore space and induce the rapid decay of the Tuffeau stone. On the contrary, Sebastopol stone has a very low water retention capacity, a higher permeability and most of its pore spaces are larger compared to those of Tuffeau. Hence, the salt can fill pore spaces of the Sebastopol stone without generating high crystallization pressures.

In case of the samples constituted by a set of Tuffeau and Sebastopol linked together, the behaviour seems to cumulate the properties of the two stones tested separately (Figs 1 and 2). Generally, the weight increases slightly because of the storage of NaCl salt. However, during immersion, some of salts accumulated near the surface can be dissolved and so the total amount of salt may reduce during one to two cycles and increase for the following cycles. At the end of the aging test (after 15 cycles with NaCl salt solution), a slight alteration can be observed to the Tuffeau cubes as some granular disintegration on the surfaces (Fig. 4). It must be mentioned that Tuffeau samples seem to maintain some water at the end of the drying cycle. The drying time (20 hours) is too short for the Tuffeau to dry completely, particularly in the case of NaCl solution.

The compatible-incompatible characteristics of the stones can be observed through the interaction between the blocks of stones linked together. Figure 5 shows the photographs of the stone blocks linked together for the 4th cycle. The stones demonstrate high efflorescence in the case of the cycles carried out with NaCl. Nevertheless, for Sebastopol stone, a zone of approximately 1 cm thickness near the Tuffeau does not have any efflorescence and seems to be humid. This behaviour is due to the interaction between the two stones. Since the pores of the Tuffeau are smaller, the capillary suction generated tends to absorb the solution in the Sebastopol stone. Tuffeau therefore absorbs more salt when it is in contact with Sebastopol stone during the first cycles. From the 5th cycle, the behaviour of the two stones with NaCl solution begins to change because the NaCl salt content becomes enough high to slow water transfer, mainly in the case of Tuffeau. Indeed, Tuffeau has a high water retention capacity and Sebastopol stone has a high water capillarity and a high permeability (Table 2). Sebastopol stone can therefore drain water whereas water transfer is lower within Tuffeau.

Moreover, the Sebastopol stone dries faster than Tuffeau stone. After the drying of Sebastopol is almost complete, part of the moisture within the pore space of the Tuffeau can be absorbed by the capillary action into the Sebastopol stone. Therefore, even if capillary suction is higher in the Tuffeau (suction generated by smaller pores), water can be transferred from the Tuffeau stone to the Sebastopol stone during the drying stage via the contact surface, since the Tuffeau is never completely dry (Table 3). Water absorbed by Sebatopol contains NaCl salts, which results in the reduction of salts in Tuffeau. This is confirmed by the measurements of salt content after the 15th cycle; the NaCl content is higher in the Sebastopol stone than in the Tuffeau (1.131 compared to 1.086) and this figure is even higher for the isolated case (1.131 compared to 1.082).

Regarding deterioration by Na_2SO_4, the Tuffeau disaggregates very quickly during the 8th cycle. The remnant of the Tuffeau cube is maintained with the cube of Sebatopol stone by the strings of connection. The Sebastopol stone continues to disaggregate uniformly on the surface (Fig. 4). After

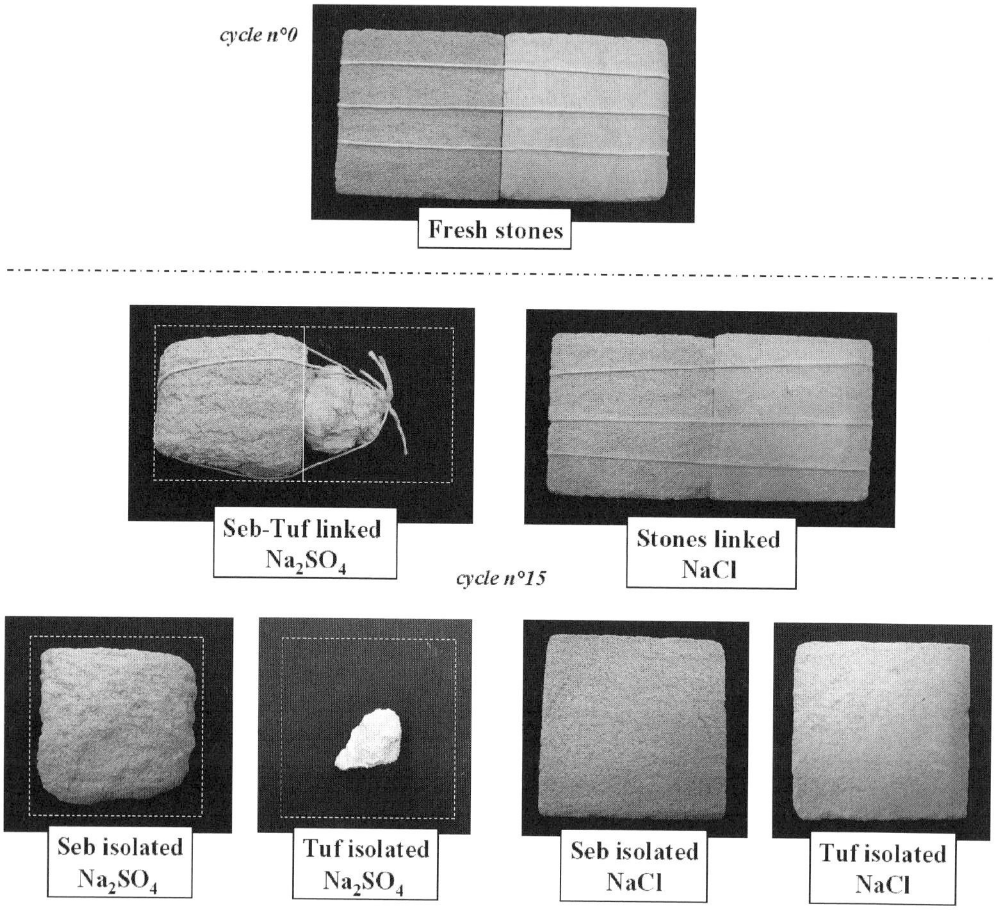

Fig. 5. 'Sebastopol/Tuffeau samples linked together' after 4 cycles of immersion/drying in the case of Na_2SO_4 and NaCl crystallization.

15 cycles, the weight loss of the Sebastopol stone linked to the Tuffeau (normalized weight: 0.785) is less than for the isolated Sebastopol stone (normalized weight: 0.688). The stone therefore deteriorates less when it is in contact with the Tuffeau. Indeed, according to Figure 4, the erosion of the surface close to the face in contact with the Tuffeau is less than for the remaining parts of the sample. The alteration of the Tuffeau is somewhat slowed and that of the Sebastopol is slightly reduced. The replacement of Tuffeau by Sebastopol can be advantageous to construction as the latter is more resistant to deterioration processes including sodium sulphate or sodium chloride.

Concluding remarks

The aim of this study was to analyse the behaviour and evaluate the durability of two limestones subjected to an accelerated aging test. As the normalized aging test is conducted with sodium sulphate (Na_2SO_4), we carried out a similar accelerated test with sodium chloride NaCl. Finally, the two previous tests were also carried out on a set of two stones (Tuffeau and Sebastopol) in order to determine the compatibility between the two stones. The main result is that the quantity of sodium chloride salt stored in the Tuffeau is higher than that of sodium sulphate. This stone, however, appears to withstand sodium chloride cycles. The durability of the Tuffeau stone face to sodium sulphate is limited as samples were completely damaged during the 8th cycle. This behaviour can be attributed to salt crystallization pressure and thenardite/mirabilite phase transition generated by sodium sulphate during immersion-drying cycles. An additional property of sodium chloride can also explain the reaction of stone: indeed' the water absorption capacity of the NaCl salt is higher than that of the Na_2SO_4. During the drying cycle, samples are not completely dry and the crystallization pressure is lower than that for sodium sulphate. A drying stage of 20 hours duration is clearly insufficient in the case of the Tuffeau, in particular with sodium chloride.

Sodium chloride salt was accumulated more than sodium sulphate in the Sebastopol stone. However, durability of Sebastopol stone seems to be better than that of Tuffeau with the two salts. Less damage was observed with sodium sulphate. This behaviour can be explained by the difference between the two stones in the mineralogical composition and in the pore space distribution, inducing different water transfer properties (permeability, capillarity and water retention). Sebastopol stone samples are always completely dry at the end of drying cycle which is not the case for the Tuffeau samples. This demonstrates that water transfer in Sebastopol stone is much faster than that in Tuffeau (Beck & Al-Mukhtar 2005). Moreover, pores in the Sebastopol stone are large and so harder to block (obstruct) by the salt. Therefore, at the end of a drying cycle, the pores of the Sebastopol stone are practically dry while the pores in Tuffeau stone are still humid.

The normalized test (EN 12370 standard) seems to be very destructive for the soft limestones over a small numbers of cycles (about only five in the case of Tuffeau stone); it therefore cannot be used to reveal the behaviour under realistic environmental conditions. The same test carried out with NaCl allows us to track more accurately the response of the stone throughout the test. Moreover, the alteration type during these ageing tests using salt crystallization is always granular disintegration. This is a natural (*in situ*) type of alteration of these stones, but not the unique type for the case of Tuffeau stone (which also experiences exfoliation, flaking, fissuring and alveolar).

The ageing tests carried out on the two types of stone gives some indication of compatibility. The use of sodium sulphate quickly induces an important alteration in the set, mainly displaying granular disintegration in Tuffeau. It is therefore not easy to study the compatibility with this normalized ageing test. The use of NaCl attenuates, or reduces slightly, the alteration of the Tuffeau stone. This can be attributed to water transfer in Sebastopol stone which is very quick. Through the contact surface between the stones, water and salt solution can be transferred from Tuffeau to Sebastopol and then ejected from the set (efflorescences and subflorescences are observed). It can be stated that replacing Tuffeau stone with Sebastopol stone is advantageous as the durability of the set is improved. However, in construction the contact between stones is usually realized by a mortar joint. If the water transfer properties of the mortar are similar to that of the stone, the conclusions presented in this study are also valid. It is necessary to consider the physicochemical compatibility between the mortar and the stone for the restoration works, however (Beck & Al-Mukhtar 2008).

Because the damage depends on the type and the quantities of salts in the stone as well as the water transfer properties (permeability, capillarity and water retention), other accelerated weathering tests must be developed to evaluate the localization of salts during immersion-drying cycles (Topal & Sozmen 2003; Angeli *et al.* 2007), imbibition-drying (Benavente *et al.* 2001; Van *et al.* 2007*a*) and relative humidity cycles (Van *et al.* 2007*b*).

More characterization tests must be planned such as X-ray diffraction, mercury intrusion porosimetry and scanning electronic microscope in order to determine the mineralogical and pore space changes during ageing tests. The microstructure and rock fabric of the two stones strongly affect the durability and the nature of the deterioration, and can influence the phenomenon of compatibility.

The authors would like to express their thanks to the quarry Lucet (St-Cyr-en-Bourg, France) for providing white Tuffeau and the quarry Rocamat (Saint-Maximin, France) for providing Sebastopol stone. Thanks also to Sami Al-Mukhtar who conducted some of experiments presented in this paper.

References

Angeli, M., Bigas, J. P., Menéndez, B., Hébert, R. & Christian, D. 2006. Influence of capillary properties and evaporation on salt weathering of sedimentary rocks. *In*: Fort, R., Alvarez de Buero, M., Gomez-Heras, M. & Vasquez-Calvo, C. (eds) *Heritage, Weathering and Conservation.* Taylor & Francis/Balkema, Leiden, 253–259.

Angeli, M., Bigas, J. P., Benavente, D., Menéndez, B., Hébert, R. & Christian, D. 2007. Salt crystallization in pores: quantification and estimation of damage. *Environmental Geology*, **52**, 205–213.

Beck, K. & Al-Mukhtar, M. 2005. Multi-scale characterisation of two French limestones used in historic constructions. *International Journal of Restoration of Buildings and Monuments*, **11**, 219–226.

Beck, K., Al-Mukhtar, M., Rozenbaum, O. & Rautureau, M. 2003. Characterization, water transfer properties and deterioration in Tuffeau: building material in the Loire valley – France. *Building and Environment*, **28**, 1151–1162.

Beck, K., Adib-Ramezani, H. & Al-Mukhtar, M. 2007. Mechanical strength and water content of porous limestones. *In*: Gdoutos, E. E. (ed.) *Experimental Analysis of Nano and Engineering Materials and Structures.* Part C, Subpart 37. Springer, Dordrecht, 963–964.

Beck, K. & Al-Mukhtar, M. 2008. Formulation and characterization of an appropriate lime-based mortar for use with a porous limestone. *Environmental Geology*, **56**, 715–727.

BENAVENTE, D., GARCIA DEL CURA, M. A., BERNABÉU, A. & ORDONEZ, S. 2001. Quantification of salt weathering in porous stones using an experimental continuous partial immersion method. *Engineering Geology*, **59**, 313–325.

EN 12370 STANDARD. 1999. Natural stone test methods – determination of resistance to salt crystallization.

ESPINOSA MARZAL, R. M. & SCHERER, G. W. 2008. Study of the pore clogging induced by salt crystallization in Indiana limestone. *In*: LUKASZEWICZ, J. & NIEMCEWICZ, P. (eds) *Proceedings of the 11th International Congress on Deterioration and Conservation of Stone*. Nicolaus Copernicus University Press, Torun, **I**, 81–88.

FLATT, R. J. 2002. Salt damage in porous materials: how high supersaturations are generated. *Journal of Crystal Growth*, **242**, 435–454.

GOUDIE, A. S. & PARKER, A. G. 1998. Experimental simulation of rapid rock blocks disintegration by sodium chloride in a foggy coastal desert. *Journal of Arid Environments*, **40**, 347–355.

PRICE, D. G. 1995. Weathering and weathering processes. *Quarterly Journal of Engineering Geology*, **28**, 243–252.

RODRIGUEZ-NAVARRO, C. & DOEHNE, E. 1999. Salt weathering: influence of evaporation rate, supersaturation and crystallization pattern. *Earth Surface Processes and Landforms*, **24**, 191–209.

TOPAL, T. & SOZMEN, B. 2003. Deterioration mechanisms of tuffs in Midas monument. *Engineering Geology*, **68**, 201–233.

VAN, T. T., BECK, K. & AL-MUKHTAR, M. 2007*a*. Accelerated weathering tests on two highly porous limestones. *Environmental Geology*, **52**, 283–292.

VAN, T.-T., BECK, K., BRUNETAUD, X. & AL-MUKHTAR, M. 2007*b*. Crystallization pressure and salt distribution in Tuffeau limestone. *In*: GDOUTOS, E. E. (ed.) *Experimental Analysis of Nano and Engineering Materials and Structures*. Part C, Subpart 37. Springer, Dordrecht, 965–966.

WEST, G. 1994. Effect of suction on the strength of rock. *Quarterly Journal of Engineering Geology*, **27**, 51–56.

Evaluation of three Italian tuffs (Neapolitan Yellow Tuff, Tufo Romano and Tufo Etrusco) as compatible replacement stone for Römer tuff in Dutch built cultural heritage

TIMO G. NIJLAND[1]*, ROB P. J. VAN HEES[1,2] & LAURA BOLONDI[1,3]

[1]*Conservation Technology group, TNO Built Environment and Geosciences, Delft, The Netherlands*

[2]*®MIT, Faculty of Architecture, Delft Universty of Technology, The Netherlands*

[3]*Present address: IMT Alti Studi Lucca, Lucca, Italy*

**Corresponding author (e-mail: timo.nijland@tno.nl)*

Abstract: Rhenish tuffs from the volcanic Eifel region in Germany, in particular the so-called Römer tuff, are among the most prominent and voluminous natural stones in Dutch monuments. The Römer tuff has been used since Roman times, and was widely used again in Romanesque (and to a lesser extent Romano-Gothic and early Gothic) architecture. The limited (or non) availability of Römer tuff for restoration purposes is posing an increasing problem. Last decennia, the availability of Römer tuff was practically limited to blocks from the lower parts of the pyroclastic flows with abundant basalt (and other) xenoliths, giving the rock a different appearance from its traditional type; the different types of Römer tuff also demonstrate different physical and hygric properties. Given the wide use of tuff stone in Italian architecture, several Italian tuffs have been evaluated in search of a compatible replacement stone for Römer tuff. The replacement stones should approach the original as much as possible, that is, in terms of authentic appearance, physical characteristics and durability. The Italian tuffs evaluated include tuffs commercially denominated as Tufo Etrusco and Tufo Romano (from the central part of Italy) and a variety of Neapolitan Yellow Tuff (Naples region). Hygric behaviour, resistance to frost-thaw cycles, petrographic characteristics and mineralogy of Italian tuffs have been determined and compared with original Römer tuff. In all three cases, resistance to frost-thaw cycles is unfortunately shown to be considerably less than that of original Römer tuff. In addition, hygric expansion of the Neapolitan Yellow Tuff appeared to be considerably larger than that of original Römer tuff. Of the tuffs evaluated, the variety of Neapolitan Yellow Tuff is a good match with the original Römer tuff in terms of visual appearance. It has already been sparsely used in the Netherlands in minor amounts. However, the durability characteristics require additional evaluation.

Volcanic tuffs from the Eifel region, Germany, constitute a major part of the Dutch built cultural heritage. These so-called Rhenish tuffs comprise several types, namely the Römer (Roman) tuff deposited by the 11 900 BP eruption of the Laacher See volcano and the Ettringer (including its variety Hasenstoppler), Weiberner (including its variety Hohenleie) and Riedener tuffs from older eruptions in the Riedener caldera. In the Netherlands, the former three tuffs have been used frequently. The amount of Römer tuff available for restoration purposes today is rather limited, if available at all. Mutual differences between the Rhenish tuffs are, however, considerable; neither Weiberner nor Ettringer tuff provide a suitable replacement stone (at least from an aesthetical point of view). For this reason, other volcanic tuffs are being considered as replacement stones. The current study reports on the compatibility of three Italian tuffs as replacement stones for Römer tuff.

Use of Rhenish tuffs in Dutch monuments

Rhenish tuffs have widely been used in the Netherlands. Nijland & van Hees (2006) provide a detailed account of their use. A brief summary will be given here. The use of Rhenish tuff as a building stone in the Netherlands dates back to Roman times, and the tuff used in that period derives its name from the people who introduced the stone in the Netherlands: Römer is German for Roman. This tuff was used again during the period of Romanesque architecture, from the 10th until the early 13th century, during which it was the commonly used natural stone in the Netherlands

From: PŘIKRYL, R. & TÖRÖK, Á. (eds) *Natural Stone Resources for Historical Monuments*.
Geological Society, London, Special Publications, **333**, 119–127.
DOI: 10.1144/SP333.12 0305-8719/10/$15.00 © The Geological Society of London 2010.

(Slinger *et al.* 1980; Nijland & van Hees 2006). The cities of Utrecht and Deventer became staple markets from which tuff, imported over the river Rhine, was distributed over large parts of the country and the most western part of Belgium (around Bruges; Slinger *et al.* 1980) and in northern Germany and western Denmark (Haiduck 1992). Important Dutch monuments (partly) constructed in Römer tuff include: the gothic Dom of Utrecht (Fig. 1) and its Romanesque predecessor; other 11th century churches from the Utrecht cross of churches (Nijland *et al.* 2007; fig. 1); many Romanesque parish churches in the northern province of Groningen (de Olde 2002) and the western provinces of Holland (den Hartog 2002); some (partly) surviving medieval houses including the oldest preserved prophane building in the Netherlands in the city of Deventer (oldest part *c.* 1130, façade second half 12th century; de Vries *et al.* 1992; fig. 1) as well as several defence buildings including the Burcht in Leiden, a 12th century chateau-en-motte. From the early 13th century onwards, Römer tuff was pushed out of the market by local fired clay brick and other types of natural stone, such as Bentheim sandstone. The latter was partly due to the levy of toll on the river Rhine, whereas Bentheim sandstone could be imported over toll-free rivers.

During the 15th and early 16th century, Weiberner tuff from the Eifel was used, especially its fine-grained variety Hohenleie (or Hohen Ley) tuff. The latter was well suited for carved and sculptured works, for example for the blind traceries of the 15th century cloister of the gothic Dom in the city of Utrecht (Nijland *et al.* 2007) and on the rampant arches of St John's cathedral, 's-Hertogenbosch. The tuff was also used for cladding the walls of several major churches, including Our Lady's Church in the city of Zwolle and the tower of the Main Church in Dordrecht. Much later, during the second part of the 19th century and first half of the 20th century, Weiberner and Hohenleie tuffs were reintroduced both for restoration purposes and newly constructed buildings. During the same period, Ettringer and Hasenstoppler tuff were introduced for the first time. These too have been used for both new buildings and restoration purposes (in the latter case not only for replacement of tuff, but also of white Belgian stone; van Hees *et al.* 2005). Examples of newly constructed buildings include the tower of the town hall of Rotterdam built in 1916, which is entirely clad with Ettringer tuff, and the KAS bank (behind the Royal Palace in Amsterdam) completed in 1932. A typical use of Ettringer tuff during the 1930s is the application to church buildings in an ecclectic style with a strong orthodox influence and as small building elements, such as corner stones, sills, etc. as accents in red clay brick façades. The latter combination was encountered again in the 1950s (Nijland & van Hees 2006; Nijland *et al.* 2007).

Characteristics of Römer tuff

The Römer tuff is obtained from the lithified ash flows and glow avalanches of the 11 900 BC eruption of the Laacher See volcano (e.g. Schmincke 1988). This type of tuff has variously been described as *duifsteen*, *trastuf*, *lapillituf* or *Andernach tuf* in older Dutch literature. The tuff has a trachytic composition. All Rhenish tuffs are macroporous rocks with varying amounts of pumice and rock fragments occurring in a fine-grained matrix that was originally composed of volcanic glass, showing considerable variation within each type. Typical igneous minerals and xenocrysts in the Römer tuff are sanidine, other feldspars, clinopyroxene (Ti-augite, diopside), olivine, amphibole, biotite, ore minerals and carbonate (Fitzner 1994; Nijland *et al.* 2003). X-ray diffraction studies of both fresh quarry material and material from Dutch monuments have shown that zeolite assemblages of analcime + chabazite as well as analcime + chabazite + phillipsite occur, the latter being most common. In weathered samples from Dutch monuments, the assemblages analcime, analcime + phillipsite and phillipsite also occur (Nijland *et al.* 2003).

Although a weather-susceptible stone (e.g. Fitzner & Lehners 1990), Römer tuff dating from the time of the original building survives on many churches, including the 11th century part of St John's Church, Utrecht. Weathering phenomena are not the same for all Rhenish tuffs. Typical weathering phenomena for the Römer tuff include spalling, alveoli, powdering and dissolution of the matrix and salt efflorescence; in addition, they are susceptible to biocolonization (Nijland *et al.* 2003). Porosity is variable, ranging from 37.7 vol. % in original building stone from St John's cathedral to 41.5 vol. % in Römer tuff from the bottom of the flows with abundant bazalt xenoliths (used in restorations over the last decades). Porosity values of up to 53.9 vol. % have been observed in Römer tuff recently obtained for the restoration of St John's cathedral. This tuff visually resembles material from the original building. Water absorption coefficients for the latter two are 0.17 and 0.32 kg m^{-2} $s^{-0.5}$, respectively (van Hees *et al.* 2003, 2004). The stones also show considerable differences in pore size distribution, wetting and drying behaviour and resistance to Na_2SO_4 crystallization. There is no damage on the variety with 53.9 vol. % porosity but spalling on the type with 41.5 vol. % porosity. The variety with 53.9 vol. % porosity also proved to be frost resistant, even under the most severe conditions.

Fig. 1. Examples of the use of Rhenish tuff in Dutch architecture: (**a**) façade of the oldest surviving secular prophane stone building in the Netherlands, the proosdij, Deventer (late 12th century); Römer tuff together with Drachenfels trachyte at the ground floor, and together with later fired clay brick at the first floor; (**b**) surviving 11th century Römer tuff on the north façade of St John's church, Utrecht, built by bishop Bernulphus; (**c**) choir of the gothic Dom of Utrecht, with Weiberner tuff in addition to Römer and Ettringer tuff as well as several other kinds of natural stone, in particular corner stones (Bentheimer and Red Weser sandstone, among others).

Italian tuffs evaluated

Volcanic tuffs have also been used in several other countries. Their use is widespread in many Italian cities, including Rome (e.g. Heiken *et al.* 2005) and Naples (e.g. Cardone 1990; de'Gennaro *et al.* 2000), as well as the Campanian region in general (Langella *et al.* 2000). Tuff from the latter area, the Neapolitan Yellow Tuff (Tufo Giallo Napoletano), matches the Römer tuff reasonably well from an aesthetic point of view and demonstrates some similarities in composition (Sersale & Aiello 1964). Neapolitan Yellow Tuff is therefore considered a potential candidate to replace Römer tuff. It has recently been applied in small amounts in the restoration of a few Dutch monuments including the Lamberti tower in the eastern village of Zelhem (Fig. 2) and the Dutch Reformed church in Winterswijk in the same area. In addition to the Neapolitan Yellow Tuff, two other Italian tuffs have been evaluated on request of the Netherlands Department of Conservation (RACM). Visually different from the Römer tuff, they are traded under the names Tufo Romano (also Fiorditufo classic) and Tufo Etrusco from the areas of Civita Castellana and Riano, respectively. To some extent, Tufo Romano visually resembles some varieties of the Weiberner tuff from the Eifel. A macroscopic overview of all three tuffs investigated is given in Figure 3.

Fig. 2. (**a**) and (**b**): Neapolitan Yellow Tuff amidst Römer tuff on the Lamberti tower, Zelhem.

Analytical methods

Mineralogy and microstructure have been studied by polarization-and-fluorescence microscopy (PFM) and X-ray diffraction analysis (XRD). Open porosity has been determined according to RILEM CPC 11.3 (1984). Resistance to crystallization by thenardite, Na_2SO_4, was determined according to internal TNO procedures (e.g. van Hees *et al.* 2003). Na_2SO_4 was chosen because it is by far the most common salt in efflorecences on tuff stone in the Netherlands (Nijland *et al.* 2003). Freeze/thaw resistance has been determined with reference to the intended use of the tuffs, that is, cladding, following the Dutch standard NEN 2879 (1989) for masonry stones.

Characterization of Italian tuffs

Petrography and mineralogical composition

The sample of Neapolitan Yellow Tuff has an open, porous matrix in which the original glass shards are easily recognized (in contrast to the Römer tuff). Pheno- and xenocrysts include sanidine, plagioclase, quartz, clinopyroxene, biotite and opaque minerals. Xenoliths are basalt fragments and pumice. Pumice has a 'thread-like' internal structure, with flattened voids and occasional large euhedral plagioclase crystals. Pumice has been zeolitized, but individual crystals are rather difficult

Fig. 3. Macroscopic impressions of (**a**) Neapolitan Yellow Tuff (10 × 10 cm); (**b**) Tufo Romano (10 × 10 cm) and (**c**) Tufo Etrusco (9 × 10 cm).

Table 1. *Water absorption coefficient and porosity. Results for Italian tuffs evaluated, in comparison to literature data for Neapolitan Yellow Tuff (Bolondi* et al. *2007) and some Rhenish tuffs (van Hees* et al. *2003, 2004; Nijland* et al. *2005)*

Tuff	Water absorption coefficient kg m^{-2} $s^{-0.5}$	Porosity vol. %
Neapolitan Yellow Tuff	0.46	51.8
Tufo Romano	0.26	50.1
Tufo Etrusco	0.27	47.8
Neapolitan Yellow Tuff		38.6–52.4
Römer tuff, original stone St John's cathedral	0.32	37.7
Römer tuff, restoration St John's cathedral		53.9
Römer tuff, recent restorations	0.17	41.5
Weiberner tuff, quarry material	0.24–0.38	47.4–45.7
Weiberner, original stone St John's cathedral		45.9

to identify in most cases. Both cubic chabazite and rosettes of prismatic phillipsite occur. X-ray diffraction confirms the presence of chabazite and phillipsite, in addition to analcime. The surrounding matrix is denser directly around pumice fragments.

Tufo Romano contains abundant fragments of limestone, sandstone/quartzite and siltstone, as well as pumice fragments with a typical size of about 5 mm. Pheno- and xenocrysts include plagioclase, sanidine, quartz, hornblende, clinopyroxene, possible leucite and non-corroded cleavage fragments of calcite. Fine grained carbonate is also dispersed locally in the matrix. Pumice shows relatively coarse-grained zeolites along its outer margin, while the surrounding matrix is denser than the pumice. Zeolites have a cubic habit or form rosettes, probably chabazite and phillipsite, respectively. X-ray diffraction confirms the presence of both zeolites, in addition to analcime. The sample shows some microcracking, with cracks passing through both matrix and rock fragments.

Tufo Etrusco has a matrix with locally limited internal cohesion, and is composed of shards of (altered) volcanic glass that have a rather open internal structure. Besides pumice, rock fragments include various types of basalt and other volcanic rocks including leucitite as well as some schist. Pheno- and xenocrysts include plagioclase, K-feldspar (sanidine, orthoclase), clinopyroxene, biotite and opaque minerals. Both matrix and pumice fragments contain relatively large zeolite crystals with either prismatic or cubic habit. The matrix is generally thread-like, with abundant rosettes and wafers of zeolites. X-ray diffraction shows the presence of chabazite and analcime. Directly around the pumice fragments, the matrix is denser. Minor microcracks occur in the matrix.

Porosity and hygric behaviour

Results for porosity and calculated water absorption coefficients are given in Table 1. Water absorption and drying curves are depicted in Figure 4. Römer tuff itself shows considerable variation in both

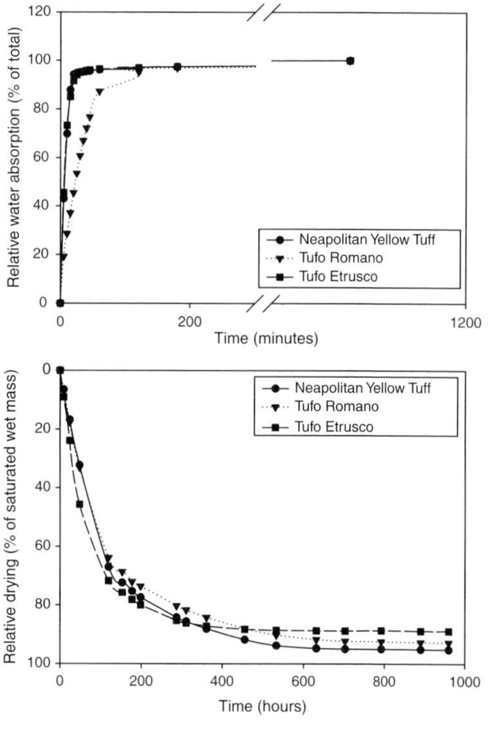

Fig. 4. (**a**) Water absorption and (**b**) drying curves.

porosity and water absorption coefficient; the range of porosity displayed by Neapolitan Yellow Tuff (Bolondi *et al.* 2007) is also large, but the presently investigated sample of Neapolitan Yellow Tuff shows a water absorption coefficient that is considerably higher than the highest observed for Römer tuff.

Freeze/thaw resistance

Results for freeze/thaw resistance are given in Figure 5 and Table 2 and compared to those for selected Römer and Weiberner tuffs which were tested according to the same procedures. As indicated by the results of Table 2, performance of the Neapolitan Yellow Tuff, Tufo Romano and Tufo Etrusco is inferior. This is particularly the case under conditions relevant to outside use, that is, preconditioning at 75% and 100% vacuum.

Fig. 5. Selected Italian tuffs after freeze/thaw test: (**a**) overview, with Neapolitan Yellow Tuff on the left, Tufo Roman in the middle and Tufo Etrusco on the right; preconditioning conditions decreasing from the front (100% vacuum) to back (50% vacuum); (**b**) same tuffs, preconditioned at 100% vacuum.

Conclusion

Römer tuff and other Rhenish tuffs show considerable variation in properties relevant to durability such as open porosity, hygric behaviour and pore size distribution (van Hees *et al.* 2003, 2004). This makes it difficult to evaluate the compatibility of any other tuff as a replacement stone. From the three Italian tuffs evaluated, namely Neapolitan Yellow Tuff, Tufo Romano and Tufo Etrusco, the former provides the best aesthetic match to the Römer tuff. Tufo Romano would, from an aesthetic point of view, provide an acceptable replacement stone for Weiberner tuff. The latter is, however, still available. In addition, freeze/thaw resistance of the Tufo Romano, an important prerequisite for outside use in Dutch climate, appears to be less than that of Weiberner tuff (Table 2).

In Neapolitan Yellow Tuff, volcanic glass has been altered to the same zeolites as encountered in Römer tuff, namely chabazite, phillipsite and analcime. However, the latter displays several assemblages (Nijland *et al.* 2003) and, although a relationship between zeolite assemblage and durability of the tuff as a building stone has not yet been established, it is likely that (as well as physical characteristics such as apparent porosity, hygric behaviour, etc.) these minerals play a role, for example in response to (acid) rain (e.g. de'Gennaro *et al.* 1984). Coupled chemical–mechanical–hydraulic processes (cf. Blacic 1993), controlled in part by the relative stability of zeolites, may influence the durability of the stone by time-dependent change of properties of tuffs. The properties include increasing water absorption with time (as observed for Rhenish tuffs; Brendle 2003) and changes in linear strain and shrinkage with time (as observed for Grey Campanian Tuff; Cioffi *et al.* 1991). It is therefore important to note that the microstructure of Neapolitan Yellow Tuff investigated is less continuous than that encountered in any Römer tuff (Ebisch 2005).

With respect to relevant physical properties such as apparent porosity (Table 1; Bolondi *et al.* 2007), Neapolitan Yellow Tuff, like Römer tuff, displays considerable variation. Compared to Römer tuff, Neapolitan Yellow Tuff shows comparable porosity to the variety of Römer tuff currently

Table 2. *Freeze/thaw resistance of evaluated Italian tuffs, compared to those of selected Rhenish tuffs (Nijland & van Hees 2003), using same test procedures*

Tuff		Wetting conditions (% vacuum)		
		100%	75%	50%
Neapolitan Yellow Tuff	1 2	↑	↑	Some scaling, not possible to break by hand
Tufo Romano	1 2	Strong exfoliation, complete disintegration on first touch	Strong exfoliation, complete disintegration on first touch	No visual damage
Tufo Etrusco	1 2	↓	↓	No visual damage, but brittle and easily broken by hand
Römer tuff, restoration St John's cathedral	1 2	No visual damage	↑	↑
Weiberner tuff, quarry material	1 2	↑		
Weiberner tuff, quarry material	1 2	Strong exfoliation	No visual damage	No visual damage
Weiberner tuff, quarry material	1 2	↓ Brittle	↓	↓

applied in the restoration of St John's cathedral, 's-Hertogenbosch. The latter also shows good freeze/thaw resistance in the laboratory. This is unfortunately not the case with the type of Neapolitan Yellow Tuff investigated in this study (Table 2). In addition to the properties presented here, Neapolitan Yellow Tuff shows a slightly larger loss of material in response to acid rain as well as considerably higher hygric expansion coefficients compared to Römer tuff (Bolondi *et al.* 2007). In summary, although Neapolitan Yellow Tuff might provide an aesthetic alternative to Römer tuff, the type currently investigated shows considerable less freeze/thaw resistance. With respect to several other properties, compatibility with the same tuff stone masonry is questionable.

Part of the research was funded by the Netherlands Department of Conservation, RACM. Discussions with RACM's H. J. Tolboom are gratefully acknowledged.

References

BLACIC, J. D. 1993. Hydration swelling effects on time-dependent deformation of zeolitized tuff. *Jorunal of Geophysical Research*, **98B**, 15909–15917.

BOLONDI, L., NIJLAND, T. G. & VAN HEES, R. P. J. 2007. Performance of stone outside its original environment: Neapolitan Yellow Tuff as replacement stone for Rhenish tuff in the Netherlands. *In*: *CITTAM Conference 'Stone' Building Between Innovation and Tradition, Naples*. Luciano, Naples, 121–126.

BRENDLE, S. 2003. *Weathering of tuff stone*. TNO report 2003-CI-R0044, TNO, Delft.

CARDONE, V. 1990. *Il Tufo Nudo Nell'architettura Napoletana*. CUEN, Naples.

CIOFFI, R., MARINO, O. & MASCOLO, G. 1991. The physical action of water on the decay of building grey-tuff stone. *Materials Engineering*, **2**, 263–275.

DE' GENNARO, M., CALCATERRA, D., CAPPELLETTI, P., LANGELLA, A. & MORRA, V. 2000. Building stone and related weathering in the architecture of the ancient city of Naples. *Journal of Cultural Heritage*, **1**, 399–414.

DE' GENNARO, M., COLELLA, C., AIELLO, R. & FRANCO, E. 1984. Italian zeolites 2. Mineralogical and technical features of Campanian tuff. *Industrial Minerals*, **204**, 97–109.

DE OLDE, H. 2002. Tufstenen kerken in Groningen. *Groninger Kerken*, **19**(1), 4–30.

DE VRIES, D. J., BLOEMINK, J. W. & PROOS, R. H. P. 1992. De Proosdij in Deventer. *Bulletin Koninklijke Nederlandse Oudheidkundige Bond*, **91**, 156–165.

DEN HARTOG, E. 2002. *De Oudste Kerken van Holland. Van kerstening tot 1300*. Matrijs, Utrecht.

EBISCH, M. 2005. *TNO atlas of tuff stone (version 1.0)*. TNO report 2005-CI-R0138, TNO, Delft.

FITZNER, B. 1994. Volcanic tuffs: The description and quantitative recording of their weathered state. *In*: CHAROLA, A. E., KOESTTLER, R. J. & LOMBARDI, G. (eds) *Lavas and Volcanic Tuffs. Proceedings of the International Meeting, Easter Island, Chile*, 1990. ICCROM, Rome, 33–51.

FITZNER, B. & LEHNERS, L. 1990. Rhenish tuff – A widespread, weathering-susceptible natural stone. *In*: PRICE, D. G. (ed.) *Proceedings of the 6th International*

Congress of the International Association of Engineering Geology. Balkema, Rotterdam, 3181–3188.

HAIDUCK, H. 1992. *Beginn und Entwicklung des Kirchenbaues im Küstengebiet zwischen Ems- und Wesermündung bis zum Anfang des 13 Jahrhunderts.* Ostfriesische Landschaft, Aurich.

HEIKEN, G., FUNICIELLO, R. & DE RITA, D. 2005. *The Seven Hills of Rome. A Geological Tour of the Eternal City*. Princeton University Press, Princeton & Oxford.

LANGELLA, A., CALCATERRA, D., CAPPELLETTI, P., COLELLA, A., DE' GENNARO, M. & DE GENNARO, R. 2000. Preliminary contribution on durability of some macroporous monumental stones used in historical towns of Campania region, southern Italy. *In*: FASSINA, V. (ed.) *Proceedings of the 9th International Congress on Deterioration and Conservation of Stone*, Venice, June 19–24 2000, Elsevier Science, **1**, 59–67.

NEN 2872 1989. *Beproeving van Steenachtige Materialen; Bepaling van de Vorstbestandheid; Eenzijdige Bevriezing in Zoetwatermilieu.* NEN, Delft.

NIJLAND, T. G. & VAN HEES, R. P. J. 2003. *Beoordeling van Weiberner en Römer tufsteen ten behoeve van de restauratie van de St. Janskathedraal te 's-Hertogenbosch.* TNO report 2003-CI-R0042, TNO, Delft.

NIJLAND, T. G. & VAN HEES, R. P. J. 2006. Use of Rhenish tuff in the Netherlands. ARKUS-Tagung Denkmalgestein Tuff, Koblenz. *Institut für Steinkonservierung Bericht*, **22**, 7–18.

NIJLAND, T. G., BRENDLE, S., VAN HEES, R. P. J. & DE HAAS, G. J. L. M. 2003. Decay of Rhenish tuff in Dutch monuments. Part 1: Use, composition and weathering. *Heron*, **48**, 149–166.

NIJLAND, T. G., DUBELAAR, W. & TOLBOOM, H. J. 2007. De historische bouwstenen van Utrecht. *In*: DUBELAAR, W., NIJLAND, T. G. & TOLBOOM, H. J. (eds) *Utrecht in Steen. Historische Bouwstenen in de Binnenstad.* Matrijs, Utrecht, 31–109.

NIJLAND, T. G., VAN HEES, R. P. J., BRENDLE, S. & GOEDEKE, H. K. 2005. Tufsteen. Deel 2: Invloed van vocht op de duurzaamheid van 'Rheinische' tuf. *Praktijkboek Instandhouding Monumenten* **21**(15).

RILEM CPC 11.3 1984. Absorption of water by concrete by immersion under vacuum. *Materials & Structures*, **17**.

SCHMINCKE, H. U. 1988. *Vulkane im Laacher See-Gebiet. Ihre Entstehung und heutige Bedeuting.* Doris Bode Verlag, Haltern.

SERSALE, R. & AIELLO, R. 1964. Costituzione e reattivà del 'trass' renano. *L'Industria Italiana del Cemento*, **34**, 747–760.

SLINGER, A., JANSE, H. & BERENDS, G. 1980. *Natuursteen in Monumenten.* Rijksdienst voor de Monumentenzorg, Zeist/Bosch & Keuning, Baarn.

VAN HEES, R. P. J., BRENDLE, S., NIJLAND, T. G., DE HAAS, G. J. L. M. & TOLBOOM, H. J. 2003. Decay of Rhenish tuff in Dutch monuments. Part 2: Laboratory experiments as basis for the choice of restoration stone. *Heron*, **48**, 167–177.

VAN HEES, R. P. J., BRENDLE, S., NIJLAND, T. G., DE HAAS, G. J. L. M. & TOLBOOM, H. J. 2004. Decay of Rhenish tuff in Dutch monuments. *In*: KWIATKOWSKI, D. & LÖFVENDAHL, R. (eds) *Proceedings of the 10th International Congress on Deterioration and Conservation of Stone*, ICOMOS Sweden, Stockholm, **1**, 91–98.

VAN HEES, R. P. J., DUBELAAR, C. W. & NIJLAND, T. G. 2005. Toepassing, verwering en vervanging an witte Belgische steen in Nederland. *Praktijkboek Instandhouding Monumenten*, **24**(16).

Petrophysical and mechanical properties of soft and porous building rocks used in Apulian monuments (south Italy)

GIOACCHINO F. ANDRIANI & NICOLA WALSH*

Dipartimento di Geologia e Geofisica, Università degli Studi di Bari, Via Orabona 4, 70125 Bari, Italy

**Corresponding author (e-mail: nwalsh@geo.uniba.it)*

Abstract: This paper brings a comprehensive review of the main petrophysical and mechanical properties of calcarenite rocks used from time immemorial in Apulia (south Italy), with load-bearing and decorative functions both in constructions of specific historic and architectonic interest and in more common buildings. These soft and porous rocks show a reduced ability to maintain their characteristics of strength, appearance and resistance to decay over a considerable period of time. Even more than other sedimentary rocks, calcarenites belonging to the same formation can change considerably in terms of physical properties and mechanical behaviour due to the complex spatial arrangement of facies strongly conditioned by depositional fabric and diagenetic processes. A number of calcarenite varieties belonging to the Calcarenite di Gravina Fm. and Pietra Leccese Fm. was selected from different parts of Apulia and characterized according to petrographical, physical and mechanical properties. These included porosity, pore size distribution, density, water absorption, degree of saturation, permeability, thermal properties as well as compressive strength and flexural strength. Particular attention was given to the relationships between rock fabric features and physico-mechanical behaviour of the calcarenites. In addition, a comparison of data for the examined varieties was also discussed. A classification of the Apulian calcarenites based on rock fabric features and uniaxial compressive strength was proposed. Critical observations regarding the durability of the Apulian calcarenites were made, taking into account other data from literature.

The Apulia region (south Italy) is essentially formed by shallow-water carbonates. Extensive deposits of fine- to coarse-grained calcarenites belonging to the Plio-Pleistocene successions of the Murge plateau (Iannone & Pieri 1982), Oligocene-Miocene and Plio-Pleistocene sequences of the Salento peninsula (Bossio *et al.* 1988) and Miocene and Pliocene sequences of the Gargano promontory (D'Alessandro *et al.* 1979; Abbazzi *et al.* 1996) characterize both the inner areas of the region, where open and underground quarries are still active today, and the coastal areas rich in small historic exploitation sites (Andriani & Walsh 2007*a*). These calcarenite deposits belong to the Calcarenite di Gravina Fm. (Middle Pliocene-Early Pleistocene), Lecce Fm. (Late Oligocene-Early Miocene) and Pietra Leccese Fm. (Late Burdigalian-early Messinian). Of lesser importance, due to a lesser extension of the outcrops and use in the course of time as building and ornamental stone, are some calcarenite varieties belonging to the Calcareniti di Porto Badisco Fm. (Late Oligocene), Calcareniti di Andrano Fm. (Late Miocene) and *Terraced marine deposits* dated from Middle Pleistocene to Late Pleistocene.

The ready availability, good workability and aesthetic appeal of the calcarenites, together with their lightness and low values of thermal diffusivity and conductivity, give them excellent insulation properties. This explains their continuing success as building and ornamental stone, despite strong competition from artificial materials that imitate their characteristics and technical properties.

As a result of the ease of excavation of the calcarenites, the areas where they outcrop have been settlement areas since ancient times. These settlements have the form of simple shelters in the rock and more complex artificial underground dwellings that are scattered over the territory of Apulia, principally along the sides of the '*Gravine*' and '*Lame*' (Parise *et al.* 2003). Subsequently, the calcarenite rock known in the Murgia area as '*Calcareous Tufa*', was widely used in the construction of modest habitations and prestigious buildings such as important churches, Romanesque cathedrals, fortified farms (*masserie*), imposing castles and mediaeval towers. In particular, some varieties belonging to the Pietra Leccese Fm. (Miocene) and Lecce Fm. (Late Oligocene-Early Miocene) are the principle stones in numerous monuments of the Lecce Baroque (17th–18th centuries) (Fig. 1).

Historically important quarry districts, some of them still active, are those of Gravina in Puglia,

From: PŘIKRYL, R. & TÖRÖK, Á. (eds) *Natural Stone Resources for Historical Monuments*.
Geological Society, London, Special Publications, **333**, 129–141.
DOI: 10.1144/SP333.13 0305-8719/10/$15.00

Fig. 1. The Basilica of Santa Croce (1548–1646), Lecce. The church was built using local stone (*Pietra leccese*).

Canosa di Puglia, Trani-Andria, Minervino Murge-Poggiorsini, Ginosa-Mottola-Massafra, Grottaglie-San Giorgio Jonico, Bari, Fasano and Polignano a Mare-Monopoli along the Murge edge, Ionian side and Adriatic coastal belt, those of Lecce, Cursi-Melpignano-Martano, Gallipoli and Cutrofiano in the Salento area and those of Apricena and San Giovanni Rotondo in the Gargano promontory (Fig. 2).

The methods of opening and excavating the pit, hole and cutting quarries differ according to the morphology of the locality. The underground quarries of Canosa di Puglia, Mottola, Gallipoli and Cutrofiano give a powerful impression of the hard work required to extract material from them. Here the tunnels have been dug on various levels and create serious problems of stability on the surface above (Bruno & Cherubini 2005). The area of underground quarries has been absorbed by rapid urban expansion and historical records of some of them have been lost (Walsh 2006).

Previous works on the geological setting, chemical and mineralogical composition, depositional environment and stratigraphy of the Apulian calcarenites have been presented by a great number of authors (Giovene 1810; Cappellini 1878; Di Stefano & Viola 1892; Sacco 1911; Gignoux 1913; De Giorgi 1922; D'Erasmo 1934; Cantelli 1960; D'Onofrio 1960; Valduga 1965; Martinis 1967; Ricchetti 1965, 1970; Azzaroli 1968; Dell'Anna *et al.* 1968, 1978; Di Geronimo 1969; Balenzano & Di Pierro 1972; Iannone & Pieri 1979; Caldara 1982; D'Alessandro & Iannone 1982; Bromley & D'Alessandro 1987; Ricchetti *et al.* 1988; Bossio *et al.* 1989, 1991; Palmentola 1989; Mazzei 1994; Tropeano & Sabato 2000; Pomar & Tropeano 2001; Margiotta & Ricchetti 2002; Margiotta & Varola 2004).

The physical and mechanical properties are described by Salvati (1932), Penta (1935), Nicotera (1953), Radina & Walsh (1972), Zezza (1974), Calò *et al.* (1985), Cotecchia *et al.* (1985), Zezza *et al.* (1989), Evangelista & Pellegrino (1990), Mongelli *et al.* (1993), Caputo *et al.* (1996) and Andriani *et al.* (2006). The influence of fabric and diagenesis on the physico-mechanical performance of the calcarenites is described by Andriani & Walsh (1998, 2000, 2002, 2003, 2007*a*).

The purpose of this paper is to emphasize and review the main petrophysical and mechanical properties of calcarenite rocks used in Apulian monuments and buildings. The petrography, porosity, pore size distribution, density, water absorption and degree of saturation of these rocks have been studied together with their permeability, thermal properties and strength in different physical states. Data were compared to determine critical observations regarding the resistence to weakening or deterioration over time of the Apulian calcarenites, taking into account other data from literature.

Material description and classification

The Apulian calcarenites are principally bioclastic dominated carbonate sediments, weakly cemented, characteristic of shallow marine temperate waters and foreshore, shoreface and offshore environments. For this study, various calcarenite varieties were sampled in areas in which old quarries of historical interest and important rock exploitation sites are located. The examined calcarenites comprise fine-, medium- and coarse-grained varieties belonging to the Calcarenite di Gravina Fm. (Middle Pliocene-Early Pleistocene) and fine-grained varieties of the Pietra Leccese Fm. (Miocene). In particular, samples of Plio-Quaternary calcarenites were taken in the areas around Gravina in Puglia (Tufare and Grotta Marallo localities), Poggiorsini (Grottellini locality), Canosa di Puglia (Pietra Caduta and Cefalicchio localities), Massafra (Caprocetta and Gravina di San Marco localities), Polignano a mare (San Vito and Santa Caterina localities) and Monopoli (Cala Corvino and Torre Cintola localities). Samples of Miocene calcarenites

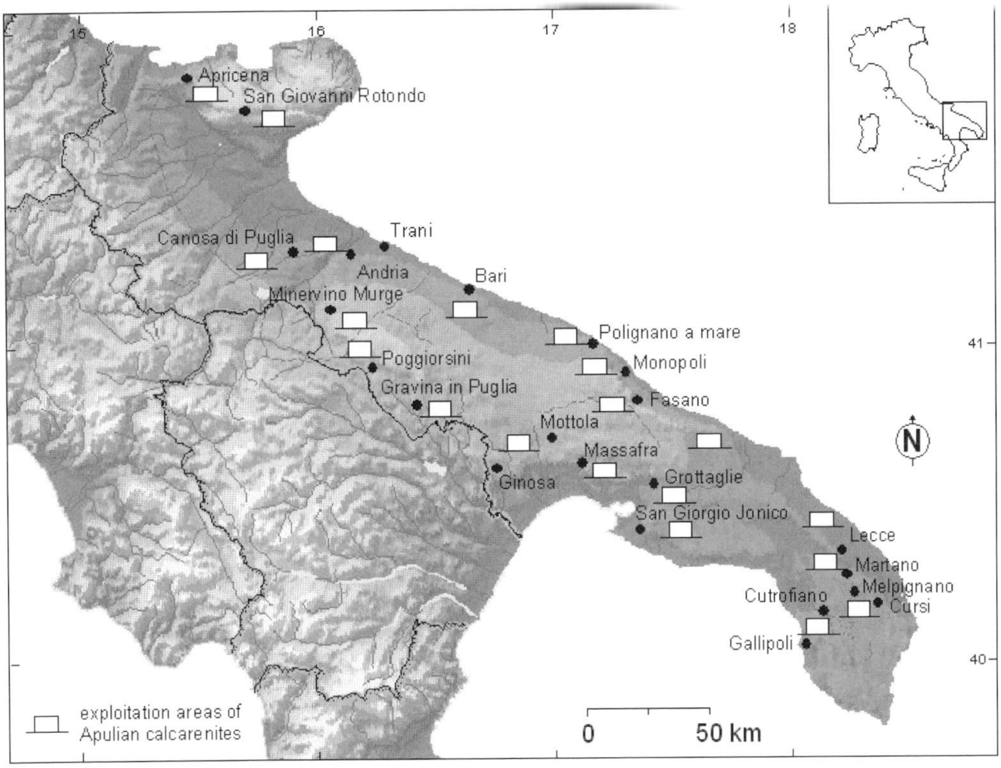

Fig. 2. Geographic location of the main extraction areas of the Apulian calcarenites.

were also taken in the Cursi-Melpignano-Martano extraction area (quarries located between Cursi and Melpignano) (Fig. 2). For the sake of simplicity, in the text we will use the term *Calcareous Tufa* for the Plio-Pleistocene lithofacies of the Murge area and *Pietra di Cursi* for those of the Miocene, found in the Salento area. The samples of *Calcareous Tufa* were classified into three categories on the basis of their grain size distribution: fine-, medium- and coarse-grained *Calcareous Tufa*. In addition, three varieties of *Pietra di Cursi* were considered, utilizing the same terminology that marks them in the commercial field. They are known as *Dura*, *Dolce* and *Gagginara* on the basis of their technical properties and principal use in the Salento area.

On a mesoscopic scale, the Apulian calcarenites are of a colour that varies between whitish and straw-yellow, tending to reddish. The homogeneity or lack of homogeneity of their appearance depends on the presence of inorganic and organic-sedimentary structures such as tabular planar or low-inclined laminations, vertical gradations and bioturbations and coarse valves of bioclasts.

Rock fabric examination was performed with transmitted light on standard thin-sections using optical polarizing microscopy. Thin-sections were taken from specimens, half of which were cut along and half across the stratification (Fig. 3).

The *Calcareous Tufa* varieties are composed of carbonates ($CaCO_3 \geq 97\%$) and (a minimal part) of clayey minerals (kaolinite, illite, chlorite, smectite and halloysite) with traces of quartz, feldspar, gibbsite and goethite. The granular framework is mainly formed by a bioclast fraction, represented by fragments of lamellibranchs, gastropods, scaphopods, brachiopods, balanis, dermal plates and prickles of echinoids, encrusting colonies of bryozoans, calcareous algae, oncolites, corals, serpulid worm tubes, benthic foraminifers and (rare) planktonic foraminifers and ostracod valves. Bioclasts exhibit micrite envelopes in places, while microboring is common especially in coarse bivalves. The lithoclasts comprise fragments of whitish-grey limestone, dolomitic limestone and havana-brown and blackish sub-rounded and, in places, sub-angular dolomites, from the erosion of the Mesozoic basement. The micritic matrix is predominantly crypto- and microcrystalline; it is mostly unresolvable with the polarizing microscope. This is carbonate mud which, within the limits of the varieties studied, is prevalently the result of the deposition

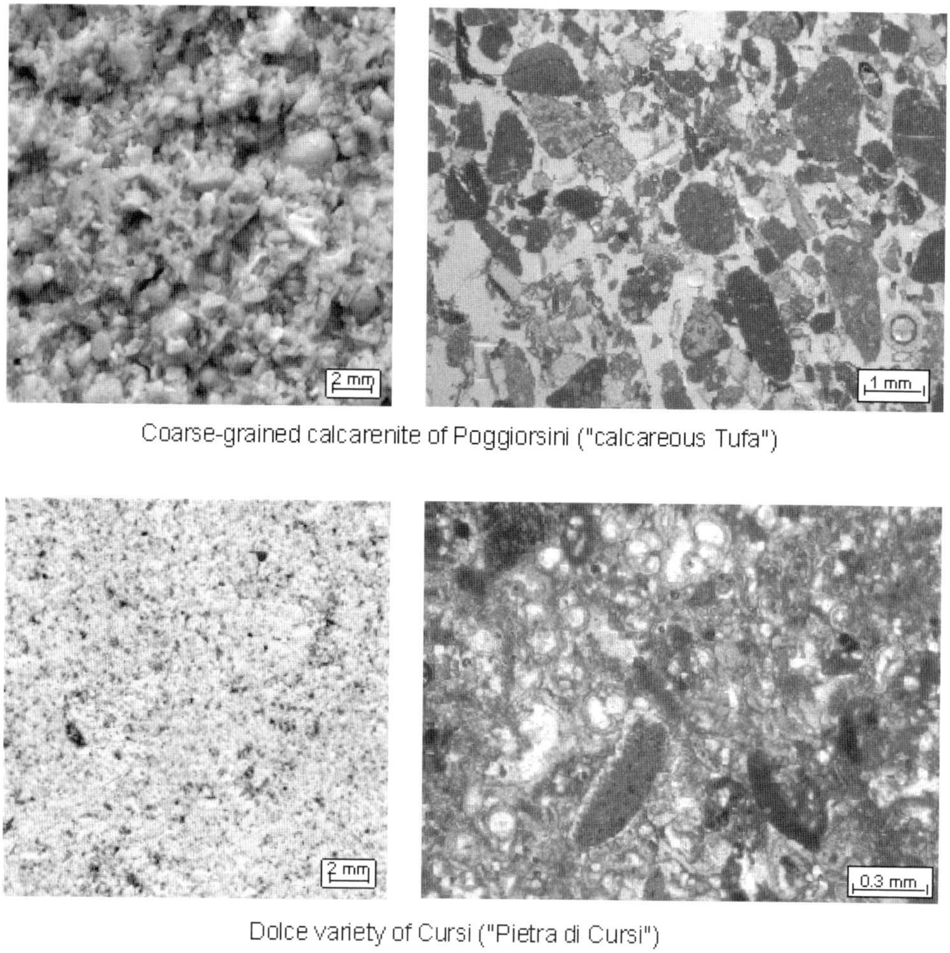

Fig. 3. Macroscopic and microscopic appearance of the coarse-grained calcarenite of Poggiorsini (*calcareous Tufa*) and the *Dolce* variety of Cursi (*Pietra di Cursi*). On the right, microphotographs in plane-polarized light; on the left, microphotographs of transversal sections of the specimens used in the experiments.

of bioclasts disintegrated by bioerosion and boring or simply by breaking off and abrasion in agitated marine waters (allomicrite). The micritic matrix is replaced in some places by microspar (aggrading neomorphism). The fabric is open and is typical of grain-supported to mud-supported bioclastic and biolithoclastic calcarenites that vary from well sorted to moderately sorted. These are principally biosparites, grainstone and packstone and, to a lesser degree, packed and sparse biomicrites, packstone and wackestone. The latter are very rare and characteristic, for instance, of a sedimentary facies which is located at the lower levels of the stratigraphic succession of the Calcarenite di Gravina Fm. observed at the *Caprocetta* quarries (Massafra).

The *Pietra di Cursi* varieties also reveal homogeneous minero-petrographic characteristics, as they are almost exclusively formed by low-Mg $CaCO_3$ (about 94%). A much lower quantity is found of glauconite, quartz grains, feldspar and rare pyroxenes (*Dolce* variety), as well as clayey minerals finely distributed in the matrix with a carbonate composition. The general fabric is one of a relatively well-packing and fine-grained calcarenite, with a self-supporting framework of skeletal grains of marine organisms (above all, planktonic foraminifers and, to a lesser extent, benthic foraminifers and rare lamellibranchs, bryozoans, echinoderms), fossil debris and pellets. The micritic matrix is not very common; it is dark coloured and predominantly forms a cryptocrystalline-based mass not resolvable by polarizing microscope or thin envelopes around skeletal grains (*Dolce* variety).

The greater part of the samples reveals grain-supported fabric, packstone in type. They are principally poorly washed biosparites and packed biomicrites (with the latter clearly subordinate in placement to the former) from very well sorted to moderately well sorted.

Unlike *Pietra di Cursi*, which generally shows good packing density, the degree of packing and spatial disposition of the grains together with the total porosity values indicate that, for the *Calcareous Tufa* varieties, the diagenesis of the carbonate sediment took place soon after deposition. The precipitation of the cement therefore occurred in the initial phases of compacting or even before experiencing increases of pressure and temperature due to burial. Not by chance, the *Calcareous Tufa* varieties studied show meniscus calcitic cement (early-stage cement) at grain contacts in many cases. This is accompanied by a border of finely crystalline calcite on their external surfaces, covered only in some places by lengthened crystals and microcrystals with a scalenohedronic or rhombohedral form (dog tooth cement).

Late stage cement (sparry calcite), which partially or totally fills pore spaces, is typical only of the more resistant varieties of *Calcareous Tufa* (e.g. medium-grained variety of *Grotte Marallo*, Gravina in Puglia). It is common in the *Pietra di Cursi* varieties in the form of moulds formed by the dissolution of bioclasts, especially those smaller in size.

A stime using a point count method on optical microscope reveals that the quantity of cement varies between 8% and 24% in the *Calcareous Tufa* varieties, and between 15% and 22% in that of *Pietra di Cursi*. Using this method, it is very difficult to obtain a reliable evaluation of the quantity of cement because of the effects of the phenomena of recrystallization and/or neomorphism in the rock. In fact, with a polarizing microscope it is not always easy to establish the difference between recrystallization, neomorphic fabric and fine sparry cement. Recrystallization and neomorphic fabric are especially evident in the varieties with a high bioclast content and those which are fine-grained with micritic matrix. Finally, in all the varieties (especially in those of *Pietra di Cursi*) it is possible to find traces of bioturbations: these are branching burrows, holes, passages and traces of locomotion left by organisms during sedimentation which were subsequently filled with non-selected material, formed by micritic carbonate mud surrounding chaotically spread lithoclasts and bioclasts. The fabric within these passages is different from that of the surrounding material as well as the amount of calcite cement.

On the basis of the pore types and porosity classification of carbonate rocks proposed by Choquette and Pray (1970), in all the varieties of *Calcareous Tufa* the greatest contribution to the total porosity is provided by the primary intergranular porosity. This is followed, according to an order that may vary from variety to variety, by intragranular porosity, mouldic porosity and fracture porosity on a microscopic and mesoscopic scale. On the contrary, mouldic porosity (generated by dissolution of aragonite bioclasts) and intragranular porosity, essentially linked to the internal structures of the skeletal shells, are especially effective in the *Pietra di Cursi* varieties. Intercrystal porosity is typical of lithofacies showing the effects of recrystallization and/or neomorphic processes. Isolated porosity, linked to the non-communicating interstices caused by the effect of irregular cementation of the grains, is uncommon. Its contribution to total porosity is always less than 6%. With the exception of the coarse-grained variety of *Pietra Caduta* (Canosa di Puglia) and the medium-grained variety of *Grotta Marallo*, all the varieties belonging to *Calcareous Tufa* and *Pietra di Cursi* are characterized by open porosity so that all the pores are interconnected and accessible.

Physical and mechanical properties

Following the standard test procedure outlined in ISRM (1979), EN 1926 (1999) and EN 12372 (1999), dry density δ_d, total porosity n, uniaxial compressive strength in the dry σ_n and saturated state σ_{sat} and after 20 freeze-thaw cycles σ_{ft} and flexural strength in the dry state σ_f were determined on 10 samples of each variety of the calcarenites considered. In particular, according to Andriani & Walsh 2003, 2007*b* water absorption w_a and degree of saturation Sr were evaluated on specimens immersed and suspended in distilled water at 20 °C for 48 h and then saturated completely under vacuum (80 kPa) without removing them from the water basket. Full saturation (Sr = 100%) was obtained for almost all the varieties studied. The degree of saturation reaches 97.4% and 94.3% only for the coarse-grained variety of *Pietra Caduta* (Canosa di Puglia) and the medium-grained variety of *Grotta Marallo* (Gravina in Puglia), respectively. It follows that porosity in the Apulian calcarenites can be considered an effective porosity.

Considering that there is no unified definition of soft or weak rock and the conventional classification schemes for intact rock appear to be inappropriate to synthesize the complex stress-strain behaviour of calcarenites, the subdivision of the calcarenite varieties in categories was based on rock fabric features and uniaxial compressive strength. Physical properties and geotechnical behaviour of a sedimentary rock are, in fact, controlled strongly by depositional and diagenetic fabric (Flügel 2004). The samples examined were therefore subdivided

Table 1. *Classification of the Apulian calcarenites (AC)*

Group	Range of uniaxial compressive strength (MPa)	General rating of rock based on strength	Rock fabric features
AC1	10–25	Moderately soft	Medium- to fine-grained packstone and grainstone; partial and total void-filling drusy and granular cement; tangent and long contacts between grains
AC2	5.0–10	Soft	Medium-grained grainstone and packstone; partial void-filling and pore-lining dog tooth cement; tangent and long contacts between grains
AC3	1.0–5.0	Very soft	Coarse-grained grainstone, medium-grained packstone; scarce cement, meniscus and microcrystalline in types; tangent contacts between grains; medium-fine wackestone with a crypto- and/or microcrystalline-based mass
AC4	0.6–1.0	Extremely soft	Coarse grainstone and medium packstone very scarce in cement, microcrystalline in type; floating and tangent contacts between grains; microsparstone as a result of complete obliterative recrystallization or replacement

in four groups: moderately soft, very soft, soft and extremely soft (Table 1). Total porosity was obtained from the classical expression

$$n = \left(1 - \frac{\delta_d}{G_S \delta_w}\right) 100\% \qquad (1)$$

using measured values of dry density δ_d, water density ($\delta_w = 1.0\ \mathrm{Mg\ m^{-3}}$) and assumed specific gravity (G_s) of 2.7 on the basis of the mineralogical composition of all the calcarenites examined.

A detailed study of pore size distribution was carried out by mercury intrusion porosimetry (MIP) technique on oven-dried samples of about 2.5 g using a Micromeritics porosimeter (Autopore IV 9500). The analyses were performed at low (3.44–345 kPa) and high pressure (0.1–228 MPa) on calcarenite fragments of irregular shape detached from fresh specimens. Considering the limitation of the operative conditions and the applied method, the pore size distribution and relative porosity (n_{MIP}) for pores with a diameter between 0.005 μm and 420 μm were evaluated.

For coarse- and medium-grained varieties, the results of the MIP were integrated with the pore size distribution obtained by image analysis on microphotographs of thin sections, according to the procedure proposed by Andriani & Walsh (2002). The Apulian calcarenites are, in fact, characterized by a wide distribution of pores which, although unimodal or bimodal, also includes coarse tails for the medium-grained variety. Cumulative curves were obtained in this way for the pore size (diameter) by a combination of both image analysis and MIP tests (Fig. 4).

Analysis of grain size distribution was also carried out. To obtain loose material for the grain size analyses, one representative saturated cylindrical sample of each calcarenite variety was subjected to numerous freeze-thaw cycles and then disaggregated by hand. The loose material thus obtained was dried in an oven at 105 °C for 24 hours and afterwards sieved using sieve sizes ranging from 2.00–0.063 mm. The remaining fine fractions (passing 230, ASTM series) were examined through sedimentation analysis. A comparison of the cumulative curves obtained for the *Calcareous Tufa* varieties and the *Pietra di Cursi* varieties is shown in Figure 5.

Water permeability tests were conducted in a purpose-built cell on cylindrical rock samples

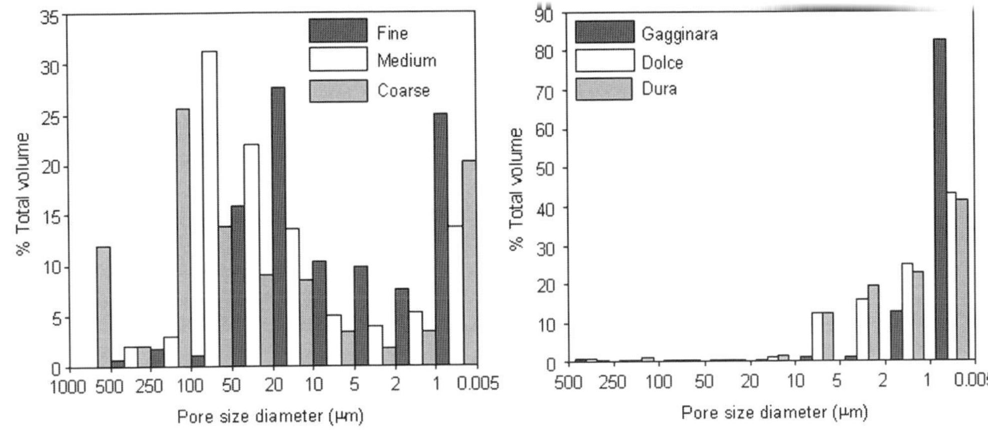

Fig. 4. Percent frequency of pore size diameter carried out by mercury intrusion porosimetry technique and image analysis. On the right, plot of the *Pietra di Cursi* varieties; on the left, plot of the *Calcareous Tufa* varieties: fine-, medium- and coarse-grained varieties from Massafra, Gravina in Puglia and Poggiorsini, respectively.

(diameter 71 mm and height 140 mm) using the constant head and falling head methods according to the procedure proposed by Andriani & Walsh (2003). The hydraulic conductivity standardized at 20 °C (k_{20}) was evaluated for a range of hydraulic gradients between 0.5 and 15.

Thermal properties of the calcarenites were obtained from the measurement of the thermal linear expansion coefficient α_l between 20 °C and 80 °C on rock bars of 350 mm × 15 mm × 15 mm, the thermal conductivity λ, specific heat Cp and thermal diffusivity D using the experimental 'cut carrot' method (Mongelli 1968), first in the dry state then in the saturated state and for different water contents.

Results and discussion

Geological factors influencing petrophysical data

Before providing any considerations about the obtained data and with reference to Tables 2 and 3, it is necessary to point out that, within each category that was proposed in the text for *Calcareous Tufa*, different calcarenite varieties were classified. On the other hand, each category of *Pietra di Cursi* corresponds to a single variety. For this reason, the ranges of values obtained from physical and mechanical tests are wider for each category of *Calcareous Tufa* than for those of *Pietra di Cursi*.

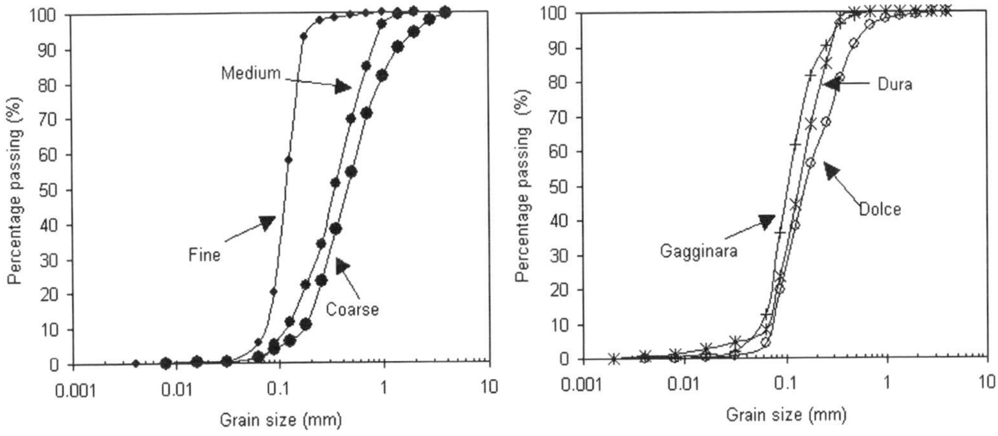

Fig. 5. Grain size distribution curves obtained using sieve and sedimentation analysis. On the right, characteristic curves of the *Pietra di Cursi* varieties; on the left, typical curves of the *Calcareous Tufa* varieties: fine-, medium- and coarse-grained varieties from Massafra, Gravina in Puglia and Poggiorsini, respectively.

Table 2. *Physical and mechanical properties of the Apulian calcarenites*

Properties	Calcareous Tufa			Pietra di Cursi		
	Fine	Medium	Coarse	Dura	Dolce	Gagginara
Specific gravity, G_s		2.70			2.70	
Dry density, δ_d (Mg m^{-3})	1.3–1.8	1.4–2.3	1.2–1.7	1.5–1.9	1.5–1.7	1.5–1.6
Sat. density, δ_{sat} (Mg m^{-3})	1.8–2.1	1.9–2.4	1.8–2.1	1.9–2.2	1.9–2.1	1.9–2.0
Porosity, n (vol.%)	33–52	15–48	37–56	30–44	37–44	41–44
Water absorption, w_a (wt.%)	18–40	6–34	21–47	16–29	22–29	26–29
Degree of saturation, Sr (%)	100	94–100	97–100	100	100	100
Compr. strength (dry), σ_n (MPa)	1.4–6.5	1.5–25.0	0.9–5.2	16.7–22.7	12.8–15.5	11.3–18.3
Compr. strength (sat), σ_{sat} (MPa)	0.9–6.0	1.1–24.0	0.7–5.0	13.0–22.1	8.1–10.1	9.4–12.3
Compr. strength (fr-th), σ_{ft} (MPa)	0.6–4.2	0.9–19.0	0.5–4.0	10.1–18.0	6.8–10.0	7.6–12.1
Flexural strength (dry), σ_f (MPa)	0.2–1.5	0.3–7.8	0.2–1.3	3.3–5.0	3.1–3.9	1.6–3.2
Hydraulic conductivity, k_{20} (10^{-5} m s^{-1}) Constant head test	0.74–2.1	0.34–8.3	7.8–12	3.5–4.9	4.1–5.6	3.1–3.6
Hydraulic conductivity, k_{20} (10^{-5} m s^{-1}) Falling head test	0.92–3.4	0.46–8.9	9.2–14	6.2–7.1	7.1–7.5	4.5–6.0

In addition, *Calcareous Tufa* constitutes wide and continuous exposure of calcarenites composed of several lithofacies from relatively well-cemented and massive to thinly laminated and irregularly cemented. The complex arrangement of facies, both vertically and laterally, is strongly conditioned by depositional fabric and diagenetic processes and derives directly from the particular depositional environment and the underlying substrate irregularities. It follows that most of *Calcareous Tufa* are anisotropic when considering rock fabric at the sample scale. An anisotropic fabric reveals an anisotropic material behaviour and this can be caused by microstructural features such as preferred grain orientation and lamination. The assumption of isotropy in terms of physical and mechanical properties can be approximately considered valid only for *Pietra di Cursi* with random distribution of allochems, microcracks and pores. The main factor of an anisotropic behaviour for *Pietra di Cursi* is the presence of bioturbations.

Starting from the specimen's preparation, it is more difficult for *Calcareous Tufa* than for *Pietra di Cursi*. Some varieties of *calcareous Tufa* are locally very friable, so that the rocks can easily break apart. This is the case for the coarse-grained varieties from Poggiorsini (*Grottelini* locality) and Canosa di Puglia (*Pietra Caduta* locality), which are irreguraly cemented with a low grain packing.

In general, as can be seen from Table 2, the *Pietra di Cursi* varieties are characterized by lower total porosity (Table 2), lower permeability and higher strength than those of *Calcareous Tufa*. Grain size, sorting and degree of packing seem to have no influence on porosity, which is greater in the calcarenite with bioclast content. The

Table 3. *Thermal properties of the Apulian calcarenites (average values)*

Varieties		Conductivity λ (W m^{-1} K^{-1})		Specific heat C_p (kJ kg^{-1} K^{-1})		Diffusivity D (10^{-7} m^2 s^{-1})		Linear expansion, α_l (10^{-6} K^{-1})
		dry	sat.	dry	sat.	dry	sat	sat
Calcareous Tufa	Fine	0.9	1.1	1.2	1.5	4.8	3.8	3.2
	Medium	0.8	1.4	1.2	1.9	3.6	3.4	3.7
	Coarse	0.7	1.0	0.9	1.4	5.4	3.7	2.1
Pietra di Cursi	Dura	1.0	1.5	1.1	1.5	5.6	5.1	5.2
	Dolce	1.0	1.4	1.3	1.4	4.7	4.2	4.6
	Gagginara	0.9	1.4	1.2	1.8	4.4	4.0	3.5

determination of porosity and degree of saturation has shown an open porosity with intercommunicating voids for almost all the varieties.

Permeability

Regarding permeability, the experimental values of the constant head text show ranges of variation of k_{20} between 0.34×10^{-5} and 12×10^{-5} m s^{-1} for *Calcareous Tufa* and between 3.1×10^{-5} and 5.6×10^{-5} m s^{-1} for *Pietra di Cursi*. In the falling head text, the hydraulic conductivity measurements vary between 0.46×10^{-5} and 14×10^{-5} m s^{-1} for *Calcareous Tufa* and between 4.5×10^{-5} and 7.5×10^{-5} m s^{-1} for *Pietra di Cursi*. With the exception of some coarse-grained varieties of *Calcareous Tufa* (Poggiorsini, Canosa di Puglia) which have shown high values of hydraulic conductivity, the ranges of data measured reveal a moderate water permeability for the Apulian calcarenites. A wider range of values was obtained for *Calcareous Tufa* due to higher variability in rock fabric features. The coarse-grained varieties show a higher water permeability than the other varieties due to their reduced degree of packing and a remarkable presence of intergranular macropores that provide rapid fluid transfer across the samples.

Grain size seems to have no direct influence on the permeability, which was lower in the calcarenites with a higher degree of packing, and on matrix and cement contents. In addition, other factors being equal, medium-grained varieties with sparry calcite (granular in type) show the lowest values of the hydraulic conductivity.

Thermal behaviour

The analytical examination of the thermal data suggests that *Calcareous Tufa* show a lower capability to conduct, propagate and accumulate heat with respect to *Pietra di Cursi* (Table 3). This is due largely to their loose degree of packing. In general, the thermal conductivity and the thermal diffusivity are higher in the calcarenites with a higher degree of packing and lower total porosity. No direct influence of grain size on the thermal properties was observed. Moreover, in all the varieties, the substitution of the air with water causes an increase in the thermal conductibility λ and a reduction of the thermal diffusivity D. In practice, considering that

$$D = \frac{\lambda}{Cp\,\delta} \tag{2}$$

the substitution of air by water leads to a more modest increase of λ than that of the product of Cp (specific heat) by δ (density of the stone), as the values of λ and Cp of the air (0.024 W m^{-1} K^{-1} and 1.01 kJ kg^{-1} K^{-1} at 25 °C, respectively) are lower than those of the water (0.60 W m^{-1} K^{-1} and 4.19 kJ kg^{-1} K^{-1} at 25 °C, respectively) while D is of two orders higher ($D_a = 1.87\ 10^{-5}$ m^2 s^{-1}; $D_w = 1.44\ 10^{-7}$ m^2 s^{-1}).

Strength

According to the strength classification system for intact rocks proposed by Deere & Miller (1966), all the investigated calcarenite varieties investigated fall into very low strength class. The uniaxial compressive strength (UCS) at the dry state is in fact less than 25 MPa; the calcarenites can therefore be considered as soft rocks. On average *Pietra di Cursi* shows higher values of UCS than those of *Calcareous Tufa*. It is characterized by a good packing density, wider distribution of cement and is richer of sparry calcite (although smaller in size). For the Apulian calcarenites, the positive correlation between UCS and dry density or grain packing (Andriani & Walsh 1998, 2000, 2003) is not always verified. Even although the influence of fabric on the behaviour of the calcarenite is difficult to quantify, it is possible to state that the strength of these soft rocks is above all controlled by type and amount of calcite cement. The highest UCS value (25 MPa) characterizes the medium-grained variety of *Grotte Marallo* (Gravina in Puglia) with widespread sparry calcite, granular in type. The lowest UCS value is of 0.9 MPa and is typical of the coarse-grained variety of Poggiorsini, with low grain packing and little early cement irregularly dispersed. Sample preparation was more complex in this latter case, and this might have influenced the UCS value. However, a strain-softening behaviour was always observed for the Apulian calcarenites. Differences in mechanical behaviour for a single variety can be attributed to the anisotropy of samples which show lamination and, in the thin section, some clusters of higher grain packing. Specimens which present the maximum strength were therefore cut with their axis practically parallel to the planes of anisotropy. This takes place more frequently for *Calcareous Tufa*; on the other hand the general behaviour of the *Pietra di Cursi* specimens is similar in all directions and can be considered approximately isotropic regarding UCS.

At the saturated state, the UCS values decrease by about 17% and 24% on average for *Calcareous Tufa* and *Pietra di Cursi*, respectively. The mechanical behaviour of the Apulian calcarenites is strongly dependent on whether the specimens are dry or saturated with water. Measuring the UCS on calcarenite samples saturated under vacuum and subjected to 15 soaking and drying cycles with distilled water, Andriani & Walsh (2007*b*) demonstrated that the negative influence of water imbibition on the overall resistance of some calcarenite varieties increases with the number the cycles, especially for fine-grained varieties. These showed a decrease in the UCS of 45% at the end of the test. In substance, the behaviour of calcarenites is affected by the presence of water in pores. Fine-grained varieties are able to hold water during the UCS test maintaining a high degree of saturation. Coarse- and medium-grained calcarenites, on the other hand, show a higher percentage of intercommunicating meso- and macropores which allow a sudden loss of water during the UCS measurements. Open porosity and pore size distribution influence the water absorption and retention of the rocks. Thus *Pietra di Cursi*, especially the *Dolce* and *Gagginara* varieties, lose water very slowly when the samples are removed from the water basket.

After 20 cycles of freeze-thaw, the UCS values can decrease to 46% for *Calcareous Tufa* (fine-grained variety) and 41% for *Pietra di Cursi* (*Dolce* variety), indicating that the Apulian calcarenites are not durable regarding freezing-thawing. It is clear that these are limit values as the sensitivity of calcarenite strength due to freezing-thawing varies between varieties and above all depends upon pore size distribution. Although it is well known that crystallization pressure is inversely related to pore size (Weyl 1959; Arnold & Zehnder 1989; Rodriguez-Navarro & Dohene 1999; Scherer 2000; Flatt 2002; Andriani 2006; Andriani & Walsh 2007*b*), samples with a high proportion of pores with diameters smaller than $<10\ \mu m$ connected to larger pores, and those weakly and irregularly cemented with higher grain size seem to be the most susceptible to frost damage.

The fine-grained varieties and some of the coarse-grained varieties (Poggiorsini, Canosa di Puglia) of *Calcareous Tufa* and the *Dura* and *Dolce* varieties of *Pietra di Cursi* therefore reveal relevant sensitivity to freezing-thawing. Detachments of coarse fragments from rock samples occur before the end of the 20th freeze-thaw cycle in loosely packed calcarenites (*Calcareous Tufa*) with a small amount of early-stage calcite cement growing irregularly at the contact between grains.

Regarding flexural strength (FS) in the dry state, for the Apulian calcarenite it is lower than UCS as expected. In particular, the ratio between UCS and FS varies on average from between 3.2 and 4.6 for *Calcareous Tufa* and between 4.1 and 6.2 for *Pietra di Cursi*. The higher values of this ratio refer to the fine-grained varieties for *Calcareous Tufa* and the *Gagginara* variety for *Pietra di Cursi*. The obtained data do not allow simple cross-correlation between fabric features and flexural behaviour of the material. Generally speaking, *Pietra di Cursi* is approximately isotropic regarding FS; on the other hand *Calcareous Tufa* shows maximum strength when loaded normally to the planes of anisotropy.

Durability

It is clear from the results that the Apulian calcarenites are sensitive to weatherability. The latter depends on both pore structure and rock strength (Benavente *et al.* 2004). In fact, the Apulian calcarenites are characterized, for the most part, by an interconnected system of pores and wide distribution of pore diameters that include micro-, meso- and macropores. The pore size distribution is bimodal for medium- and coarse-grained varieties, and unimodal for fine-grained varieties. With the exception of the *Gagginara* variety, which almost exclusively presents micropores, the connection of a large number of micropores to meso- and/or macropores in the rock pore system is responsible for the potential of the stone to take in and hold water solutions, and hence to weather. In other words, open fabric and the interconnection of intergranular and mouldic pores to intragranular and intercrystal pores determine the hydraulic behaviour of the stone in terms of sorptivity, hygroscopicity, water absorption and retention and provide a qualitative evaluation of the potential weatherability for the stone (Andriani & Walsh 2003; Benavente *et al.* 2007).

The presence of interstitial water plays a significant role in reducing the strength of the calcarenites, especially in weak cemented and fine-grained varieties. An increase in water content and saturation persisting over time tends to decrease the range of elastic behaviour of the calcarenites. The negative influence of water on the overall resistance of the Apulia calcarenites is also evident from the uniaxial compressive strength data obtained after freeze-thaw cycles. By analogy with the growth of salt crystals in porous systems, it is possible to confirm that these rocks are very susceptible to the process and mechanism of salt weathering (Everet 1961; Fitzner & Snethlage 1982; Goudie & Viles 1997; Scherer 2000). Salt damage by hydration, crystallization and thermal expansion are the most common deterioration processes in Apulia, especially, in coastal areas (Zezza & Macrì 1995; Andriani & Walsh 2007*a*, 2007*b*).

Conclusions

Many historic buildings and monuments in Apulia have been built with soft and porous calcarenites due to their ready availability, easy workability and aesthetic appeal together with their lightness and good thermal performances in terms of thermal diffusivity and conductivity. At the same time, these rocks are particularly susceptible to weathering by environmental pollution, marine aerosols and meteoric precipitations as a consequence of their low overall resistance and hydraulic behaviour, closely linked to the geometry and topology of the pore network.

Different calcarenite varieties belonging to *Calcareous Tufa* and *Pietra di Cursi* were classified into three categories and submitted to the same petrophysical and mechanical tests. The results obtained allowed further classification into four groups according to their rock fabric features and uniaxial compressive strength: moderately soft (10–25 MPa), soft (5–10 MPa), very soft (1–5 MPa) and extremely soft (below 1 MPa).

Special importance was given to the rock fabrics influencing the anisotropy of the technical properties. Determination of porosity and degree of saturation has shown an open porosity with intercommunicating voids for almost all the varieties. In general, the *Pietra di Cursi* varieties were characterized by lower total porosity, lower permeability and higher strength than those of *Calcareous Tufa*. Regarding thermal properties, *Calcareous Tufa* has shown a lower capability to conduct, propagate and accumulate heat with respect to *Pietra di Cursi*. The mechanical behaviour of all the varieties was strongly controlled by the presence or absence of water in pores. The Apulian calcarenites have shown high sensitivity to the freeze-thaw cycles. The stone varieties with wide pore size distribution (including micro- and macropores) and those weakly and irregularly cemented are the most sensitive to frost damage and, by analogy, to salt deterioration.

Thanks are due to Joann Cassar and an anonymous referee for their helpful comments on the preliminary version of this paper. This research was supported by the 2008 MURST 60% project 'Analisi dei caratteri geologico-tecnici e idrogeologici per la tutela e la valorizzazione delle risorse naturali, ambientali e culturali' (Resp.: Prof. Nicola Walsh).

References

ABBAZZI, L., BENVENUTI, M. *ET AL.* 1996. Revision of the Neogene and Pleistocene of the Gargano Region (Apulia, Italy). The marine and continental successions and the mammal faunal assemblages in an area between Apricena and Poggio Imperiale (Foggia). *Memorie della Società Geologica Italiana*, **51**, 383–402.

ANDRIANI, G. F. 2006. Considerazioni sugli effetti del weathering in rocce calcarenitiche utilizzate per manufatti di interesse storico e architettonico. *In*: BALDASSARRE, G. & BADINO, V. (eds) *Le Risorse Lapidee dall'Antichità ad Oggi in Area Mediterranea.* GEAM, Torino, 233–238.

ANDRIANI, G. F. & WALSH, N. 1998. Caratteri tessiturali e resistenza al taglio diretto di calcareniti tenere e porose. GEAM, **93**, 35–42.

ANDRIANI, G. F. & WALSH, N. 2000. Thermal properties and their influence on strength and deformability of calcareous rocks. *In*: CALVI, G. & ZEZZA, U. (eds.) *Proceedings of the International Congress Quarry-Laboratory-Monument.* September 26–30, 2000, **1**, 81–90.

ANDRIANI, G. F. & WALSH, N. 2002. Physical properties and textural parameters of calcarenitic rocks: qualitative and quantitative evaluations. *Engineering Geology*, **67**, 5–15.

ANDRIANI, G. F. & WALSH, N. 2003. Fabric, porosity and water permeability of calcarenites from Apulia (SE Italy) used as building and ornamental stone. *Bulletin of Engineering Geology and Environment*, **62**, 77–84.

ANDRIANI, G. F., BALDASSARRE, G. & WALSH, N. 2006. Porosità e permeabilità delle calcareniti di Pietra Caduta. *In*: BALDASSARRE, G. & BADINO, V. (eds) *Le Risorse Lapidee Dall'Antichità ad Oggi in Area Mediterranea.* GEAM, Torino, 113–116.

ANDRIANI, G. F. & WALSH, N. 2007*a*. Rocky coast geomorphology and erosional processes: A case study along the Murgia coastline South of Bari, Apulia, SE Italy. *Geomorphology*, **87**, 224–238.

ANDRIANI, G. F. & WALSH, N. 2007*b*. The effects of wetting and drying, and marine salt crystallization on calcarenite rocks used as building material in historic monuments. *In*: PŘIKRYL, R. & SMITH, B. J. (eds) *Building Stone Decay: From Diagnosis to Conservation.* Geological Society, London, Special Publications, **271**, 179–188.

ARNOLD, A. & ZEHNDER, K. 1989. Salt weathering on monuments. *In:* ZEZZA, F. (ed.) *The Conservation of Monuments in the Mediterranean Basin.* Grafo, Bari, 31–58.

AZZAROLI, A. 1968. Studi illustrativi della Carta Geologica d'Italia. Formazioni Geologiche. *Servizio Geologico d'Italia*, **1**, 183–185.

BALENZANO, F. & DI PIERRO, M. 1972. Le Calcareniti calabriane della zona di Trani. Osservazioni mineralogiche e chimiche. *Geologia Applicata ed Idrogeologia*, **7**, 126–172.

BENAVENTE, D., GARCÍA DEL CURA, M. A., FORT, R. & ORDÓÑEZ, S. 2004. Durability estimation of porous building stones from pore structure and strength. *Engineering Geology*, **74**, 113–127.

BENAVENTE, D., CUETO, N., MARTÍNEZ-MARTÍNEZ, J., GARCÍA DEL CURA, M. A. & CAÑAVERAS, J. C. 2007. The influence of petrophysical properties on the salt weathering of porous building rocks. *Environmental Geology*, **52**, 215–224.

BOSSIO, A., MAZZEI, R., MONTEFORTI, B. & SALVATORINI, G. 1988. Nuovo modello stratigrafico del Miocene-Pleistocene inferiore del Salento in chiave geodinamica. *In*: *Proceedings of 74th National Congress Società Geologica Italiana*, Sorrento, Italia, 35–38.

BOSSIO, A., GUELFI, F., MAZZEI, R., MONTEFORTI, B. & SALVATORINI, G. 1989. Studi sul Neogene e

Quaternario della Penisola salentina. III – Stratigrafia del pozzo Poggiardo (N. 54, PS 140/3). *Quaderni Ricerche del Centro Studi di Geotecnica e di Ingegneria*, Lecce, **11**, 55–88.

BOSSIO, A., GUELFI, F., MAZZEI, R., MONTEFORTI, B. & SALVATORINI, G. 1991. Note geologiche e stratigrafiche sull'area di Palmariggi (Lecce, Puglia). *Rivista Italiana di Paleontologia e Stratigrafia*, **97**, 175–234.

BROMLEY, R. G. & D'ALESSANDRO, A. 1987. Bioerosion of the Plio-Pleistocene transgression of Southern Italy. *Rivista Italiana di Paleontologia e Stratigrafia*, **93**, 379–442.

BRUNO, G. & CHERUBINI, C. 2005. Subsidence induced by the instability of weak rock underground quarries in Apulia. *Giornale di Geologia Applicata*, **1**, 33–39.

CALDARA, M. 1982. Segnalazione di Pliocene medio e Superiore nelle Murge settentrionali (Puglia). *Bollettino della Società Geologica Italiana*, **106**, 153–162.

CALÒ, G., DI PIERRO, M., FEDERICO, A. & MONGELLI, G. 1985. Caratteri geologici petrografici mineralogici e meccanici dei "Carpari" della Provincia di Lecce. *Quarry and Construction*, **7**, 45–51.

CANTELLI, C. 1960. Sul Quaternario di Gravina di Puglia. *Giornale di Geologia, serie 2*, **28**, 211–226.

CAPPELLINI, G. 1878. Della pietra leccese e di alcuni suoi fossili. *Memorie Academia delle Scienze di Bologna*, **9**, 227–258.

CAPUTO, M. C., QUADRATO, E. & WALSH, N. 1996. Influenza dello shock termico sui parametri fisicomeccanici del "Tufo Calcareo" del bordo occidentale delle Murge. *Memorie della Società Geologica Italiana*, **51**, 813–822.

CHOQUETTE, P. W. & PRAY, L. C. 1970. Geologic nomenclature and classification of porosity in sedimentary carbonates. *AAPG Bulletin*, **54**, 207–250.

COTECCHIA, V., CALÒ, G. & SPILOTRO, G. 1985. Caratterizzazione geolitologica e tecnica delle calcareniti pugliesi. *In:* III Convegno Nazionale *Attività estrattiva dei minerali di 2° categoria, Bari 17–19 gennaio 1985*, PEI, Parma, **1**, 209–216.

D'ALESSANDRO, A., LAVIANO, A., RICCHETTI, G. & SARDELLA, A. 1979. Il Neogene del Monte Gargano. *Bollettino della Società Paleontologica Italiana*, **18**, 1, 9–116.

D'ALESSANDRO, A. & IANNONE, A. 1982. Pleistocene carbonate deposits in the area of Monopoli (Bari province): sedimentology and paleoecology. *Geologica Romana*, **21**, 603–653.

DE GIORGI, C. 1922. *Descrizione geologica ed idrografica della Provincia di Lecce*. Tipografia Salomi, Lecce, 1–50.

DELL'ANNA, L., GARAVELLI, C. L. & NUOVO, G. 1968. Sui cosiddetti "tufi calcarei" della Regione Pugliese-Lucana. *Periodico di Mineralogia*, **37**, 657–715.

DELL'ANNA, L., DE MARCO, A., RICCHETTI, G. & DI PIERRO, M. 1978. Ricerche geologiche e mineralogiche sulle Calcareniti di Monte Castiglione. *Bollettino della Società Geologica Italiana*, **97**, 451–474.

DEERE, D. U. & MILLER, R. P. 1966. *Engineering classification and index properties for intact rocks*. Technical Report. Air Force Weapons Laboratory, New Mexico, No. AFNL-TR, 65–116.

D'ERASMO, G. 1934. Il mare pliocenico nella Puglia. *Memorie Geologiche e Geografiche di G. Dainelli, Firenze*, **4**, 45–138.

DI GERONIMO, I. 1969. I depositi quaternari della costa fra Brindisi e Torre Canne. *Atti dell'Accademia Gioenia di Scienze naturali in Catania*, serie VI, **20**, 195–224.

DI STEFANO, G. & VIOLA, C. 1892. L'età dei tufi calcarei di Matera e di Gravina di Puglia e il sottopiano "Materino" M. E. *Bollettino del Regio Comitato Geologico d'Italia*, **23**, 125–149.

D'ONOFRIO, S. 1960. Sui foraminiferi di alcuni campioni raccolti presso Gravina di Puglia. *Gionale di Geologia, Serie 2*, **28**, 203–210.

EN 1926, 1999. *Natural stone test methods – Determination of compressive strength*. European Committee for Standardization, Brussels.

EN 12372, 1999. *Natural stone test methods – Determination of flexural strength under concentrated load*. European Committee for Standardization, Brussels.

EVANGELISTA, A. & PELLEGRINO, A. 1990. Caratteristiche geotecniche di alcune rocce tenere italiane *In*: BARLA, G. (ed.) *Le rocce tenere – Terzo ciclo di Conferenze di Meccanica e Ingegneria delle Rocce – MIR 90*, SGE, Padova, **2**, 2–32.

EVERET, D. H. 1961. The themodynemics of frost damage to porous solid. *Transactions of the Faraday Society*, **57**, 1541–1551.

FITZNER, B. & SNETHLAGE, R. 1982. Über Zusammenhange zwischen salzkristallisationsdruck und Porenradienverteilung. *Group Petrography Newsletter*, **3**, 13–24.

FLATT, R. J. 2002. Salt damage in porous material: how high supersaturations are generated. *Journal of Crystal Growth*, **242**, 435–454.

FLÜGEL, E. 2004. *Microfacies of Carbonate Rocks, Analysis, Interpretation and Application*. Springer-Verlag, Berlin.

GIGNOUX, M. 1913. Les forlations marines Pliocenes et Quaternaries de l'Italie du Sud et de la Sicilie. *Annales de l'Université de Lyon*, **1**, 36, 24–693.

GIOVENE, G. M. 1810. Notizie geologiche e meteorologiche sulla Japigia, ossia della provincia di Terra d'Otranto nel Regno di Napoli. *Memorie di Matematica e di Scienze Fisiche e Naturali della Società Italiana delle Scienze*, **15**, 274–290.

GOUDIE, A. S. & VILES, H. 1997. *Salt Weathering Hazards*. Wiley, Chichester.

IANNONE, A. & PIERI, P. 1979. Considerazioni critiche sui "Tufi Calcarei" delle Murge. Nuovi dati litostratigrafici e paleoambientali. *Geografia Fisica e Dinamica Quaternaria*, **2**, 173–186.

IANNONE, A. & PIERI, P. 1982. Caratteri neotettonici delle Murge. *Geologia Applicata e Idrogeologia*, **18**, 147–159.

ISRM 1979. Suggested methods for determining water content, porosity, density, absorption and related properties and swelling and slake-durability index properties. *International Journal of Rock Mechanichs and Mining Sciences and Geomechanics Abstracts*, **16**, 141–156.

MARGIOTTA, S. & RICCHETTI, G. 2002. Stratigrafia dei deposti oligomiocenici del Sdalento (Puglia). *Bollettino della Società Geologica Italiana*, **121**, 243–252.

Margiotta, S. & Varola, A. 2004. Nuovi dati geologici e paleontologici su alcuni affioramenti nel Territorio di Lecce. *Atti della Società Toscana di Scienze Naturali, Memorie, Serie A*, **109**, 1–12.

Martinis, B. 1967. Note Geologiche sui dintorni di Casarano e Castro (Lecce). *Rivista Italiana di Paleontologia e Stratigrafia*, **73**(4), 1297–1380.

Mazzei, R. 1994. Età della Pietra leccese nell'area di Cursi-Melpignano (a sud di Lecce, Puglia). *Bollettino della Società Paleontologica It.*, **33**, 2, 243–248.

Mongelli, F. 1968. Un metodo per la determinazione in laboratorio della conducibilità termica delle rocce. *Bollettino di Geofisia Teorica ed Applicata*, Trieste, **10**, 51–58.

Mongelli, F., Sciruicchio, V. & Walsh, N. 1993. Proprietà termiche del Tufo calcareo pugliese. *In*: *Proceedings of the International Congress "La Pietra da costruzione: Il Tufo calcareo e la Pietra leccese"*, CNR-IRIS, Bari, 329–349.

Nicotera, P. 1953. La Pietra leccese. *L'Industria Mineraria, Faenza*, **4**, 1–12.

Palmentola, G. 1989. Lineamenti geologici e morfologici del Salento leccese. *Quaderni Ricerche del Centro Studi di Geotecnica e di Ingegneria, Lecce*, **11**, 7–30.

Parise, M., Federico, A., Delle Rose, M. & Sammarco, M. 2003. Karst terminology in Apulia (Southern Italy). *Acta Carsologica*, **6**, 65–82.

Penta, F. 1935. I materiali da costruzione dell'Italia meridionale. Molise, Puglia, Lucania e Calabria. *Fondazione Politecnica del Mezzogiorno d'Italia, Napoli*, **2**, 1–672.

Pomar, L. & Tropeano, M. 2001. The calcarenite di gravina formation in Matera (Southern Italy): New insights for coarse-grained, large-scale, cross-bedded bodies encased in offshore deposits. *AAPG Bulletin*, **85**, 661–689.

Radina, B. & Walsh, N. 1972. Proprietà tecniche delle Calcareniti (Tufi Calcarei) di Gravina in Puglia. *Geologia Applicata e Idrogeologia*, **7**, 56–75.

Ricchetti, G. 1965. Alcune osservazioni sulla serie della Fossa Bradanica. "Le Calcareniti di Monte Castiglione". *Bollettino della Società dei Naturalisti in Napoli*, **74**, 3–11.

Ricchetti, G. 1970. Nuove osservazioni sui depositi plio-pleistocenici nei dintorni di Taranto. I "Tufi calcarei con Arctica islandica". *Bollettino della Società Geologica Italiana*, **89**, 3–10.

Ricchetti, G., Ciaranfi, N., Luperto Sinni, E., Mongelli, F. & Pieri, P. 1988. Geodinamica ed evoluzione sedimentaria e tettonica dell'Avampaese apulo. *Memorie della Società Geologica Italiana*, **41**, 57–82.

Rodriguez-Navarro, C. & Doehne, E. 1999. Salt weathering: influence of evaporation rate, supersaturation and crystallisation pattern. *Earth Surface Processes and Landforms*, **24**, 191–209.

Sacco, F. 1911. La Puglia. Schema geologico. *Bollettino della Società Geologica Italiana*, **30**, 529–638.

Salvati, M. 1932. *I Tufi di Terra di Bari.* Giuseppe Laterza & Figli, Bari.

Scherer, G. W. 2000. Stress from crystallisation of salt in pores. *In*: Fassina, V. (ed.) *Proceedings of the 9th International Congress on Deterioration and Conservation of Stone, Venice*. Elsevier, 187–194.

Tropeano, M. & Sabato, L. 2000. Response of Plio-Pleistocene mixed bioclastic-lithoclastic temperate-water carbonate systems to forced regressions: the Calcarenite di Gravina Formation, Puglia, SE Italy. *In*: Hunt, D. & Gawthorpe, R. L. (eds) *Sedimentary Responses to Forced Regressions.* Geological Society, London, Special Publications, **172**, 217–243.

Valduga, A. 1965. Contributo alla conoscenza geologica delle Murge baresi. *Istituto di Geologia e Paleontologia Universita degli studi di Bari, Adriatica Bari*, 1–15.

Walsh, N. 2006. Caratteri petrofisici e meccanici di calcareniti pugliesi e lucane. *In*: Baldassarre, G. & Badino, V. (eds) *Le Risorse Lapidee Dall'Antichità ad Oggi in Area Mediterranea*. GEAM, Torino, 225–230.

Weyl, P. K. 1959. Pressure solution and the force of crystallization, a phenomenological theory. *Jounal of Geophysical Research*, **64** (11), 2001–2025.

Zezza, F. 1974. Le pietre da costruzione e ornamentali della Puglia. Caratteristiche sedimentologico-petrografiche, proprietà fisico-meccaniche e problemi geologico-tecnici relativi all'attività estrattiva. *Rassegna Tecnica Pugliese Continuità, Bari*, **8** (3–4), 3–51.

Zezza, F. & Macrì, F. 1995. Marine aerosol and stone decay. *The Science of the Total Environment*, **167**, 123–143.

Zezza, U., Veniale, F., Zezza, F. & Moggi, G. 1989. Effetti dell'imbibizione sul decadimento meccanico della pietra leccese. *In*: Zezza, F. (ed.) *Proceedings of the First International Symposium on the Conservation of Monuments in the Mediterranean Basin*, 7–10 June, 1989, Grafo, Brescia, **1**, 263–269.

Petrophysical properties of selected Quaternary building stones in western Austria

MICHAEL UNTERWURZACHER[1,2*], ULRICH OBOJES[1], ROLAND HOFER[1] & PETER W. MIRWALD[1]

[1]*Institute of Mineralogy and Petrography, University of Innsbruck, Innrain 52, 6020 Innsbruck, Austria*

[2]*Now: Archaeometry and Cultural Heritage Computing Working Group, Institute of Geography and Geology, University of Salzburg, Hellbrunnerstrasse 34, 5020 Salzburg, Austria*

**Corresponding author (e-mail: Michael.Unterwurzacher@sbg.ac.at)*

Abstract: In west Austria Quaternary building stones, such as lithic breccias of alluvial fans and talus slopes or calcareous spring tufa, have been frequently used as building stones since Roman times. Spring tufas are a widely used building material of historical objects in west Austria. This porous calcareous rock, formed by carbonate precipitation from calcium carbonate supersaturated spring waters, is an appreciated building stone: easy to quarry, lightweight, easily workable and relatively resistant to weathering. The Hötting Breccia, a lithified talus and alluvial breccia, has only been extracted in a few quarries near Innsbruck/Tyrol, however. Many of the mediaeval buildings of the towns of Innsbruck and Hall are built of this decorative type of stone. Petrography, mineralogical composition, porosity parameters and hygric properties have been investigated in this study from two tufas and one breccia occurrence. The results obtained reveal that these Quaternary stones, being formed at the Earth's surface, exhibit pore properties and hygric behaviour which differ considerably from other stone materials which have been subjected to the physical-chemical formation conditions of the upper Earth crust. This has implications for their workability, internal stability and weathering behaviour.

In the Alpine region, the masonry of many historical objects such as churches, castles and monasteries, is composed of natural stone that (in most cases) was taken from local sources. The geological setting of the Alps of west Austria implies that carbonate rocks play a very important role as building stones. As will as marbles, different calcareous and dolomitic rocks, the so-called calcareous tufa, are a prominent material. Compared to dense marbles and calcareous-dolomitic rocks the tufa, which is a highly porous freshwater carbonate, differs significantly in appearance and petrophysical properties from the other carbonates.

Calcareous tufas '*sensu lato*' make very important building stones in non-Alpine regions such as Italy (e.g. Rome), Spain (e.g. Albacete; Sancho *et al.* 1997; Pena *et al.* 2000), Croatia (Plitvice area; Emeis *et al.* 1987) and/or Bosnia & Herzegovina (e.g. Una river area). While the formation of travertine limestone of Middle Italy (*lapis tiburtinum*) is associated with the activity of hot springs related to volcanism, the calcareous tufa of the east Alps are formed by precipitation of calcium carbonate from 'cool' spring and river waters supersaturated by a factor of ten. Supersaturation is mostly attained, or is highest, some distance downstream from the emergence of the spring or, in large tufa precipitating systems (e.g. Plitvice), also within-stream to downstream of rapids and water falls (Sanders *et al.* 2006). There is general agreement that the degassing of CO_2 out of the water is the most important single process in producing the necessary degrees of supersaturation for precipitation (Viles & Goudie 1990).

Another very important factor for tufa formation is the geological setting because it directly causes the supersaturation of waters with calcium carbonate: in the Alpine region calcareous micashist and/or till and debris sediments (lixivation of Quaternary sediments seems to be of highest importance) (e.g. Kieslinger 1964). Tufa formation often occurs in association with karst formation phenomena (Bögli 1978; Ford & Pedley 1996). The property which all tufa have in common is that they are of very young geological age (Viles & Goudie 1990).

Tufa occur in many locations in west Austria (and also in adjacent regions) and are of very different size and quality. The most prominent occurrence in Tyrol is that of Thiersee/Kufstein which served

From: PŘIKRYL, R. & TÖRÖK, Á. (eds) *Natural Stone Resources for Historical Monuments*. Geological Society, London, Special Publications, **333**, 143–152.
DOI: 10.1144/SP333.14 0305-8719/10/$15.00

as the material source for the Josefsburg of the Kufstein fortress. The tufas have been preferred materials because of their favourable properties: lightweight, good workability, considerable strength and weathering durability. In the case of the fortress of Kufstein, this material was presumably preferred to hard limestone because of its 'soft behaviour', assumed to resist a gun bombardment better.

Archaeological excavations show that tufa have already been used in Roman times (e.g. Aguntum/Eastern Tyrol and Roman villa of Rankweil/Vorarlberg). Its use peaked in the Middle Ages (e.g. the castle ruins of Kropfsberg/Northern Tyrol and of Neuburg/Vorarlberg; Unterwurzacher *et al.* 2006), when even very small occurrences were exploited (e.g. St Martin near Ludesch/Vorarlberg).

The Hötting Breccia, occurring above Innsbruck/Northern Tyrol, is interpreted as an interglacial succession that accumulated from alluvial fans and talus slopes (Penck 1921; Obojes 2003). In fractures of this Pleistocene breccia Spötl & Mangini (2006) found calcitic flowstones and dated them (U/Th ages) between 101 500 $\pm$ 1500 and 70 300 $\pm$ 1800 years. Two main types of this breccia exist: the so-called 'White Hötting Breccia' which mainly contains carbonate components and the 'Red Hötting Breccia' which contains dyeing particles of red-coloured Triassic sandstone ('Alpiner Buntsandstein') in the matrix as well as sandstone clasts.

The Hötting Breccia was quarried in two larger and several smaller quarries until the beginning of the 20th century. Most of the stone material has been employed for the buildings of the towns of Innsbruck and Hall in Tyrol (Hofer 2004). However, despite its local occurrence, the material has been used regionally in the Inn valley.

Despite hundreds of years of exposure to weathering, it is a striking observation that the extent of damage on these two building stones is relatively limited compared to other stones (marbles, sandstones) used as historical building materials. The weathering behaviour of tufa is apparently a complex process. Geochemical processes of solution and precipitation take place during formation and early diagenesis (e.g. Sanders *et al.* 2006). During the use as building stone, secondary consolidation processes seem to take place within the tufa. These processes are probably due to solution and precipitation, caused by humidity and/or fluids (mainly rainwater). A detailed investigation of these processes is the subject of a current project.

Intensive weathering phenomena can only be detected in cases of severe weathering impact: dissolution due to rain, pollution impacts and crumbling. A striking feature is that in humid micro-environments tufas tend to be wet, which may lead to all sorts of biogenic activities on the stone. The Hötting Breccia shows two forms of weathering: dark surface depositions due to pollution by particulate matter and algae growth and material loss mainly due to matrix failure. The matrix disintegration takes place preferentially in the contact areas with clasts, which ultimately leads to the loss of the mostly unaffected clasts. Sandstone clasts ('Alpiner Buntsandstein') are the only exception in this respect; their low weathering resistance results in the formation of holes.

The starting point of this investigation was the fact that both Quaternary materials have recently been formed and solidified on the Earth's surface. They are both porous carbonate rock types, however, of specific character. A number of projects conducted recently, which were concerned with mapping the stone inventory of historic objects in south and north Tyrol (Bidner *et al.* 2003; Franzen *et al.* 2005) showed that these two stone types have been frequently employed in historical constructions. The search for their properties revealed that almost no petrophysical-chemical data were available on these stone types. This obvious lack of and the increasing need for regional monument conservation to develop sustainable maintenance and conservation concepts gave rise to this study, aiming at a first overview of material properties.

Methods

As a first step, the occurrences of the stone materials were mapped in order to obtain an idea of the size and sedimentology of the units and to undertake representative sampling. The materials studied in more detail are from the two tufa sites Andelsbuch and Thiersee and from the Hötting Breccia quarry Mayr above Innsbruck (Fig. 1). These sites were chosen due to their local or regional importance. Tufa from Andelsbuch was a common building stone of local importance in east Vorarlberg. Thiersee tufa is best known as the main building component of the 'Josefsburg', the western outlier of the Kufstein fortress. Breccia from the quarry Mayr is expected to be the main building material of the historical town of Innsbruck.

The petrographic fabric and the mineral assemblages of the stones were studied by optical microscopy. In addition, X-ray powder diffraction (XRD: Siemens D500, Siemens D5005 and AXS-Bruker D8) provided a semi-quantitative determination of the mineral composition of these rocks.

Determination of hygric properties was conducted using three representative specimens of each rock of 50 $\times$ 50 $\times$ 50 mm in size, cut from stone blocks and dried for 48 hours at 40 °C. Capillary water uptake for determination of the A-coefficient (water uptake coefficient in $kg\,m^{-2}\,h^{-1/2}$) and water capacity were first determined according to

Fig. 1. The Quaternary building stones investigated in Vorarlberg and Tyrol. (**a**) wall of Thiersee tufa at the Josefsburg of Kufstein fortress; note the considerable pores and voids; (**b**) brick of Andelsbuch spring tufa in a farmers house; (**c**) polished specimen of Hötting Breccia; (**d**) Josefsburg of the Kufstein fortress, Tyrol, seen from south (most of this fortification is built from calcareous tufa from Thiersee); (**e**) part of the Andelsbuch tufa quarry and (**f**) part of a façade of the Cathedral of Innsbruck built from Hötting Breccia.

the recommendations of EN 1925. The samples were mounted on an automatic weighting device, with the base area in a water bath at a relative air humidity of 70%. Measurements were taken automatically every 5 seconds. After the capillary water uptake measurements, samples were taken out of the water bath and dried.

Subsequently, the water uptake of the materials was determined by (i) submersing the samples in a water bath for 24 hours and then by (ii) placing the cubes into a water-filled container (desiccator) which was evacuated.

Monitoring of the drying behaviour by weighing was the last of the hygric determinations. The specimens were dried at room temperature after 24 hours of water uptake. The determinations were made either manually or automatically by means of a logger system.

The characterization of the radii was performed by mercury-intrusion porosimetry (PASCAL140 and Porosimeter2000). The measurement range of this mercury porosimeter is from 0.0032 μm to 316.2278 μm. These measurements offer also – as an option – the calculation of the specific surface to be compared with values directly determined by the nitrogen adsorption (BET) method. However, the data retrieved are not presented here because of obvious inconsistencies: it seems that due to the variable pore size the consumption of Nitrogen is variable and the automatic calculation function of the machine does not yield consistent calculations for the specific surface.

Fragments of the samples allowed the determination of the specific inner surface by a Nitrogen gas adsorption (BET) analyser Quantachrome Nova2200.

Rock locations studied

Spring tufa from Thiersee

Thiersee-Mitterland (Fig. 1a) is situated near the border of Tyrol/Bavaria (Germany). Quaternary deposits (till, gravel terraces) overlie the basement rocks of Cretaceous sediments. The tufa consists mainly of fossil phytoclastic waterfall tufa and recent moss tufa. This kind of tufa is common to and characteristic of most of the east alpine occurrences (Sanders *et al.* 2006). In addition, a further type of tufa was observed: it consists of subspherical to ellipsoidal vadoids and cyanoids of a few millimetres to more than a centimetre in size (Sanders *et al.* 2006).

The spring tufa occurrence near Thiersee probably had an original lateral extension of a few hundred metres. The thickness today reaches 5–10 m. Except for minor formations of moss tufa and thin calcareous crusts on rock surfaces in waterfalls of local creeks, no precipitation of calcium carbonate takes place today. The spring tufa had been quarried mainly for local use. Its major exploitation was during the 18th century when the material was employed for the Josefsburg, the western outlier of the fortress of Kufstein (Fig. 1a & d). This enormous construction highlights the originally huge capacity of this occurrence. Today the occurrence is nearly completely exploited.

Spring tufa from Andelsbuch

The location of Andelsbuch (Fig. 1b) is situated in Vorarlberg/west Austria some 15 km east of the city of Dornbirn on the east side of the Bregenzerach river. The basement rocks are limestones, clays and marls of the Helveticum and Flysch units (Rüf 2006), covered by Quaternary sediments (till, gravel terraces). Moss tufa and phytoclastic tufa built up this spring tufa. At the tufa deposit of Andelsbuch, active deposition of tufa still takes place.

The occurrence of Andelsbuch has a lateral extension of at least one hundred metres. The thickness of the deposits is estimated to reach a minimum of 5 m at least. Today, precipitation of moss tufa is still observed. The occurrence has been quarried as local building stone over centuries (Fig. 1b, c).

Hötting Breccia

The Hötting Breccia is exposed on the south-facing slopes of the Inn valley above Innsbruck/Tyrol (Fig. 1c). Two different types of breccia exist which reach up to 40 m of thickness:

(i) the White Breccia, which contains Triassic limestones embedded in a calcitic matrix, and
(ii) the Red Breccia, which additionally contains Permo-Triassic sandstones ('Alpiner Buntsandstein', Obojes 2003).

With the exception of very rare crystalline rock clasts, which are remnants of glacial deposits from the preceding Riss-glaciation, all the clastic components are derived from the underlying Triassic substrate. The stratigraphically lower Red Breccia consists mainly of limestone components and a carbonate matrix (silty grain size: 2–63 μm), red-coloured due to abundant Permo-Triassic sandstone fragments. The ratio of clasts to matrix is very variable ranging from 30:70 to 70:30. The Red Breccia is well cemented and compact. Sedimentologically, this sequence is built up by debris-flow breccias and, subordinately, stratified breccias as well as beds of unlithified material of loessic character.

The White Breccia which overlies the Red Breccia is less consolidated, contains less matrix and more isopachous cements. The presence of conglomerates and breccias deposited from aqueous flows and their vertical association with debris-flow deposits suggests quasi-perennial to ephemeral surface runoff. Due to the low degree of lithification of the White Breccia, only the lower part of the sequence (the well-lithified Red Breccia) was used as a building stone (Fig. 1c, f).

Results

Spring tufa from Thiersee

The tufas of Thiersee are highly variable in their appearance. The Thiersee tufa most probably

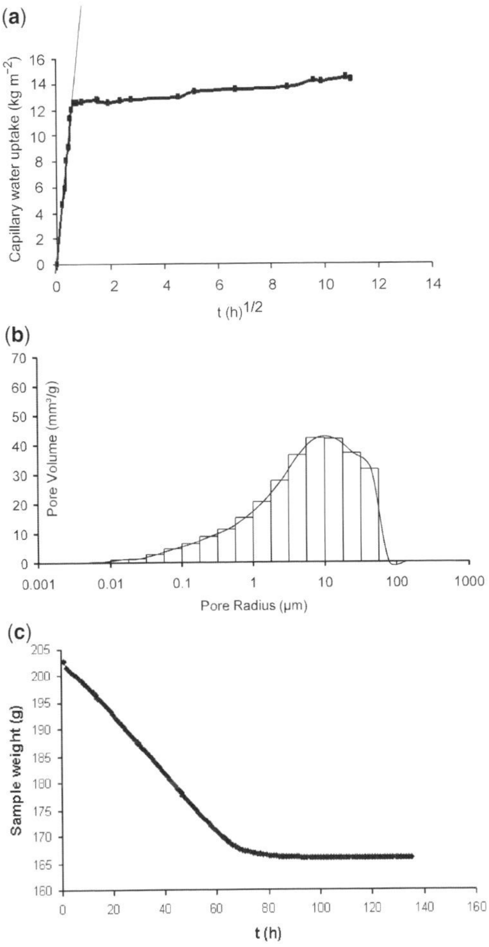

Fig. 2. Petrophysical properties of Thiersee tufa: (**a**) capillary water uptake; (**b**) pore size distribution and (**c**) drying behaviour.

pertained to the fluvial-barrage type of tufa (cf. Ford & Pedley 1996). Common tufa facies include various types of phytoclastic tufas as well as vadolithic tufas and, subordinately, moss tufas.

Three representative samples were petrophysically investigated in more detail. The capillary water uptake yielded a considerable A-coefficient of 26 kg m^{-2} $h^{-1/2}$ (Fig. 2a and Table 1) which, however, seems comparatively low for tufa (see below). The water capacity point at 12.5 kg m^{-2} was reached within 30 minutes with the cube samples (weight increase 32 g; initial dry weight 170 g). This corresponds to a water-filled pore volume of 19% (Fig. 2a). Submersed in water for 24 hours, the water uptake amounted to 22% (weight increase 38 g; initial dry weight 170 g). The saturation value obtained under vacuum conditions yielded a total pore volume of 24% (weight increase 41 g; initial dry weight 170 g). The drying behaviour documented by continuous weighing is shown in Fig. 2c. The course of the drying curve practically consists of two linear branches. The water loss of the 5 × 5 × 5 cm samples reaches over an extended period of 70 hours under room temperature, with a room humidity content of about 70%.

The pore radii spectrum of a typical sample is displayed in Fig. 2b, showing an asymmetric shape with a maximum in the distribution at 10 µm. The specific inner surface, determined on the two samples, ranges between 0.8 m^2 g^{-1} and 1.0 m^2 g^{-1}.

Spring tufa from Andelsbuch

The tufa depositional system pertains to the perched-springline type according to the classification of Ford & Pedley (1996) and consists mainly of moss tufa and phytoclastic tufa.

Table 1. *Petrophysical data of the Thiersee and Andelsbuch tufa, the Hötting Breccia and the Gröden Sandstone (Franzen 2002)*

	Thiersee tufa	Andelsbuch tufa	Hötting Breccia	Gröden Sandstone (Franzen 2002)
Density (g cm^{-3})	2.0 ± 0.2	1.4 ± 0.2	2.3 ± 0.2	2.57
Average pore radius (µm)	10	30	0.6	0.5
Shape of the pore radii spectrum	asymmetric	asymmetric	symmetric	asymmetric bimodal
Specific surface area (BET) (m^2 g^{-1})	0.9 ± 0.2	1.05 ± 0.2	1.6 ± 0.3	1.5 ± 0.3
A-coefficient (kg m^{-2} $h^{-1/2}$)	26 ± 5	355 ± 20	1.9 ± 0.2	0.9
Water capacity (vol. %) attained for	19 ± 7	42 ± 10	4 ± 3	6 ± 3
50 × 50 × 50 mm cubic samples	0.5 h	70 seconds	3.5 h	6.4 h
Water saturation in 24 h (vol. %)	22 ± 6	44 ± 8	8 ± 1	6.6 ± 3
Water saturation under vacuum (vol. %)	24 ± 6	48 ± 9	9.5 ± 1	8 ± 4
Drying (h) to 'dry' equilibrium state at room temp. and humidity	70 ± 8	110 ± 10	36 ± 5	35 ± 5
Ratio of drying /soaking duration	1:88	1:37	1:2.7	1:1.2

The spring tufas from Andelsbuch are highly variable in their petrophysical data. On average, they show a very high capillary coefficient of 355 kg m^{-2} $h^{-1/2}$; the water capacity point at 24 kg m^{-2} is reached within 70 seconds (Fig. 3a and Table 1) which corresponds to a water filling of 42%. The water uptake of submersed samples was 44% after 24 hours, and reached a total porosity of about 48% under vacuum condition. The Hg-porosimetry yielded a pore radii spectrum displayed in Fig. 3b with a maximum at 30 μm and, again, of asymmetric shape. The specific surface area, determined by BET, is about 1.05 m^2 g^{-1}.

Drying experiments (Fig. 3c) starting from the 24 hour water uptake value shows a similar behaviour to the Thiersee tufa, only attaining room moisture equilibrium after 110 hours.

Hötting Breccia

Petrographically as well as petrophysically, the Red Breccia shows considerable differences between the clastic components (mainly Triassic limestones, size between 0.05 and 0.4 m) and the red matrix (silty grain size of 2–63 μm; Hofer 2004). The major mineral phases of the matrix are calcite, dolomite and quartz in varying proportions; illite is found in minor amounts. A striking rock fabric characteristic to be observed in thin sections is that millimetre-sized pores exhibit isopachous seams of cements of calcite crystals. The ratio between clasts and matrix varies in a wide range (30:70 to 70:30), implying that all data scatter considerably. While the clastic components show specific inner surfaces of less than 0.2 m^2 g^{-1}, the matrix exhibits up to 2 m^2 g^{-1}. The capillary

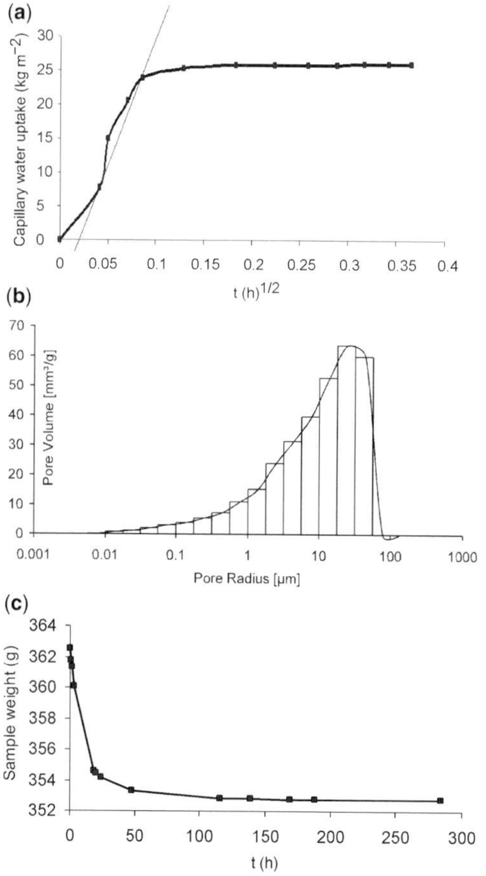

Fig. 3. Petrophysical properties of Andelsbuch tufa: (**a**) capillary water uptake; (**b**) pore size distribution and (**c**) drying behaviour.

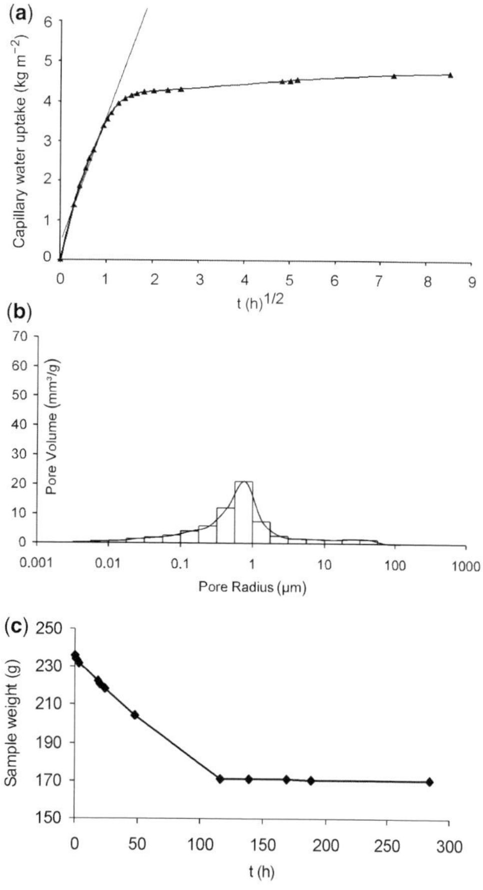

Fig. 4. Petrophysical properties of Hötting Breccia: (**a**) capillary water uptake; (**b**) pore size distribution and (**c**) drying behaviour.

A-coefficient of 1.9 kg m^{-2} h$^{-1/2}$ (Fig. 4a and Table 1) is therefore governed by the matrix. The average porosity determined from the water capacity value amounts to about 4.2 vol.%, which is reached for the cube-shaped samples after about 200 minutes. The water uptake within 24 hours by submersion amounts to 8 wt%. The water uptake under vacuum conditions reached a saturation value of 9.5%, which is representative of the total porosity volume. Drying experiments starting from the 24 hours water uptake value approach after the 'dry' equilibrium state at ambient conditions is attained (Fig. 4c).

The pore radii spectrum is of fairly symmetrical shape. It has a maximum at 0.6 μm (Fig. 4b); this value holds for the matrix and the contributions of the clasts can almost be neglected.

Discussion

The data determined from three localities give the first outline of the basic petrophysical properties of calcareous tufa and the Hötting Breccia being widely as used building stones in west Austria. In contrast to many other stone materials, they are young Quaternary stones which were formed on the Earth's surface. Their properties reflect the frame of thermodynamic conditions of the Earth's surface and diagenetic processes related to elevated *P-T* conditions can be ruled out. The average data of the two different kinds of stones are compiled in Table 1. In particular, those of the Hötting Breccia are characterized by a considerable scatter due to their pronounced heterogeneity. For comparison, data of the Gröden Sandstone (Franzen 2002) have also been included in Table 1. This Permo-Triassic sandstone from South Tyrol/Italy represents a fairly consolidated, widely used clastic material which is similar in properties to the Hötting Breccia (see below).

The inspection of the data presented in the figures and compiled in Table 1 show that the two tufas differ distinctly from the Hötting Breccia.

The most striking difference is the density due to the dramatic discrepancy in porosity which is half a magnitude. Surprisingly, this very porous structure of the tufa does not lead to strength problems. Exhibiting a strength of 50 ± 20 N mm^{-2}, this material serves in historical structures as a lightweight but appropriate material for statically sophisticated components, for example, for vaults and archways.

The porosity of both materials implies unusual features. The tufa porosity can only be determined by water saturation measurements since the fraction of macropores, which often exceed millimetre size, can be substantial. The pore radius spectrum as determined by Hg-porosimetry (Fig. 5a) shows the situation in the range of micropores <1 μm and in the 1–100 μm section of the capillary pores (1–1000 μm). Both tufa have an asymmetric pore radii spectrum with a maximum in the very active 1–100 μm capillary range. A mesopore range (0.05–0.002 μm) of the tufa is not really developed.

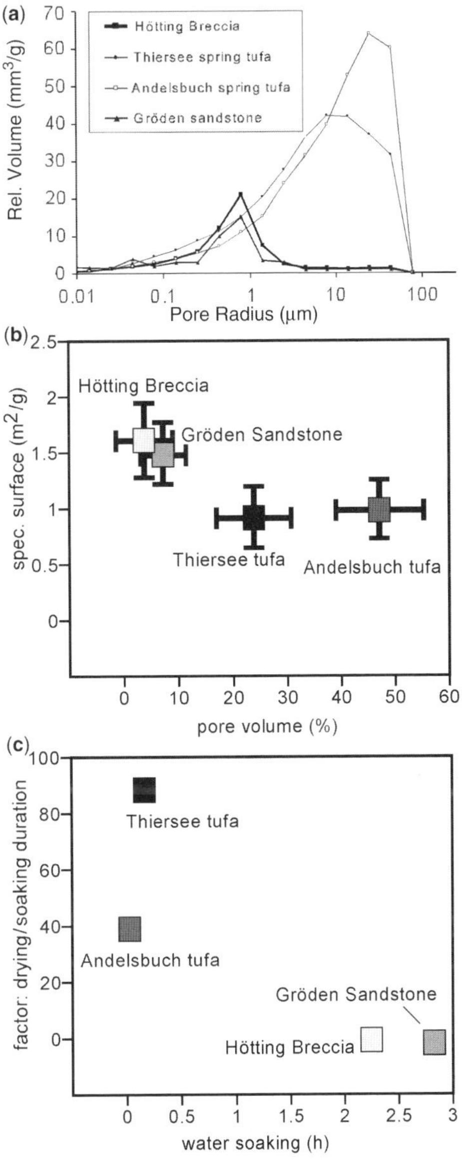

Fig. 5. (**a**) Pore size distribution of the three stone types investigated compared to Gröden Sandstone; (**b**) correlation between specific surface and pore volume of the three stone types investigated compared to Gröden Sandstone (bars symbolize the variance of the data) and (**c**) correlation between the drying/soaking duration and the water soaking duration of the three stone types investigated, compared to Gröden Sandstone.

With respect to the A-coefficient, the almost free access of water via the abundant macropores into the tufa stone seems mainly responsible for the extremely high value and the wide scattering of this parameter. With values of about 26–355 kg m^{-2} $h^{-1/2}$ this parameter exceeds that of other stone materials by 1–2 orders of magnitude.

The porosity of the Hötting Breccia (10 vol.%) is determined by the middle-fine-grained matrix, since one of the clasts – mostly carbonates – has less than 1 vol.% of porosity (Hofer 2004). This circumstance implies that the properties of the Hötting Breccia are related in a specific way to the whole rock. Due to this, the pore radius spectrum shows a maximum in the 1 μm range and is symmetric. A comparison with the pore radius spectrum data of the fine to middle-grained Gröden Sandstone (Table 1 and Fig. 5a) initially shows a similarity to that of the Hötting Breccia. The main pore radius maximum of the Gröden Sandstone also lies at 1 μm. However, the spectrum of the Gröden Sandstone indicates an asymmetry with a bimodality in the range <1 μm. This might be related to the diagenetic processes to which this material was subjected, and which tend to reduce the porosity and to modify the pore radius distribution (Franzen 2002). A comparative representation of the pore radii spectra of all four materials is depicted in Figure 5a; their average pore radii are listed in Table 1.

The specific surface of the four materials does not differ substantially at first sight. The tufa dominated by their macroporosity exhibit only 60% of the Hötting Breccia and the Gröden Sandstone. Taking into account that the clasts of the Hötting Breccia contribute only very little to the specific surface, the effective value of the specific surface of the matrix may actually be slightly higher at about 1.9–2.0 m^2 g^{-1} (Hofer 2004). A graphical correlation of the specific surface and the pore volume of the four materials is given in Figure 5b. The macroporous tufa are characterized by a lower specific surface than the Hötting Breccia and the Gröden Sandstone, which exhibits a pore radii maximum below 1 μm. This also proves that the specific surface of rocks may be proportionally related to pore size.

With respect to the hygric properties (the A-coefficient, the water capacity and the water saturation as well as the drying behaviour), the A-coefficients of the investigated stones show the biggest discrepancy (see Table 1). The speed of water uptake of the tufa is faster by 1.5–2.5 orders of magnitude than that of the Hötting Breccia (and Gröden Sandstone). A qualitative explanation is provided by comparison of the pore radii spectra and the pore volume of the tufa and the Hötting Breccia (and Gröden Sandstone), respectively (cf. Fig. 5a). The most effective capillary water transport takes place in the pore radii range below 1000 μm (Klopfer 1985). The Andelsbuch tufa exhibits a maximum in the radii spectrum at 30 μm and a pore volume of 48% related to the capillarity coefficient of 355 kg m^{-2} $h^{-1/2}$; the Thiersee tufa has a radii maximum of 10 μm and a pore volume of 24%, which yields an A-coefficient of 26 kg m^{-2} $h^{-1/2}$. The corresponding values of the Hötting Breccia are a maximum of pore radii distribution at 0.6 μm, a pore volume of 9.5% and an A-coefficient of 1.9 kg m^{-2} $h^{-1/2}$. A similar comparison would hold for the Gröden Sandstone.

Investigating the drying behaviour of the materials reveals further differences. In particular, the time-dependent water loss for the tufa is 0.55 ± 0.05 g h^{-1} and for the Hötting Breccia is 0.33 g h^{-1}. Comparing these time intervals from a soaked to dry state with the water uptake time up to water capacity point (soaking duration), the ratios of drying to soaking duration for the tufa are about 1:90 (Thiersee) to 1:40 (Andelsbuch), for the Hötting Breccia 1:2.7 and for the Gröden Sandstone 1:1.2 (see Table 1). A plot of the factor of drying/soaking duration v. soaking duration (hours) given in Fig. 5c demonstrates this.

These ratios show that the speed of the uptake of water, which is usually 2–5 times faster than the drying process for stone materials, is larger by more than one order of magnitude in the case of the tufa. This qualitatively explains why tufa tends to be 'wet' under atmospheric conditions and may serve as a good nutrient medium for organisms. A crucial role is probably taken here by the numerous millimetre-centimetre sized pores of the tufa which may function as internal water reservoirs. We believe that the observed almost linear drying behaviour of the tufa is due to this reservoir effect (Figs 2c & 3c) combined with the huge capillarily active micropores.

As to the weatherability of the tufa and the Hötting Breccia, it was noted above that they are fairly resistant materials. The effect of solubility of carbonates, which represents the main constituent in both cases, is not significant under the conditions of the Alpine climate. This seems to apply in particular for the tufa that may retain much of the driving rain and stormwater by their huge capillary pore capacity. The solute may re-precipitate during following drying periods. However, this is difficult to prove microscopically in the very variable fabrics of tufa. Possibly, this may be correlated to the mentioned secondary consolidation effect of tufa. Such a phenomenon is also shown by the Hötting Breccia to some extent, where calcite, the most soluble mineral component of this rock, is deposited in seams of isopachous cement in macropores. Based on this observation, one may assume that, on the stone surface, frequent interchange processes in water uptake and drying lead to a partial solution of the calcitic matrix and consequently to disintegration of the fabric. Apparently, these processes are especially efficacious on the fabric discontinuities between matrix and clasts, which give rise to losses of clasts.

Frost/thaw events are another important climatic impact factor within the Alps. According to Hirschwald (1912), one may assume a critical frost parameter S for building stones which is defined by the ratio of the water saturation obtained after 24 hours of submersion at normal atmosphere condition (WS_{24}) and the water saturation under vacuum conditions (WS_{vac}). The critical regime of S is at a value of 0.9. The values retrieved for the studied tufa lie at 0.9 and for the Hötting Breccia at 0.85, which indicate that they are located in that range. Despite this fact, both materials demonstrate relatively frost-resistant behaviour. This is confirmed by very recent unpublished results from frost/thaw cycle experiments carried out by Unterwurzacker & Obojes. However, detailed investigations of this are still lacking.

Conclusion

Quaternary stones are common building stones in the Alpine regions. In contrast to many other stone materials, they are young, formed recently on the Earth's surface. Their properties reflect the frame of thermodynamic conditions of the Earth's surface; diagenetic processes related to elevated P-T conditions can be ruled out. The geochemical processes of solution and precipitation that already take place during formation and sedimentation, as well as during their use as building stone, lead to secondary consolidation within the stones. The relatively good weathering resistance of the two materials is that they were formed under conditions similar to those they are now exposed to; in this fact, they differ significantly from most other building stone materials.

We would like to thank U. Griesser for making available the Hg-porometer of the Institute of Technical Pharmacy of the University of Innsbruck. The manuscript was greatly improved by discussions with D. Sanders, C. Spötl and T. Bidner, as well as by valuable comments on a previous draft of this manuscript by K. Germann and an unknown reviewer.

References

BIDNER, T., LANZ, B., MITTERER, S., RECHEIS, A., HAUSER, W., FRANZEN, C. & MIRWALD, P. W. 2003. Material properties, room-climate measurements and building history. An interdisciplinary project to set the conditions of restoration and re-use for the fortress of Kufstein, Tyrol, Austria. *Building and Environment*, **38**, 1133–1141.

BÖGLI, A. 1978. *Karsthydrographie und Physische Speläologie*. Springer Verlag, Berlin.

EMEIS, K.-C., RICHNOW, H.-H. & KEMPE, S. 1987. Travertine formation in Plitvice National Park, Yugoslavia: chemical versus biological control. *Sedimentology*, **34**, 595–609.

EN 1925, 1999. *Natural stone test methods – Determination of water absorption coefficient by capillarity*. European Committee for Standardization, Brussels.

FORD, T. D. & PEDLEY, H. M. 1996. A review of tufa and travertine deposits of the world. *Earth Science Reviews*, **41**, 117–175.

FRANZEN, C. 2002. *Historische Bauwerksteine in SüdTyrol, Verteilung und Verwitterungsverhalten*. Unpublished M.Sc. Thesis, University of Innsbruck, 111 pp.

FRANZEN, C., DIEKAMP, A., OBOJES, U., UNTERWURZACHER, M. & MIRWALD, P.-W. 2005. Lithologische Kartierung der Kapelle in der Burgruine Kropfsberg, Reith im Alpachtal/Tyrol. *Zeitschrift der Deutschen Gesellschaft für Geowissenschaften*, **156**, 197–203.

HIRSCHWALD, J. 1912. *Handbuch der Bautechnischen Gesteinsprüfung*. Gebr. Bornträger, Berlin.

HOFER, R. 2004. *Das Naturbausteininventar der Altstadt von Hall in Tyrol einschließlich einer materialkundlichen Charakterisierung der wichtigsten Gesteinsarten*. Unpublished M.Sc. thesis, University of Innsbruck, 86 pp.

KIESLINGER, A. 1964. *Die Nutzbaren Gesteine Salzburgs*. Verlag, Das Bergland-Buch, Salzburg/Stuttgart.

KLOPFER, H. 1985. *Wassertransport durch Diffusion in Feststoffen*. Bauverlag, Wiesbaden and Berlin.

OBOJES, U. 2003. *Quartärgeologische Untersuchungen an den Hängen der Innsbrucker Nordkette (Hötting Breccia)*. Unpublished M.Sc. thesis, University of Innsbruck, 91 pp.

PENA, J. L., SANCHO, C. & LOZANO, M. V. 2000. Climatic and tectonic significance of Late Pleistocene and Holocene tufa deposits in the Mijares river canyon, Eastern Iberian Range, Northeast Spain. *Earth Surface Processes and Landforms*, **25**, 1403–1417.

PENCK, A. 1921. *Die Höttinger Breccie und die Inntalterrasse nördlich Innsbruck*. Abhandlungen Preussische Akademie Wissenschaften, Berlin.

RÜF, B. 2006. *Typen und Bildung von Kalktuff in ausgesuchten aktiven Vorkommen Vorarlbergs (Österreich), und deren Bedeutung für historische Bauwerke*. Unpublished M. Sc. thesis, University of Innsbruck.

SANCHO, C., PENA, J. L. & MELENDEZ, A. 1997. Controls on Holocene and present-day travertine formation in the Guadalaviar River (Iberian Chain, NE Spain). *Zeitschrift für Geomorphologie*, **41**, 289–307.

SANDERS, D., UNTERWURZACHER, M. & RÜF, B. 2006. Microbially-induced calcium carbonate in tufa of the western Eastern Alps: a first overview. *Geo.Alp*, **3**, 167–189.

SPÖTL, C. & MANGINI, A. 2006. U/Th age constraints on the absence of ice in the central Inn Valley (eastern Alps, Austria) during Marine Isotope Stages 5c to 5a. *Quaternary Research*, **66**, 167–175.

UNTERWURZACHER, M., RÜF, B. & SANDERS, D. 2006. Quelltuff in Vorarlberg – Bildung, verwendung, materialtechnische eigenschaften. *Forschen und Entdecken*, **19**, 207–224.

VILES, H. A. & GOUDIE, A. S. 1990. Tufa, travertines and allied carbonate deposits. *Progress in Physical Geography*, **14**, 19–41.

Contribution to the technological characterization of two widely used Portuguese dimension stones: the 'Semi-rijo' and 'Moca Creme' stones

CARLOS FIGUEIREDO[1]*, RITA FOLHA[1], ANTÓNIO MAURÍCIO[1], CARLOS ALVES[2] & LUÍS AIRES-BARROS[1]

[1]*Center of Petrology and Geochemistry, CEPGIST, IST, Av. Rovisco Pais, 1049-001, Lisbon, Portugal*

[2]*University of Minho, Centro de Investigação Geológica e Valorização de Recursos, DCT, Gualtar, 4710-057 Braga, Portugal*

**Corresponding author (e-mail: carlos.m.figueiredo@ist.utl.pt)*

Abstract: Stone weathering and durability are two major concerns widely recognized within construction, cultural heritage and monument stone decay assessment and conservation works. This paper aims to complement the technological data on the two commercial varieties of Portuguese dimension stones ('Semi-rijo' and 'Moca Creme'), widely used in pavement and cladding inside and outside buildings. New data on the pore structures (fluid transport–related properties) and the durability (salt crystallization tests) of these limestones are presented. The pore structure was studied by a combined application of optical microscopy, scanning electron microscopy and mercury injection porosimetry. Fluid migration physical tests (open and free porosity, capillary imbibition and Hirschwald coefficient) were also performed, according to Portuguese and French standards. The resistance to salt crystallization was determined using the Portuguese standard NP EN 12370. An integrated analysis of all data has allowed a comparison of the results of the durability tests with characteristics of the pore network and properties related to fluid transport as well as petrographical features of the stones. The open porosity and freely-interconnected pores accessible to water after 48 hours (N48) seem to be the main features controlling their durability.

Natural stone has been used in the construction of buildings and structures for centuries. Although initially used as a convenient, local construction material as well as for its natural qualities of fire resistance and strength, natural stone came to be appreciated in time for its aesthetic value (Hora & Miller 1994). However, the aesthetic characteristics of a stone together with its commercial name is not be enough to determine its suitability for a particular application (Bradley 1998; Yavuz 2006). Due to the complexity of the geological processes generating the rocks, their properties can vary within apparently similar rocks: in a single quarry, from quarry to quarry, and even within a single quarried block (Farmer 1968; Purcell 1969; Goldstrand & Shevenell 1997; Meng *et al.* 2006). This stresses the need for a close scrutiny and control of the quality constancy of any natural stone to enable informed selection and appropriate use in very specific applications. Stones used successfully on a particular project in the past may not have the same physical or mechanical properties as the stone used for a current project. Moreover, stone weathering and durability are two major concerns widely recognized within construction, cultural heritage and monument stone decay assessment and conservation works (Amoroso & Fassina 1983; Winkler 1997; Přikryl 2007). Stone deterioration usually involves several interacting agents and processes, and moisture and salts are among the most damaging factors in stone decay (Arnold & Zehnder 1989; Chabas & Jeannette 2001; Doehne 2002; Franzen & Mirwald 2004; Vásárhelyi & Ván 2006; Rothert *et al.* 2007; Sumner & Loubser 2008). As well as rock-forming minerals and their textural relations, pores and fissures are the most significant petrographical component governing the weathering and durability of natural stones (Fitzner 1993; Houck & Scherer 2006; Angeli *et al.* 2007; Benavente *et al.* 2007; Török *et al.* 2007). The pore space represents the preferred area for physical, chemical and biological weathering processes (Leith *et al.* 1996; Smith & Kennedy 1998; Pérez-Bernal & Bello 2002; Ruedrich & Siegesmund 2007; Shahidzadeh-Bonn *et al.* 2007).

This paper aims to complement the technological data on the two commercial varieties of Portuguese dimension stones ('Semi-rijo' and 'Moca Creme'), widely used in pavements and cladding inside and outside buildings. In addition, similar

From: PŘIKRYL, R. & TÖRÖK, Á. (eds) *Natural Stone Resources for Historical Monuments.* Geological Society, London, Special Publications, **333**, 153–163.
DOI: 10.1144/SP333.15 0305-8719/10/$15.00

Portuguese limestones from other Jurassic formations were used in the past to build some of the most important monuments located in the central region of Portugal such as, the 'Igreja de Santa Cruz' in Coimbra (12th century) and the 'Mosteiro da Batalha' in Batalha (14th–16th centuries) (Aires-Barros 2001). The research carried out in this paper is focused on the evaluation of the role of pore structure in controlling the durability of these limestones. Some new data have been obtained for each stone type regarding fluid migration physical tests (including open and free porosity, capillary absorption and Hirschwald coefficient), optical and scanning electron microscopy, mercury porosimetry and salt crystallization tests.

Methodology

To characterize the pore structure geometry a combined application of classical and modern methods and techniques usually used in the context of fluid flow studies in engineering, hydrology, sedimentology and petroleum industry such as optical microscopy, scanning electron microscopy (SEM) and mercury injection porosimetry (MIP) (Bear 1972; Basan *et al.* 1997) was used.

Two large samples of 15 × 15 × 100 cm were obtained for the 'Semi-rijo' (SR) and 'Moca Creme' (MC) Portuguese dimension stones, from Germánio & Cordeiro and Pedramoca quarries, respectively, in 2006. From these large samples, an equal number of testing specimens was cut parallel (||, H) and perpendicularly (⊥, V) to the bedding planes and prepared according to the standards used for experimental tests. Thin sections were prepared for a detailed petrographic characterization by optical microscopy. For each stone type, twelve cylindrical specimens (50 mm diameter by 50 mm long, with six of them cut perpendicular to the bedding planes) were used for the determination of open and free porosity and capillary absorption, according to Portuguese standards NP EN 1925 (2000) (capillary absorption) and NP EN 1936 (2001) (open porosity, measured by imbibition under vacuum conditions) and also the French standard N FB 10-504 (1973) (freely accessible porosity under atmospheric pressure after 48 hours, measured as N48 porosity).

According to the standard test procedure outlined in NP EN 1936 (2001), the open porosity (the total interconnected pore volume that intercommunicates with the stone surface and is available to water under vacuum condition (Carmichael 1989; Fitzner 1993) was determined using the buoyancy or hydrostatic method. After drying for 24 ± 2 hours in a ventilated oven at a temperature of 70 ± 5 °C to constant mass and cooled at room temperature 20 ± 5 °C, the samples were weighed in air and their dry mass were determined. Samples were then completely saturated with distilled water under vacuum condition (2.0 ± 0.7 kPa), weighed surface-dry in air (water-saturated mass) and finally weighed suspended by a fine-diameter wire in water (immersed in water or hydrostatic mass).

Following Mertz (1991), Hammecker (1993) and the French standard N FB 10-504 (1973), the N48 porosity (the freely water-accessible porosity under atmospheric pressure after 48 hours) was determined by imbibition of the samples in distilled water under atmospheric pressure for 48 hours. During the first 24 hours of imbibition, the samples were only partially immersed in distilled water: up to one-quarter of sample height for 1 hour followed by 23 hours at half of their height. The samples were then completely immersed in distilled water for 24 hours and their 48 hours water-saturated mass was determined. The Hirschwald coefficient (S48 water-saturation coefficient, %) was calculated as the ratio of N48 porosity to open porosity.

A visual picture of the pore space and its structural relations with the rock-forming minerals and allochemical components were obtained by SEM of several horizontal, polished sections cut through the cylindrical specimens previously used for petrophysical characterization. Fourteen cylindrical-shaped specimens (25 mm diameter by 25 mm long) for both stone types were also studied by MIP with a Micromeritics 33 000 psia (228 MPa) mercury porosimeter AutoPore III 9400. Using this equipment, the pore diameter ranging from approximately 360 μm to 0.005 μm could be analysed (Micromeritics 1997). According to the Portuguese Standard NP EN 12370 (2001), twelve cubic samples (40 mm side) were used to evaluate the resistance to salt crystallization. Fifteen cycles were performed for each sample. After they have been soaked in a 14% solution of sodium sulphate decahydrate ($Na_2SO_4 \cdot 10H_2O$) for two hours, the samples were dried up to 105 ± 5 °C over sixteen hours in an ARALAB climatic chamber (FITO-CLIMA 300EDTU).

Finally, the study of statistical correlations contributed to an integrated discussion of results and comparison of durability tests with pore size, geometry and structure of pore network, properties related to fluid transport as well as petrographical features of the stones.

'Semi-rijo' (SR) and 'Moca Creme' (MC) stones

Geological and geographical setting

These stones are both calciclastic limestones representing different facies exploited from the same Bathonian age (Middle Jurassic) 'Valverde'

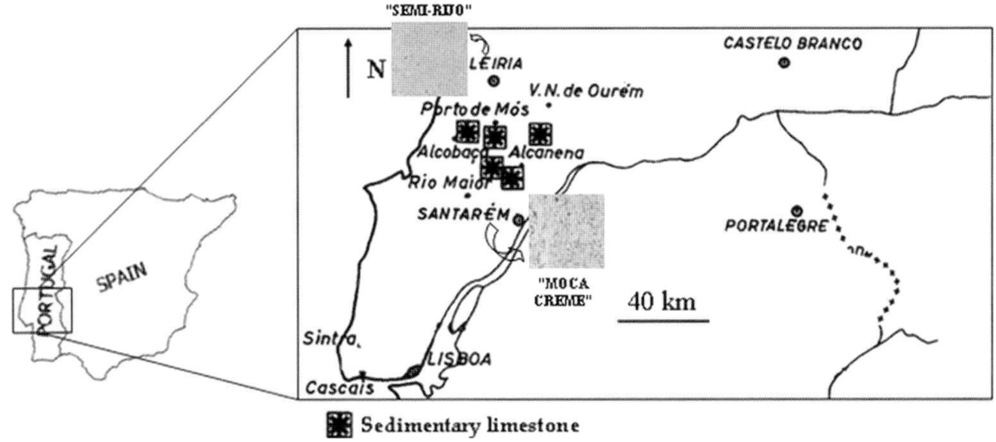

Fig. 1. Map of Portugal showing the location of some active quarries of SR and MC stones. Modified after Catálogo de Rochas Ornamentais Portuguesas (1983, 1984, 1985) and Portuguese Natural Stones (1995).

formation that belongs to the Calcareous Massif of Estremadura in the Mesozoic/Cenozoic occidental border of Portugal (Catálogo de Rochas Ornamentais Portuguesas 1983, 1984, 1985; Manuppella *et al.* 1985; Costa *et al.* 1988; Portuguese Natural Stones 1995).

The Calcareous Massif of Estremadura is the main Portuguese centre of ornamental limestones with a production of 280 000 tons/annum (Carvalho *et al.* 1998). It covers an area of 900 km^2 in the centre of Portugal, about 100 km north of Lisbon. It is a thick sequence of carbonate Mesozoic rocks, structurally elevated, that were deposited in a sedimentary basin along the continental Portuguese margin (Carvalho *et al.* 2000; Moura 2001).

Some active quarries of both stone types are located in the central region of Portugal (Fig. 1): district of Leiria, Municipality of Porto de Mós for the SR and district of Santarém, Alcanede parish, for the MC. The SR is a white oolitic, calciclastic limestone, scarcely bioclastic. According to the area of exploitation (Arrimal and Çodaçal parishes), two commercial varieties are respectively known as 'Semi-rijo' Arrimal (SR_a) and 'Semi-rijo' Codaçal (SR_c) (Carvalho *et al.* 2000). Blocks of medium size are usually available and no preferential plane is taken into account to achieve a good ornamental pattern. MC is a beige limestone, generally coarsely calciclastic and abundantly bioclastic. However, three grain size varieties are usually identified when taking into account the size of its allochemical components: coarse (MC_c), medium (MC_m) and fine (MC_f) 'Moca Creme' (Carvalho 1995; PNS 1995). Its best ornamental aspect is achieved by cutting the blocks normal to the bedding planes (CROP 1983, 1984, 1985; Costa *et al.* 1988).

The large samples obtained from Germánio and Cordeiro and Pedramoca quarries are respectively similar to the commercial varieties SR_c and MC_f.

Chemical composition

Some average values referring to the chemical composition of these stones (concentration in weight percent obtained for major chemical elements in terms of oxides) are summarized in Table 1, as given in CROP (1983, 1984, 1985), PNS (1995) and Carvalho *et al.* (2000).

Table 1. *Chemical composition in weight percent of studied rocks (average values)*

Stones types	SiO_2	TiO_2	Al_2O_3	$Fe_2O_3^T$	MnO	MgO	CaO	Na_2O	K_2O	L.O.I.	Total	CO_2
SR_a*†‡	0.06	trace	0.49	0.06	trace	0.27	55.46	0.06	0.01	43.85	100.26	43.81
SR_c*‡	0.07	0.00	0.25	0.05	n.a.	0.34	55.37	0.07	trace	43.88	100.03	43.82
MC_c*	0.09	0.00	0.20	0.06	trace	0.32	55.24	0.08	0.04	43.85	99.88	43.70
MC_m*†‡	0.07	0.00	0.54	0.06	0.00	0.10	55.86	0.06	0.07	43.60	100.36	43.95
MC_f*	0.03	0.00	0.00	0.12	n.a.	0.44	54.85	0.07	trace	43.97	99.48	43.94

n.a., not analysed; *, values from PNS (1995); †, values from CROP (1983, 1984, 1985); ‡, Values from Carvalho *et al.* (2000).

According to these results, both stones are pure limestones with values of CaO generally higher than 55% and silica content lower than 1% (Pettijohn 1975; Mason & Moore 1982).

Physical and mechanical properties

In Table 2 some values of physical and mechanical tests are summarized, compiled from CROP (1983, 1984, 1985), PNS (1995) and Carvalho *et al.* (2000).

The results show that these stones are low (SR) to medium (MC) resistant to mechanical stress, with compressive strength values varying from 48–54 MPa and 75–92 MPa, respectively (Carmichael 1989). They demonstrate a reduction in compressive strength after freeze-thaw tests from 6.5 to 11.4%, for both stones. As they also have values of water absorption higher than 0.5% they can be considered vulnerable to decay processes due to water-rock interaction (CROP 1983, 1984, 1985; Carvalho *et al.* 2000).

Results and discussion

Macroscopic description and thin section analysis

The commercial varieties SR and MC are both calciclastic limestones. SR is a light-beige pelloidal, oolitic, calciclastic limestone, with rare rounded bioclasts (foraminifera, arthropods, echinoderms, etc.) (Fig. 2a). The allochemical components generally have a wide range of sizes from 0.08 mm to 1.6 mm. MC is a beige pelloidal, oolitic, calciclastic limestone, rich in bioclasts (brachiopods, foraminifera, echinoderms, molluscs, arthropods, etc.) (Fig. 2b). These are usually less rounded and generally have a centimeteric size (Fig. 2b). The pelloids show better uniformity of size, ranging from 0.03 mm–0.08 mm. In thin sections, SR (Fig. 2c) could be classified as fossiliferous pelmicrosparite/grainstone and the MC as biopelintrasparite/grainstone (Fig. 2d) (Pettijohn 1975). Compared to SR, MC has a more heterogeneous fabric due to the presence of very thin bedding planes (usually not larger than 0.5 cm), mainly related to the alignment of bioclasts and a coarsely crystalline calcite.

As the porosity is not clearly visible in the thin section by optical petrography, it is likely that very fine pores exist in these stones as corroborated by the pore entry sizes obtained by MIP and scanning electron microscopy (see below).

Scanning electron microscopy (SEM)

Very fine pores (smaller than 5 μm, see Figs 3a and b) may occur for both stones. They are associated with sparry calcite cement, being intergranular

Table 2. *Physical and mechanical properties of studied rocks (average values; footnotes as for Table 1)*

Stones types	Compressive strength (MPa)	Compressive strength* (MPa)	Bending strength (MPa)	Apparent density (kg m^{-3})	Open porosity (%)	Water absorption (%)	Abrasion test (mm)	Impact test: minimum fall height (cm)	Linear thermal expansion coefficient ($\times 10^{-6}$ per °C)
SR_a*†‡	54	55	11	2346	12.03	5.12	5.9	35	5.7
SR_c*‡	48	45	12	2305	12.61	5.47	9.3	25	3.3
MC_c*	92	82	9	2511	6.19	2.46	2.4	35	3.9
MC_m*†‡	91	85	19	2515	5.90	2.34	4.0	40	5.3
MC_f*	75	78	17	2433	8.92	3.60	4.2	30–35	3.4

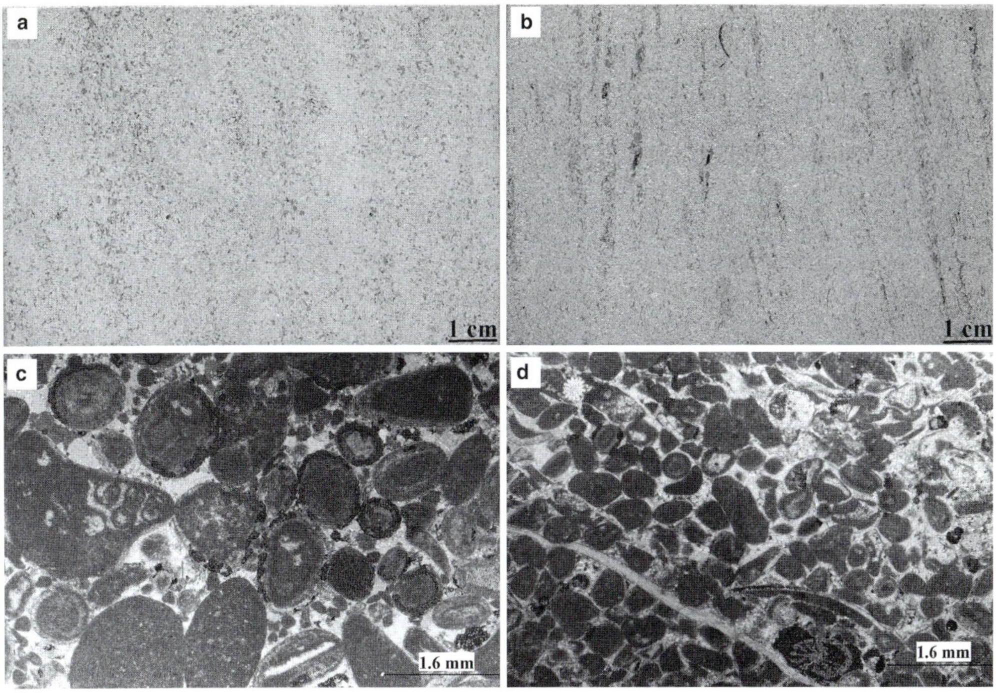

Fig. 2. (**a, b**) Cutting planes and (**c, d**) photomicrographs (crossed nicols) of thin sections, showing the most common aesthetic characteristics and fabric of the stones. (a, c) SR (fossiliferous pelmicrosparite/grainstone) and (b, d) MC biopelintrasparite/grainstone).

with fissure-like shape (Fig. 3a) or showing a more irregular shape (Fig. 3b) within the pelloids.

Open and free porosity, capillary absorption and Hirschwald coefficient

Both varieties are moderately porous stones and have an open porosity accessible to water ranging from 11.2 vol.% to 12.8 vol.% and from 12.1 vol.% to 12.9 vol.%, respectively (see Table 3). Hirschwald coefficients (S48%) are generally, higher than 80%, suggesting freely interconnected (N48%) and uniformly distributed pores (Mertz 1991; Hammecker 1993; Thomachot & Jeannette 2002). For both stones, the free porosity (N48%) is usually higher than 10.6 vol.%, corresponding to a Hirschwald coefficient (S48%) usually higher than 93.3%. There is a good correlation ($r = +0.95$, Table 4) between open (%) and free (N48%) porosity, suggesting pore networks with similar general characteristics in the two types of limestones. This agrees with the weak correlation found between porosity (%) and Hirschwald coefficient (see Table 4), mainly related to the narrow range of variation of the latter parameter regardless of porosity values.

MC shows lower values for the two capillary kinetic parameters considered, A (mass increase by area by square root of time) and B (height of capillary rise by square root of time) which range from 0.143–0.256 g cm^{-2} $h^{-1/2}$ and from 0.7872–1.9029 cm $h^{-1/2}$, respectively. SR samples have values of A ranging from 0.1812–0.2782 g cm^{-2} $h^{-1/2}$ and B ranging from 1.5721–2.0938 cm $h^{-1/2}$ (Figs 4a, b and c). This could be related to the more heterogeneous fabric and the presence of very thin bedding planes in MC (the lowest values were found for the samples cut perpendicularly to these petrographical heterogeneities). However, there is a good correlation ($r = +0.81$, Table 4) between both parameters related to capillary absorption for both limestones. The values of the capillary rise parameter B are relatively small when compared to the results from other carbonate stone types, such as those reported by Hammecker & Jeannette (1994): 2.4 cm $h^{-1/2}$ for 'Laspra limestone' (fine dolomicrite with open porosity of 30.0 vol.%), 5.5 cm $h^{-1/2}$ for 'Lourdines limestone' (calcareous micrite with open porosity of 25.8 vol.%) and 7.1 cm $h^{-1/2}$ for 'Hontoria limestone' (bioclastic limestone with open porosity of 19.8 vol.%) or compared to results reported for

Fig. 3. SEM image of sparry calcite cement (**a**) with intergranular fissure-like shape pores and (**b**) of a pellet interior with pores of irregular shape.

less porous rocks such as granites (Alves *et al.* 1996; Begonha & Sequeira Braga 2002). These reported results are in agreement with the weak correlations found between porosity and capillary absorption parameters in the stones considered in the present work (Table 4), suggesting the influence of the pore network geometrical characteristics in the capillary kinetics (Mertz 1991; Hammecker 1993; Benavente *et al.* 2007).

Mercury injection porosimetry (MIP)

According to MIP results (Fig. 5a, b and Table 5), the pore network structure is, for both stones, composed essentially by meso- and macropores ($0.002\ \mu m \leq$ pore radius $\leq 7.5\ \mu m$; IUPAC 1976) comprising more than 90% of the total space invaded by mercury injection. Both stones usually have pores with radii ranging in size from to 0.0033–168.29 µm. However, the curves obtained are typical of unimodal pore structures (Bear 1972; Mertz 1991; Hammecker 1993; Micromeritics 1997; Thomachot & Jeannette 2002).

Salt crystallization tests

Regarding the durability tests, MC seems to be less resistant to salt crystallization than SR (see Fig. 6a–d and Table 3). These results seem to be mainly related with the open (%) and freely interconnected porosity (N48%) accessible to water (Table 4). The higher correlation coefficients

Table 3. *Porosity, water transport properties and weight loss by salt crystallization tests*

	SR		MC	
	Samples ∥ to bedding	Samples ⊥ to bedding	Samples ∥ to bedding	Samples ⊥ to bedding
Open porosity (%)	11.84 ± 0.61	11.77 ± 0.53	12.60 ± 0.26	12.34 ± 0.21
N48* (%)	11.58 ± 0.52	11.40 ± 0.66	12.24 ± 0.33	11.95 ± 0.37
S48† (%)	97.86 ± 0.95	96.84 ± 1.78	97.09 ± 1.48	96.84 ± 1.69
A‡ ($g\ cm^{-2}\ h^{-1/2}$)	0.223 ± 0.023	0.218 ± 0.027	0.216 ± 0.020	0.173 ± 0.017
B‡ ($cm\ h^{-1/2}$)	1.919 ± 0.139	1.823 ± 0.179	1.682 ± 0.158	1.224 ± 0.205
Weight loss (%)	9.88–13.03	9.67–12.90	18.62–22.64	11.34–19.98

*, open porosity at 48 hours; †, Hirschwald coefficient; ‡, Capillary kinetic parameters.

Table 4 . *Correlation matrix of fluid transport-related properties and weight loss by salt crystallization tests*

	(a)	(b)	(c)	(d)	(e)	(f)
(a) Open porosity (%)	1					
(b) Open porosity at 48 hours (%) (N48)	0.95	1				
(c) Hirschwald coefficient (%) (S48)	0.05	0.36	1			
(d) Capillary kinetic parameter A ($g\ cm^{-2}\ h^{-1/2}$)	0.15	0.27	0.39	1		
(e) Capillary kinetic parameter B ($cm\ h^{-1/2}$)	− 0.11	0.01	0.37	0.83	1	
(f) Weight loss by salt crystallization tests (%)	0.77	0.76	0.44	0.26	−0.09	1

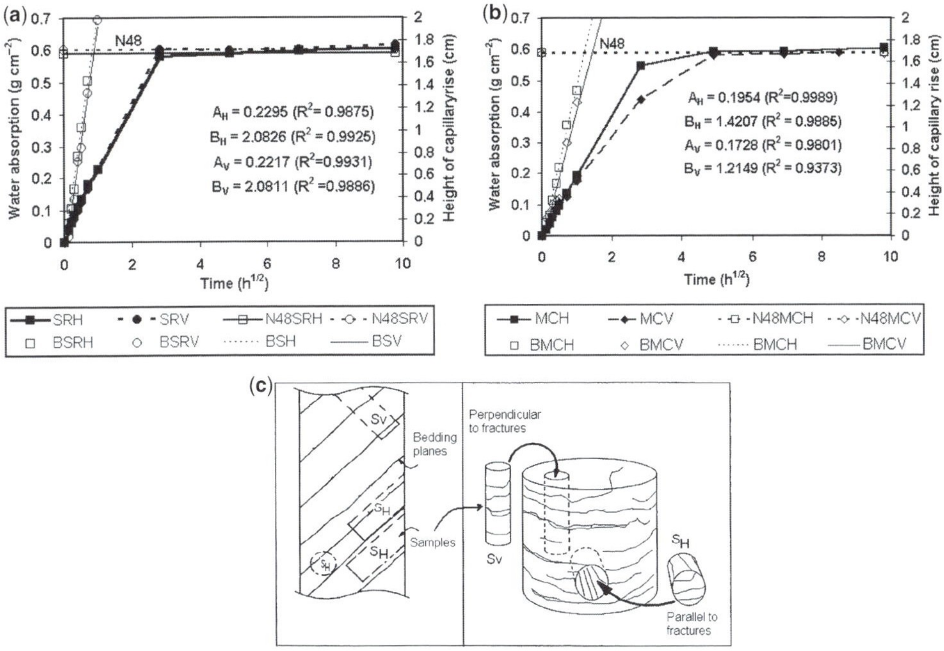

Fig. 4. Water absorption by capillary action on samples cut ∥ (H) and ⊥ (V) to the bedding planes for (**a**) SR and (**b**) MC. (**c**) Example of representative sampling of parallel (∥, H) and perpendicular (⊥, V) samples, regarding the heterogeneities (bedding, fractures, etc.) of natural stones (modified after Tiab & Donaldson 2004).

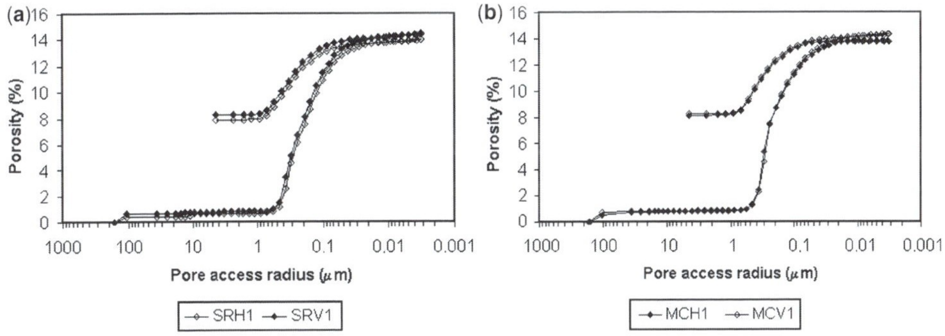

Fig. 5. Mercury porosimetry curves for samples cut ∥ (H) and ⊥ (V) to the bedding planes: (**a**) (SR) and (**b**) (MC).

Table 5. *Pore system characteristics obtained by mercury injection porosimetry (MIP)*

	SR		MC	
	Samples ∥ to bedding	Samples ⊥ to bedding	Samples ∥ to bedding	Samples ⊥ to bedding
NHg* (%)	14.20 ± 0.38	14.18 ± 0.75	14.39 ± 0.70	14.62 ± 0.43
NHg (r† ≤ 7.5 μm) (%)	13.65 ± 0.39	12.69 ± 2.07	12.40 ± 2.39	12.92 ± 1.27
NHg (r ≤ 7.5 μm)/NHg (%)	96.13 ± 0.74	89.33 ± 12.19	86.66 ± 18.24	68.41 ± 8.21
Mprv‡ (μm)	0.193 ± 0.007	0.195 ± 0.009	0.260 ± 0.089	0.235 ± 0.014
Apr§ (μm)	0.090 ± 0.006	0.091 ± 0.004	0.099 ± 0.016	0.094 ± 0.007

*, porosity by MIP; †, pore access radius (μm); ‡, median pore radius by volume; §, average pore radius (Bear 1972; Micromeritics 1997).

Fig. 6. 40 × 40 × 40 mm cubic samples (**a, c**) before and (**b, d, e**) after salt crystallization tests: (a, b) SR and (c, d, e) MC. (e) Details of salt decay features (fracturing and pelloid detachment) on an MC specimen obtained by stereomicroscope observations after three cycles.

(r = +0.77 and r = +0.76) found between weight loss (%) and open (%) and free (N48%) porosity, respectively, suggest that the volume of accessible pore space plays a decisive role in the salt decay process (Chabas & Jeannette 2001; Doehne 2002; Benavente *et al.* 2007; Rothert *et al.* 2007; Ruedrich & Siegesmund 2007). The degradation of the stones is mainly due to the pelloids detachment, surface flaking and scaling, sparry calcite cement fracturing and granular disintegration (Fig. 6a–e), causing the rounding of the edges and corners and an unevenly distributed loss of mass and volume, hence changing the shape of the cubic test samples (Langella *et al.* 2000; Kouzeli & Pavelis 2004; Rothert *et al.* 2007). The higher susceptibility of corners and edges of samples to salt decay, as usually found in field observations of several stones, can be explained by the effects of specimen geometry on water vapour pressure that would be higher on convex surfaces (relative to flat surfaces) at the same temperature (Amoroso & Fassina 1983). As well as the geometry factor, the more heterogeneous fabric and the presence of very thin bedding planes in MR seem to contribute to its degradation mainly by a very thin surface flaking and scaling when compared to SR (see Fig. 6a–e). The influence of fabric characteristics on the degradation forms due to salt decay is reported in numerous studies (Langella *et al.* 2000; Kouzeli & Pavelis 2004; Angeli *et al.* 2007; Rothert *et al.* 2007).

Final considerations

Based on this study, both SR and MC stone types have freely interconnected pore networks with uniformly distributed small pore radii. These characteristics can explain the high Hirschwald coefficients as well as the relatively small values of capillary rise parameter.

Results from salt crystallization tests highlight the importance of open porosity and freely interconnected pores accessible to water (48 hours porosity, N48) to the susceptibility to salt weathering of these stones. The effect of petrographical heterogeneities, such as the presence of stratification, pellets, oolites and fossils is also highlighted. These heterogeneities could help to explain the difference in porosity observed between samples cut parallel and perpendicular to bedding.

Despite being preliminary, these results complement the technical data available in the Portuguese Catalogue of Dimension Stones in aspects related to pore network characterization. These data could be relevant to the assessment of stone susceptibility to weathering, conservation treatments and specification establishment.

Comparing the new data and the data already published, the major difference lies in the new larger values of open porosity obtained for the MC variety. A likely explanation for these results seems to be related to the fact that new sites from the same region are being quarried to provide the commercial variety MC, a stone well known on the market (Carvalho 1995). A close scrutiny and control of the quality standards of any natural stone should be considered an essential requirement in all types of building projects, weathering studies and restoration works, given the well-known fact that stones can be conceived as random heterogeneous porous media.

A better understanding of the role of pore structure characteristics in controlling the durability of this type of materials could be obtained through an improvement of the methodological approach used so far, by using 2D or 3D computer-aided image analysis (CAIA) and non-destructive-high-resolution X-ray computed tomography (CT).

This study was partially financed by Portuguese Fundação para a Ciência e a Tecnologia (FCT), project No. POCTI/CTA/44940/2002_PORENET, with financial support from European Union FEDER and the state budget of the Portuguese Republic and Centro de Petrologia e Geoquímica/IST/TU Lisbon subproject DECASTONE.

References

AIRES-BARROS, L. 2001. *As rochas dos monumentos Portugueses: tipologias e patologías.* Cadernos do Instituto Português do Património Arquitectónico (IPPAR). Ministério da Cultura, (**I**), (**II**).

ALVES, C., SEQUEIRA BRAGA, M. A. & HAMMECKER, C. 1996. Water transfer and decay of granitic stones in monuments. *Comptes Rendus de l'Académie des Sciences*, **323**, 397–402.

AMOROSO, G. G. & FASSINA, V. 1983. *Stone decay and conservation: atmospheric pollution, cleaning, consolidation and protection.* Materials Science Monographs, **11**, Elsevier, Amsterdam.

ANGELI, M., BIGAS, J. P., BENAVENTE, D., MENÉNDEZ, B., HÉBERT, R. & DAVID, C. 2007. Salt crystallization in pore: quantification and estimation of damage. *Environmental Geology*, **52**, 187–195.

ARNOLD, A. & ZEHNDER, K. 1989. Salt weathering on monuments. *In:* ZEZZA, F. (ed) *The Conservation of Monuments in the Mediterranean Basin. Proceedings of the 1st International Symposium on The Influence of Coastal Environment and Salt Spray on Limestone and Marble*, Bari, Grafo Edizioni, 31–58.

BASAN, P. B., LOWDEN, B. D., WHATTLER, P. R. & ATTARD, J. J. 1997. Pore-size data in petrophysics: a perspective on the measurement of pore geometry. *In*: LOVELL, M. A. & HARVEY, P. K. (eds) *Developments in Petrophysics.* Geological Society, London, Special Publication, **122**, 47–67.

BEAR, J. 1972. *Dynamics of Fluids in Porous Media.* American Elsevier Publishing Company, Inc., Dover Publications, Inc., New York.

BEGONHA, A. & SEQUEIRA BRAGA, M. A. 2002. Weathering of the Oporto granite: geotechnical and physical properties. *Catena*, **49**, 57–76.

BENAVENTE, D., CUETO, N., MARTÍNEZ-MARTÍNEZ, J., CURA, M. A. G. & CAÑAVERAS, J. C. 2007. The influence of petrophysical properties on the salt weathering of porous building rocks. *Environmental Geology*, **52**, 197–206.

BRADLEY, F. 1998. *Natural Stone: A Guide to Selection.* W.W. Norton & Company, Inc. New York, London.

CARMICHAEL, R. S. 1989. *Practical Handbook of Physical Properties of Rocks and Mineral*, CRC Press, Inc. Boca Raton, Florida.

CARVALHO, J. 1995. Calcários ornamentais e industriais na área de pé da pedreira (Maciço Calcário Estremenho). *Boletim de Minas*, **32**, 25–39.

CARVALHO, J., MANUPPELLA, G. & MOURA, A. C. 1998. Contribution to the geological knowledge of the Portuguese Ornamental Limestone. *Comunicações do Instituto Geológico e Mineiro*, **84**, 74–77.

CARVALHO, J., MANUPPELLA, G. & MOURA, A. C. 2000. Calcários ornamentais Portugueses. *Boletim de Minas*, **37**, 223–232.

CROP (*Catálogo de Rochas Ornamentais Portuguesas*). 1983, 1984, 1985. Volumes **I**, **II**, **III**, Direcção-Geral de Geologia e Minas (DGGM), Ministério da Indústria, Energia e Exportação.

CHABAS, A. & JEANNETTE, D. 2001. Weathering of marbles and granites in marine environment: petrophysical properties and special role of atmospheric salts. *Environmental Geology*, **40**, 359–368.

COSTA, J. R. G., MOREIRA, J. C. B. & MANUPPELLA, G. 1988. Calcários ornamentais do maciço calcário estremenho. *Estudos, Notas e Trabalhos, Lisboa*, DGGM, **30**, 51–88.

DOEHNE, E. 2002. Salt weathering: a selective review. *In:* SIEGESMUND, S., WEISS, T. & VOLLBRECHT, A. (eds) *Natural Stone, Weathering Phenomena, Conservation Strategies and Case Studies.* Geological Society, London, Special Publications, **205**, 51–64.

FARMER, I. W. 1968. *Engineering Properties of Rocks.* E. & F. N. Spon Ltd, London.

FITZNER, B. 1993. Porosity properties and weathering behaviour of natural stones. Methodology and examples. *In:* ZEZZA, F. (ed.) *Stone Material in Monuments: Diagnosis and Conservation (Second Course)*, Heraklion – Crete, C.U.M. University School of Monument Conservation, Mario Adda Editore, Bari, 43–54.

FRANZEN, C. & MIRWALD, P. W. 2004. Moisture content of natural stone: static and dynamic equilibrium with atmospheric humidity. *Environmental Geology*, **46**, 391–401.

GOLDSTRAND, P. M. & SHEVENELL, L. A. 1997. Geologic controls on porosity development in the Maynardville Limestone, Oak Ridge, Tennessee. *Environmental Geology*, **31**, 248–258.

HAMMECKER, C. 1993. *Importance des transferts d'eau dans la dégradation des pierres en oeuvre.* PhD Thesis, University Louis Pasteur, Strasbourg.

HAMMECKER, C. & JEANNETTE, D. 1994. Modelling the capillary imbibition kinetics in sedimentary rocks: role of petrographical features. *Transport in Porous Media*, **17**, 285–303.

HORA, Z. D. & MILLER, L. B. 1994. *Dimension Stone in Victoria, B. C. A City Guide & Walking Tour.* British Columbia, Ministry of Energy, Mines and Petroluem Resource.

HOUCK, J. & SCHERER, G. W. 2006. Controlling stress from salt crystallization. *In:* KOURKOULIS, S. K. (ed) *Fracture and Failure of Natural Building Stones. Applications in the Restoration of Ancient Monuments.* Springer, Berlin, 299–312.

IUPAC. 1976. *Manual of Symbols and Terminology for Physicochemical Quantities and Units* - Appendix II. Definitions, Terminology and Symbols in Colloid and Surface Chemistry. Part II: Heterogeneous Catalysis. *Pure and Applied Chemistry*, **46**, 71–90.

KOUZELI, K. & PAVELIS, C. 2004. Study of the building materials of the "Demossion Sema" monuments in Athens. *In*: KWIATKOWSKI, D. & LÖFVENDAHL, R. (eds) *Proceedings of the 10th International Congress on Deterioration and Conservation of Stone, Stockholm*, **1**, 407–414.

LANGELLA, A., CALCATERRA, D., CAPPELLETTI, P., COLELLA, A., GENNARO, M. & GENNARO, R. 2000. Preliminary contribution on durability of some macroporous monumental stones in historical towns of Campania region, Southern Italy. *In:* FASSINA, V. (ed.) *Proceedings of the 9th International Congress on Deterioration and Conservation of Stone, Venice*, **1**, 59–67.

LEITH, S. D., REDDY, M. M., RAMIREZ, W. F. & HEYMANS, M. J. 1996. Limestone characterization to model damage from acidic precipitation: Effect of pore structure on mass transfer. *Environmental Science & Technology*, **30**, 2202–2210.

MANUPPELLA, G., MOREIRA, J. C. B., COSTA, J. R. G. & CRISPIM, J. A. 1985. Calcários e dolomitos do Maciço Calcário Estremenho. *Estudos, Notas e Trabalhos, Lisboa*, DGGM, **27**, 3–48.

MASON, B. & MOORE, C. B. 1982. *Principles of Geochemistry*, 4th edn. John Wiley & Sons, New York.

MENG, Z., ZHANG, J. & PENG, S. 2006. Influence of sedimentary environments on mechanical properties of clastic rocks. *Environmental Geology*, **51**, 113–120.

MERTZ, J. D. 1991. *Structures de porosité et propriétés de transport dans les grès.* Sciences Géologiques, **90**. PhD Thesis, University Louis Pasteur, Strasbourg.

MICROMERITICS. 1997. *AutoPore III Operator's Manual V1.xx.* Micromeritics Instrument Corporation. Norcross, GA, USA.

MOURA, A. C. 2001. A pedra natural ornamental em Portugal – Nota Breve. *Boletim de Minas, Lisboa*, **38**, 161–177.

N FB 10-504. 1973. *Pierres calcaires: mesure du coefficient d'absorption d'eau.* Association Française de Normalisation (AFNOR), Paris.

NP EN 1925. 2000. *Methods of test for natural stone units – Determination of water absorption coefficient due to capillary action.* Portuguese Institute for Quality (IPQ), Caparica, Portugal.

NP EN 1936. 2001. *Methods of test for natural stone units- Determination of real density and apparent density and of total and open porosity*. Portuguese Institute for Quality (IPQ), Caparica, Portugal.

NP EN 12370. 2001. *Natural stone test methods: Determination of resistance to salt crystallisation*. Portuguese Institute for Quality (IPQ), Caparica, Portugal.

PÉREZ-BERNAL, J. L. & BELLO, M. A. 2002. Weathering effects on stone pore size distributions. *In:* GALÁN, E. & ZEZZA, F. (eds) *Protection and Conservation of the Cultural Heritage of the Mediterranean Cities*, Swets & Zeitlinger, Lisse, 203–207.

PETTIJOHN, F. J. 1975. *Sedimentary Rocks*, 3rd edn. Harper & Row Publishers, New York.

PNS (*Portuguese Natural Stones*). 1995. CEVALOR (Centro Tecnológico para o Aproveitamento e Valorização das Rochas Ornamentais e Industriais).

PŘIKRYL, R. 2007. Understanding the Earth scientist's role in the pre-restoration research of monuments: an overview. *In:* PŘIKRYL, R. & SMITH, B. J. (eds) *Building Stone Decay: From Diagnosis to Conservation*. Geological Society, London, Special Publications, **271**, 9–21.

PURCELL, D. 1969. The repair and preservation of stonework. *Garden History Society*, Occasional paper, **1**, 11–14.

ROTHERT, E., EGGERS, T., CASSAR, J., RUEDRICH, J., FITZNER, B. & SIEGESMUND, S. 2007. Stones properties and weathering induced by salt crystallization of Maltese Globigerina Limestone. *In:* PŘIKRYL, R. & SMITH, B. J. (eds) *Building Stone Decay: From Diagnosis to Conservation*. Geological Society, London, Special Publications, **271**, 189–198.

RUEDRICH, J. & SIEGESMUND, S. 2007. Salt and ice crystallization in porous sandstones. *Environmental Geology*, **52**, 343–367.

SHAHIDZADEH-BONN, N., AZOUNI, A. & COUSSOT, P. 2007. Effect of wetting properties on the kinetics of drying of porous media. *Journal of Physics: Condensed Matter*, **19**, 1–7.

SUMNER, P. D. & LOUBSER, M. J. 2008. Experimental sandstone weathering using different wetting and drying moisture amplitudes. *Earth Surface Processes and Landforms*, **33**, 985–990.

SMITH, B. J. & KENNEDY, E. M. 1998. Moisture loss from stone influenced by salt accumulation. *In:* JONES, M. S. & WAKEFIELD, R. D. (eds) *Aspects of Stone Weathering, Decay and Conservation*. Imperial College Press, London, 55–64.

THOMACHOT, C. & JEANNETTE, D. 2002. Evolution of the petrophysical properties of two types of Alsatian sandstone subjected to simulated freeze-thaw conditions. *In:* SIEGESMUND, S., WEISS, T. & VOLLBRECHT, A. (eds) *Natural Stone, Weathering Phenomena, Conservation Strategies and Case Studies*. Geological Society, London, Special Publications, **205**, 19–32.

TIAB, D. & DONALDSON, E. C. 2004. *Petrophysics. Theory and Practice of Measuring Reservoir Rock and Fluid Transport Properties*, 2nd edn. Elsevier, Amsterdam.

TÖRÖK, Á., FORGÓ, L. Z., VOGT, T., LÖBENS, S., SIEGESMUND, S. & WEISS, T. 2007. The influence of lithology and pore-size distribution on the durability of acid volcanic tuffs, Hungary. *In:* PŘIKRYL, R. & SMITH, B. J. (eds) *Building Stone Decay: From Diagnosis to Conservation*. Geological Society, London, Special Publications, **271**, 251–260.

VÁSÁRHELYI, B. & VÁN, P. 2006. Influence of water content on the strength of rock. *Engineering Geology*, **84**, 70–74.

WINKLER, E. M. 1997. *Stone in Architecture. Properties, Durability*, 3rd edn. Springer-Verlag, Berlin.

YAVUZ, A. B. 2006. Deterioration of the volcanic kerb and pavement stones in a humid environment in the city centre of Izmir, Turkey. *Environmental Geology*, **51**, 211–227.

Pore and hygric properties of porous limestones: a case study from Bratislava, Slovakia

MAREK LAHO[1], CHRISTOPH FRANZEN[2,3]*, RUDOLF HOLZER[1] & PETER W. MIRWALD[2]

[1]*Department of Engineering Geology, Comenius University, Bratislava, Slovakia*

[2]*Institut für Mineralogie & Petrographie, Leopold-Franzens-Universität Innsbruck, Innrain 52, A-6020 Innsbruck, Austria*

[3]*Present address: Institut für Diagnostik und Konservierung an Denkmalen in Sachsen und Sachsen-Anhalt e. V., Schloßplatz 1, 01067 Dresden, Germany*

**Corresponding author (e-mail: franzen@idk-info.de)*

Abstract: Most historic stone monuments in Bratislava (Slovakia) are built with various types of porous, light and weakly cemented sedimentary rocks. These Neogene sandstones and limestones, also known as Leitha limestones, were quarried in the Vienna Basin. The various sedimentary environments are reflected in the heterogeneity of lithotypes and their cementation. The distinctly different pore structure of these rocks is reflected in their very variable physical properties and consequently in the distinct durability. Petrographic and petrophysical properties were determined on main ashlar types of the two most important historical monuments of Bratislava, the Castle and St Martin's Cathedral. Fresh samples of Leitha limestones were also obtained from the existing or abandoned quarries. The study includes a detailed petrographic examination of major rocks types and an assessment of their mineralogical composition by X-ray diffraction and microscopy. Hygric properties and porosity influence the weathering stability and the chances to apply conservation treatment. Parameters such as pore volume, pore size distribution by mercury porosimetry, specific surface of pores by nitrogen adsorption (BET) analysis as well as water sorption, capillary water uptake and drying behaviour were measured and compared for the six studied lithotypes.

Weathering of natural stone is induced by the interaction of material properties with a given regime of environmental factors (Winkler 1994). Thermodynamically, it is characterized by a non-equilibrium situation: the chemical and physical state of the stone does not adjust to the temperature and chemical conditions on the Earth's surface in most cases, but to those under which it was formed at the depth within the Earth's crust. Weathering is a complex system of interrelated processes during which material properties accommodate external conditions. It is bound to the presence of water, which acts as a transport and reaction medium to help reach the governing thermodynamic equilibrium conditions (Franzen & Mirwald 2004). These processes manifest themselves in the first intimation by changes to the property of the material which finally develop to various macroscopic forms and stages of decay (Smith & Přikryl 2007). The spectrum of stone properties such as strength, mineralogical-chemical composition, pore properties and hygric behaviour represent a frame of behaviour (Přikryl *et al.* 2003). The environmental factors are essentially related to different external parameters, climate, construction constraints, use and maintenance and anthropogenic pollution sources (Török 2002). Control of these external factors is mostly limited (Laue 2005). Detailed investigations on the different stone properties give basic information for the interpretation of weathering forms and the responding processes. This also may be used to outline possible conservation and restoration strategies and measures. In some cases, such investigations may also provide a prognosis of the future weathering behaviour.

Several monuments in Bratislava are built from natural stone from the east Austrian Danube region including the Leitha Mountains. These monuments are now showing various forms of deterioration and severe signs of decay. Consequently, extensive conservation and reconstruction campaigns on Bratislava monuments are necessary. A basic prerequisite for effective preservation activities is that – beyond the knowledge about the cultural-historical significance of the object – the materials must be characterized with respect to their material properties in order to understand their interactions in damage processes. Substantial

From: PŘIKRYL, R. & TÖRÖK, Á. (eds) *Natural Stone Resources for Historical Monuments.* Geological Society, London, Special Publications, **333**, 165–174.
DOI: 10.1144/SP333.16 0305-8719/10/$15.00

material science research must be integrated in a general methodological approach of conservation and renovation concepts. The present weathering forms and extent of damage must be related to the specific material properties and to the local environmental conditions (Fitzner & Heinrichs 2002).

In a continuation of previous work (Holzer *et al.* 2004*a*, *b*), this study focuses on the investigations of the different stone types as pristine building materials. Additionally, the other aim is to identify the potential source quarries and gather information on the provenance of the material. As well as the identification of lithotypes and their weathering forms, the mineralogical composition, sedimentary features, pore-size distribution and hygric properties were used as baselines of the comparison. Data of the pore and hygric properties of porous building rocks are key parameters in understanding weathering processes (Benavente *et al.* 2007) as well as in stone protection, conservation and reconstruction measures.

The dataset of this paper provides additional information for the restoration of the stone materials of Bratislava. The limestones used in Bratislava are very similar to the porous limestones used in the eastern part of Central Europe (Török 2002, 2004; Holzer *et al.* 2004*b*; Přikryl & Přikrylová 2004; Török *et al.* 2004). Recognition of their material properties and long-term behaviour can be compared to similar stones used in other countries, for example, Italy (Andriani & Walsh 2007).

Historic use of Leitha limestone

The Neogene Leitha limestones *s.l.* (prevailingly sandy limestones, carbonate sandstones and conglomerates) quarried in the areas of Hundsheim-Wolfsthal and in the Leitha Mountains (St Margarethen, Oggau, Mannersdorf) have been utilized as building and ornamental stones in the eastern part of Central Europe in the region between Vienna, Brno, Bratislava and Budapest. The name Leithakalk or Leitha limestone is derived from the river Leitha and the Hainburg Mountains. in the Hundsheim-Wolfsthal area.

As well as its use as building stone, the Leitha limestone was also an appreciated architectural and sculptural stone for the Viennese and the neighbouring Slovakian, Moravian and Hungarian areas since Roman times. The Roman Empire had already used the sandy limestones, calcareous sandstones and carbonate conglomerates, but more extensive application of these materials as dimension stones has been documented from 13th/14th until the late 19th century. The most prominent stone objects of the Gothic up to the late Baroque times were mostly constructed from this stone.

Fig. 1. St Martin's Cathedral as viewed from the Bratislava castle.

A further characteristic is that, after demolishing older structures, the stone was often re-used as a secondary material.

Case studies of Bratislava castle and St Martin's Cathedral (Holzer *et al.* 2004*a*) as well as a number of other monuments in Bratislava have shown that specific Leitha limestone varieties have been employed during specific times, but simultaneous use of different materials is also quite common (Holzer *et al.* 2004*b*). Documents indicate that the stones for the upper castle structure had been transported along the Danube river most likely from quarries in the Hundsheim–Wolfsthal area and from the Leitha Mountains, both for the Castle of Bratislava and St Martin's Cathedral (13th century) (Fig. 1).

Geological setting

Geologically, the Leitha limestone *s.l.* had been deposited on the margins of the Vienna basins in the early Cenozoic period. The Middle Miocene Leitha limestone was widespread sediment in the Central Paratethys. The depositional environment is interpreted to reflect tropical/subtropical conditions (Piller 2003). In evaluating the biota of these limestones, the following inventory is considered to be climatically important: zooxanthellate corals (>10 genera), coralline algae (including *Sporolithon*) and larger foraminifera (7 genera, including *Borelis*, *Amphistegina* and *Planostegina*) (Piller 2003). Corals and coralline algae formed bioconstructions in shallow-water environments representing tropical/subtropical climates. In many places, however, coralline algae are the dominating rock-forming biota of the Leitha limestone. The main reasons for this prevalence are the generally very high terrigenous input during this time. Additionally, palaeotopography did not provide adequate shallow water areas for reef growth (Piller 2003).

Methodology

Field sampling and observation

All basic rock types of Leitha limestone which have been determined at the monuments of Bratislava were sampled from potential quarries found in Hundsheim and Wolfsthal, some 10 km west of Bratislava, and in St Margarethen, Oggau, and Mannersdorf which are part of the Leitha Mountains (Fig. 2). The following main types of Leitha limestone were distinguished: carbonate conglomerate, oolitic limestones, sandy limestones, calcareous sandstones and lumachelle limestones. A number of samples of the rock material from several quarries were used for the laboratory tests. The weathering forms at the monuments are described by Smith *et al.* (2003).

Petrographic study

Mineralogical and petrographical analysis of samples was performed to identify the major lithotypes and to describe deterioration processes. After a macroscopic description, thin sections were prepared from selected samples and studied with respect to mineralogical and sedimentological characteristics. The mineralogical composition was also determined by powder X-ray diffraction (D500, Bruker AXS).

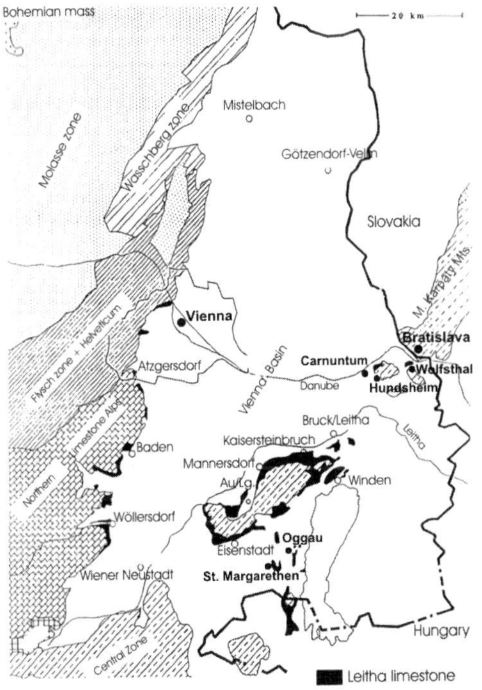

Fig. 2. Geological map of the Vienna basin showing position of quarries (Wolfsthal, Hundsheim, Oggau and St Margarethen) where samples were taken from.

Pore space analysis

The following parameters were investigated: porosity, pore-size distribution and specific surface area. The porosity and pore size distribution were measured by mercury porosimetry in two stages: low pressure to 400 kPa (Pascal 140) and high pressure to 200 MPa (Porosimeter 2000 Fision Instruments). Samples crushed into grains were assessed by mercury porosimetry. Weight of samples was measured to be 3.9–5.4 g. The specific surface area of the materials was determined by the nitrogen adsorption BET method (NOVA 2200 Quantachrome). Samples for assessment of BET were crushed into grains and outgassed at 110 °C for two hours. The samples were found to weigh 5.7–9.4 g. In addition, the specific surface was also calculated from the mercury porosimetry data for comparison.

Hygric properties

Hygric parameters are crucial properties for the weathering behaviour of a stone material, since they are responsible for the H_2O transport in to and out of the stone. In addition, they govern the internal H_2O transport processes and are a major factor of most decay processes. The following parameters were determined: vapour sorption/desorption, the capillary uptake, the total water uptake within 24 hours and the drying behaviour.

The water vapour sorption isotherm characterizes the ability of a material to interact with the air moisture. This property is most accurately determined by monitoring the mass changes of the sample equilibrated at various humidity conditions by a moisture sorption analyser (SPS 11) at constant temperature. For the experiments, the samples were crushed into grains. Samples of about 14–20 g were equilibrated in a humid atmosphere.

Capillary water uptake experiments of up to 24 hours were conducted by using cylindrical samples of 34 mm diameter and 28 mm height. The absorptive water saturation (WA_c) was determined from these measurements. Finally, the drying behaviour of the samples (cylinders of 34 mm in diameter and of 28 mm in height) was also monitored under ambient conditions until the constant weight was reached.

Petrographic character of studied limestones

Oolitic limestone, Wolfsthal

Light-coloured rock of Sarmatian age consists of calcareous ooids. In the nuclei of ooids, foraminifera fragments or quartz can be observed. The extraclasts are represented by quartz, muscovite, granite and sandstone fragments, but many of these are covered by calcitic envelopes. The extraclasts were driven from the nearby granitic source area. Foraminifers (*Miliolina, Rotaliina, Nubecularia sp.*) and bivalves are the most common fossils.

This limestone type was very often used as a building stone in the construction epoch of St Martin's Cathedral and the Bratislava castle. Analogous rocks were found in abandoned quarries on northern slopes of the Wangheimer forest (Hainburg Mountains) near the village Wolfsthal (Mišík 1997). Oolitic limestones are found as intercalations between the layers of lumachelle limestone and sandy limestone. The rock was exploited and used for its very easy workability, suitable colour and short transportation distance.

Granular disintegration of individual ooids due to the weak cementation represents the typical decay feature.

Sandy limestone, Wolfsthal

Macroscopically, the rock is light coloured, porous and fine-grained. Major rock components are ooids, foraminifers (*Miliolina, Rotaliina, Nubecularia sp., Quinqueloculina div. sp., Triloculina sp., Elphidium sp.*) and bivalves. The sand-sized carbonates, fragments of granite, quartz, biotite and plagioclase are cemented by calcitic cement. Clastic material originated from deterioration processes of the granitic bedrock, which represents the core body spreading from the territory of Malé Karpaty Mountains to the south east range of Western Carpathians.

This lithotype belongs to the oldest constructional material and is often found in the castle façade. The second period of its use dates back to about 1760, which is the so-called rococo rebuilding phase of the castle. In that time, the material was used on the western platform of the castle as exterior window frames (personal communication from Semanko & Špánik 2002).

Damage forms are flaking or crumbling. The formation of dark surface layers enriched in gypsum is also common.

Lumachelle limestone, Wolfsthal

This type of material was only sampled from the quarry in Wolfsthal. The lumachelle limestone from Wolfsthal is white coloured, with coarse grains and moderate cements responsible for high porosity. It contains foraminifers (*Miliolina, Rotaliina, Nubecularia sp., Quinqueloculina div. sp., Triloculina sp., Elphidium sp.*) and bivalves.

Some dimension stone blocks of this lithotype have been found in the socle part of St Martin's Cathedral, but have not been identified in the lower parts of the castle wall. It's use in the higher parts of the wall cannot be ruled out, in particular where parts of the walls are covered by the mortar.

Selective dissolution of cement is the most common weathering form, resulting in a positive morphology of shell fragments.

Carbonate conglomerate, Hundsheim

Two distinct lithotypes can be distinguished macroscopically: the light-grey conglomerates (beach sediments) and the slightly pink well-sorted limestone congolmerates (sublittoral zone). Microscopically, the main amount of organic substance is attributed to red coral algae (genera *Lithophyllum sp.* and *Lithothamnium sp.*, tubes of worms, bryozoans and foraminifers) Less significant but very variegated associations are represented by foraminifers of genuses *Elphidium sp., Textularia sp., Spiroloculina sp., Heterolepa sp.* and *Triloculina sp.* Terrigeneous admixture consists of quartz clasts, granitic and carbonate fragments. Significant amount of very fine, silt-sized light-grey bioclastic and non-fossiliferous carbonates provide the grey colour of the rock. Bioclasts and lithoclasts are cemented by calcite. Calcitic coating is also very common. Both fine- and coarse-grained varieties of conglomerates occur.

This type of building stone was subsequently excavated from higher beds of the old Roman quarry walls (Hainburg Mountains). The Roman settlement near Bad Deutsch-Altenburg it is well documented (Czijzek 1852). The building stone was shipped on the river Danube to the castle hill in Bratislava. Both fine- and coarse-grained types were distinguished in the entire castle wall system of Bratislava. The same rock material was also identified during the excavation works for remedial measures of the NW castle tower, where such rock blocks were found at about 3.0 m below the present surface. Conglomerates were also used during the so-called Pállfy castle reconstruction (1616–1635). The main advantage in using this type of material was the relatively short distance to the quarry sites. The fact that these materials were less workable than other rock types was apparently of minor importance.

Soiling and growth of dark gypsum crust, along with flaking and scaling, represent the most common damage forms of this lithotype. The latter forms are mostly observed at the edges of dimension stones and at the stone plinths of objects. The coarse-grained conglomerate shows detachment of larger compact stone pieces.

Sandy limestone, St Margarethen

The yellowish sandy limestone was formed in a shallow-water environment where sessile foraminifer species of *Textularia sp.*, *Triloculina sp.* and *Elphidium sp.* were present. As well as foraminifers, bivalves and ostracods are also common in these limestones. Fossils and clasts are weakly cemented and aragonitic shells are often leached and replaced by drusy calcite cement. This lithotype was quarried in St Margarethen from the Roman period. It has often been used because of the good workability and easy extraction. This material has been used in different construction periods of the St Stephan Cathedral in Vienna (Müller *et al.* 1993) and in St Martin's Cathedral in Bratislava (Holzer *et al.* 2004*a*). However, there is speculation that at least some stones have been re-used and were taken from the Petronell/Carnuntum Roman settlements in the mediaeval period.

Discolouration, efflorescence, partial leaching of the weak carbonate cement and following granular disintegration are the most common decay features. These mechanisms are triggered by air pollution, humidity and changes in temperature.

Leitha limestone, Oggau and Mannersdorf

Although localized on the opposite sides of the outcropping gneiss core of the Leitha Mountains, these two limestones are fairly similar. Macroscopically, the light-coloured fine-grained rock possesses low porosity. It is mainly composed of bioclasts belonging to the red coralline algae (*Lithophyllum sp.* and *Lithothamnium sp.*), bryozoans and foraminifers.

This lithotype was used as plinth at St Martin's Cathedral. The very good workability and its white colour enable the rock to be used as construction material.

Leitha limestone from Oggau and Mannersdorf is highly durable. A few weathering phenomena are manifested, mainly by colour changes and the formation of dark gypsum crusts.

Interpretation of mineralogical and petrophysical data

Mineralogical composition

The petrographic study on thin sections shows that all studied rocks can be classified as calcitic limestones due to the dominance of calcite (Table 1). X-ray diffraction analyses revealed that some minor admixtures of quartz and feldspar are also present. In a few lithotypes, ankerite, gypsum and dolomite were also detected. Higher amounts of quartz, ankerite, feldspar and muscovite were found in the limestones from the Hundsheim-Wolfsthal area (see lithotypes 1, 3, 4 in Table 1).

Porosity and pore-size distribution

Mercury porosimetry yielded a porosity range of 5 to 19 vol. % (Table 2). A good linear correlation between the bulk density and porosity suggests a real density value of 2.65–2.70 g cm^{-3} (at zero porosity) which corresponds to that of marble and/or calcite single crystal.

With the exception of a few samples, the limestones display a bimodal pore size distribution (Fig. 3). While the first maximum is located in the range 10–100 μm, the second maximum at smaller radii varies slightly but is usually found in the range 0.1–0.005 μm. While the intensity of the first maximum shows no systematic trend, a closer inspection of the second maximum indicates that the samples can be divided into two groups: (1) the limestones from Oggau, St Margerethen and

Table 1. *Results of mineral composition of different types of Leitha limestone determined by X-ray diffraction analysis*

No.	Rock/Locality	Calcite	Dolomite	Quartz	Feldspar	Mica
1.	Oolitic limestone/Wolfsthal	++		+++	++	
2.	Leitha limestone/Oggau	+++			(−)	
3.	Sandy limestone/Wolfsthal	++		++	+	+
4.	Oolitic limestone (weathered)	++		++	++	(+)
5.	Sandy limestone/St Margarethen	+++			(−)	
6.	Lumachelle limestone/Wolfsthal	+++		(+)	+	
7.	Leitha limestone/Mannersdorf	+++			(−)	
8.	Fine-grained conglomerate/Hundsheim	+++	+++	(+)	(+)	+
9.	Coarse conglomerate/Hundsheim	+++	+++	(+)	+	+

[+++ very high; ++ high; + few; (+) accessory; (−) unsure]

Table 2. *Pore properties of different Leitha limestones as determined by nitrogen adsorption (BET) and mercury porosimetry (Hg)*

No.	Lithology	Location	Specific surface (m^2 g^{-1}) (BET)	Spec.surf. (m^2 g^{-1}) (Hg)	Total porosity (%) (Hg)	Density (g cm^{-3}) (Hg)
1	Oolitic limestone	Wolfsthal	0.53	0.36	14.4	2.32
2	Leitha limestone	Oggau	1.16	0.79	6.3	2.57
3	Sandy limestone	Wolfsthal	0.38	0.40	16.7	2.24
4	Oolitic limestone	Wolfsthal	0.19	0.21	11.1	2.40
5	Sandy limestone	St Margarethen	1.18	1.97	18.5	2.21
6	Lumachelle limestone	Wolfsthal	0.31	0.20	11.1	2.38
7	Leitha limestone	Mannersdorf	1.11	1.82	16.9	2.21
8	Fine-grained conglomerate	Hundsheim	0.67	0.26	12.2	2.38
9	Coarse conglomerate	Hundsheim	0.36	0.28	5.0	2.60

Mannersdorf showing a pronounced second maximum, and (2) the limestones from Hundsheim-Wolfsthal area where this second maximum is less visible. The relatively high porosities of 11–19 vol. % correlate with an average pore radius range of 25–45 μm.

The specific surface area values indicate a general consistency, except for the samples of the sandy limestone from St Margarethen (No. 5) and the Leitha limestone from Mannersdorf (No. 7) (Table 2). For these samples, the specific surface area calculated from mercury porosimetry data provides values in the range 1.8–1.9 m^2 g^{-1} that is much higher than the specific surface area measured by BET analysis (1.1–1.2 m^2 g^{-1}). The comparatively high portion of pores lower than approximately 0.01 μm pore entrance radii, where mercury porosimetry has its high pressure end, serve as explanation. The techniques overlap here but, in this dimension, the molecule sorption-based measurement technology is the more dependable.

Hygric parameters and capillary water uptake

Although the shape of the sorption hysteresis is fairly similar for all samples, the final weight change attained at high relative humidity (80–89% RH) is significantly different (Fig. 4). The Leitha limestones *s.l.* show quite similar sorption behaviour compared to that of other stone material with comparable specific surface area (Franzen 2003).

The water uptake within 24 hours ranges from 2 to 15 wt. %. The weight difference between the dry and soaked sample multiplied by the density allows a rough calculation of the bulk porosity. In comparison with the Hg-porosimetry data presented in Table 2, the bulk porosity values are in agreement with the values of samples 1, 4, 5, 6 and 9. Sample 2 yields a higher bulk porosity which may indicate the presence of open voids or cracks, which are not gathered by the mercury measurement. In contrast, the calculations for samples 3, 7 and 8 resulted in a lower bulk porosity compared to the mercury porosimetry data. This is probably indicative of the existence of voids that are hardly accessible by water during the 24 hour test period.

The main portion of the capillary water uptake data are depicted in Figure 5. The data suggest that three different groups of limestones exist in terms of capillary behaviour. The first group includes samples 3, 4 and 5 with capillary absorptions of around 0.9 g cm^{-2}. Samples 1 and 2 represent an intermediate group, while samples 6, 7, 8 and 9 are characterized by low capillary uptake.

As well as the water uptake, the drying behaviour of the material during certain time periods was also monitored (Figs 6a, b). Within 24 hours, all samples reached a dry state under ambient conditions. For most of the samples the drying follows a linear to quasi-linear trend. The oolitic and the lumachelle limestones show a specific kink in the drying curve while the drying follows a smooth function of time in all other cases (Fig. 6a). However, a closer inspection of the data using a semi-logarithmic plot (Fig. 6b) reveals that samples 5 and 2 exhibit an almost linear drying behaviour implying an almost constant rate. Drying of samples 8, 7, 3 and 9 seems to follow a process governed by a power law. Intermediately positioned curves of samples 1, 6 and 4 show a kink mentioned above (Fig. 6b). We assume that the curves of these samples indicate a change in the drying mechanism. The capillary support of water to the drying interface near the surface collapses after several hours. Subsequent drying is controlled by diffusion through the pore system.

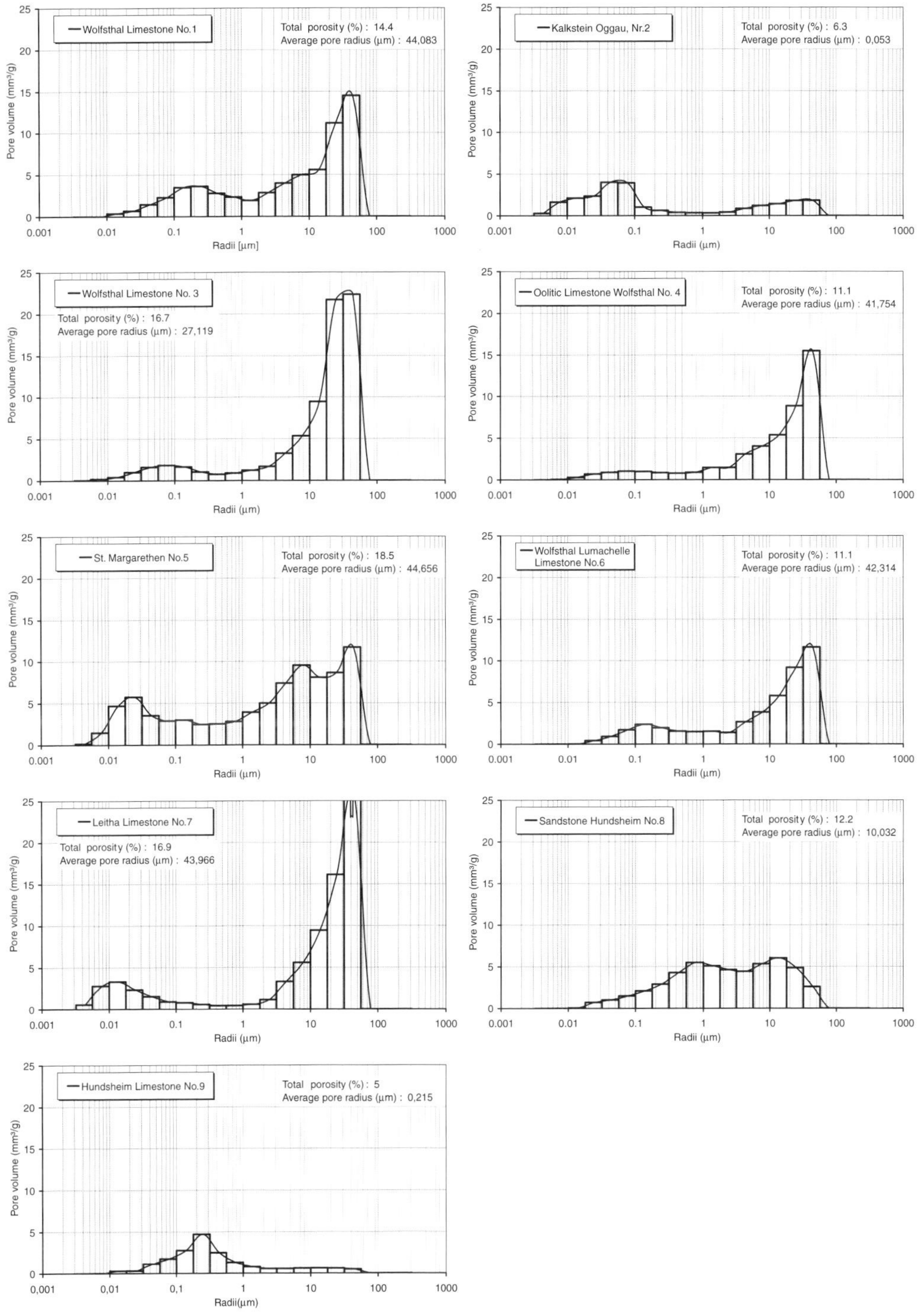

Fig. 3. Pore size distribution of studied Leitha limestones (samples taken from quarries).

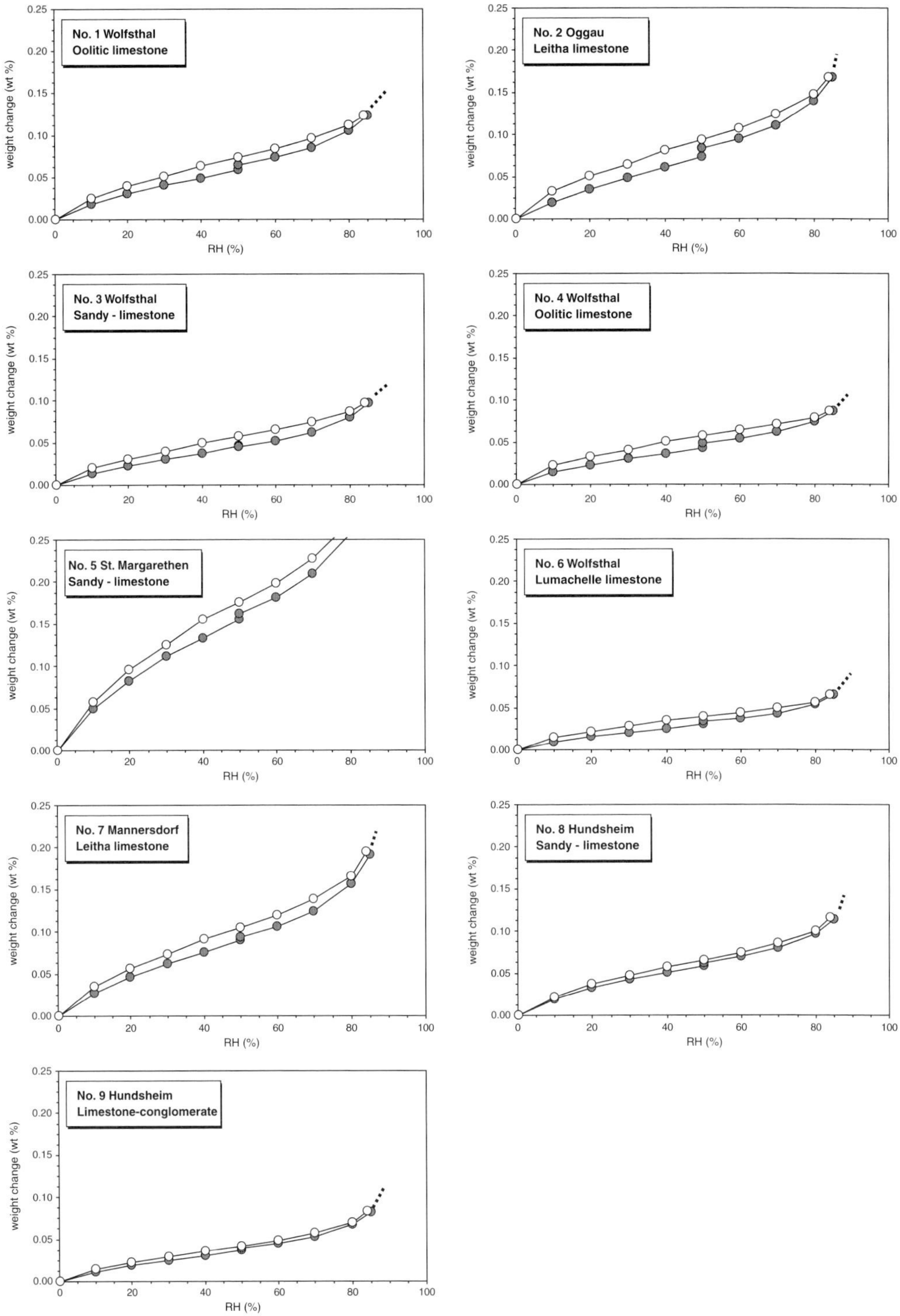

Fig. 4. Water sorption isotherms determined by SPS-11. Closed symbols, sorption; open symbols, desorption. The maximum value of water sorption of limestone from St Margarethen quarry (sample 5) exceeds 0.3%.

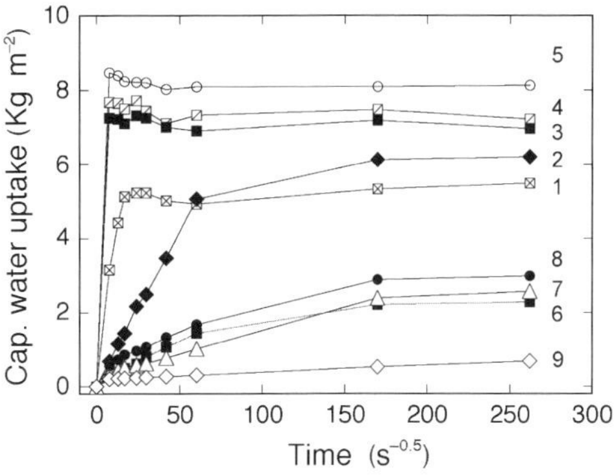

Fig. 5. Capillary water uptake of studied Leitha limestones. The numbers refer to the different stone types in Table 1.

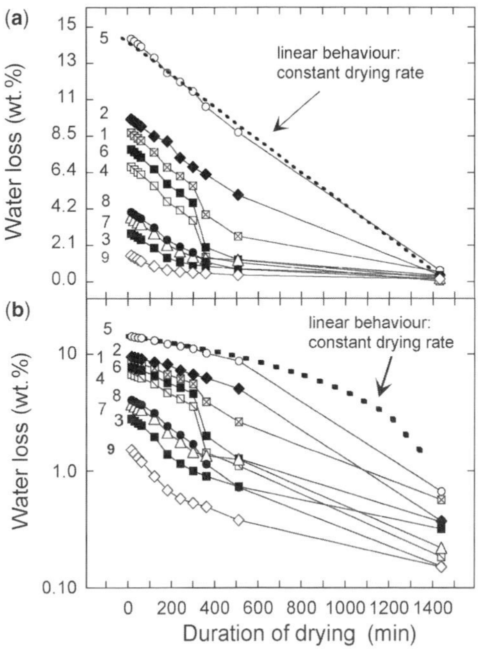

Fig. 6. The drying behaviour versus time of the studied Leitha limestones: **(a)** the linear representation of the data suggests a non-uniform drying behaviour; **(b)** the logarithmic presentation of the ordinate reveals three different patterns of behaviour: (i) a linear drying process at a constant rate (5, 2); (ii) samples 1, 6 and 4 indicate a change in the drying mechanism; (iii) the drying process of samples 8, 7, 3 and 9 follows a power-law trend. The numbers refer to the different stone types in Table 1.

Conclusions

The Neogene sedimentary rocks (limestones, sandstones and conglomerates) of the Leitha Mountains and Hainburg Mountains presented a principal source of natural building stone used for historical monuments in Bratislava. These sources were favoured both due to their close proximity to construction sites and their excellent workability. Structures such as St Martin's Cathedral and the Bratislava castle were built from several lithological types that were also identified in the quarries of Hundsheim, Wolfsthal, St Margarethen, Oggau and Mannersdorf. The provenance studies were based on test results of petrographic and petrophysical properties. Accordingly, six major lithotypes were identified displaying ranges of partly overlapping physical properties. By using pore-size distribution analyses and water sorption tests, it was possible to determine the differences between similar materials and group them accordingly. The data of these different Leitha limestones show remarkable variations, especially in terms of pore radii spectra. These properties can be used to assess the durability of materials and provide information for conservation measures.

This study was partly conducted in the frame of CEEPUS program No. 22262. We are grateful to M. Noistering who conducted the mercury porosimetry measurements and to U. Griesser (both University of Innsbruck), who granted access to the SPS 11 sorption apparatus.

References

ANDRIANI, G. F. & WALSH, N. 2007. The effects of wetting and drying, and marine salt crystallization on calcarenite rocks used as building material in historic monuments. *In*: PŘIKRYL, R. & SMITH, B. J. (eds) *Building Stone Decay: From Diagnosis to Conservation*. Geological Society, London, Special Publications, **271**, 179–188.

BENAVENTE, D., CUETO, N., MARTINEZ-MARTINEZ, J., GARCIA DEL CURA, M. A. & CANAVERAS, J. C. 2007. The influence of petrophysical properties on the salt weathering of porous building rocks. *Environmental Geology*, **52**(2), 215–224.

CZIJZEK, J. 1852. Geologische verhältnisse der umgebungen von hainburg, des leithagebirges und der ruster berge. *Jahrbuch der Kaiserlich-Königlichen Geologischen Reichsanstalt*, **1852**, 35–55.

FITZNER, B. & HEINRICHS, K. 2002. Damage diagnosis on stone monuments – weathering forms, damage categories and damage indices. *In*: PŘIKRYL, R. (ed.) *Understanding and Managing Stone Decay*. The Karolinum Press, Prague, 11–51.

FRANZEN, C. 2003. *Historische Bauwerksteine in Süd-Tyrol. Verteilung und Verwitterungsverhalten*. Unveröffentl. Dissertation, University of Innsbruck.

FRANZEN, C. & MIRWALD, P. W. 2004. Moisture content of natural stone – static and dynamic equilibrium with atmospheric humidity. *Environmental Geology*, **46**(3–4), 391–401.

HOLZER, R., LAHO, M., DURMEKOVÁ, T. & GREIF, V. 2004*a*. Inžinierskogeologický výskum Dómu sv.

Martina v Bratislave. *In*: ŠTEFANOVIČOVÁ, T., BEŇUŠ, R. *ET AL.* (eds) *Dóm sv. Martina v Bratislave*. Archeological investigation 2002–2003 (in Slovak). Elán, Bratislava, 62–70.

HOLZER, R., LAHO, M. & DURMEKOVÁ, T. 2004*b*. Ancient building stone sources of Bratislava's monuments. *In*: PŘIKRYL, R. (ed.) *Dimension Stone 2004*. A.A.Balkema Publishers, Leiden, 51–56.

LAUE, S. 2005. Salt weathering of porous structures related to climate changes. *Restoration of Buildings and Monuments - Bauinstandsetzen und Baudenkmalpflege*, **11**(6), 1–10.

MIŠÍK, M. 1997. Stratigrafické a priestorové rozmiestnenie vápencov s kalcitovými, chamositovými, hematitovými a illitovými ooidmi v Západných Karpatoch. *Mineralia Slovaca*, **29**, 83–112.

MÜLLER, H. W., ROHATSCH, A., SCHWAIGHOFER, B., OTTNER, F. & THINSCHMIDT, A. 1993. Gesteinsbestand in der Bausubstanz der Westfassade und des Albertinischen Chores von St. Stephan. *Österreichische Zeitschrift für Kunst und Denkmalpflege, Sonderdruck*, 106–116.

PILLER, W. E. 2003. The Middle Miocene Badenian Leitha Limestone of the Central Paratethys - Not a climate controversy institute for geology and paleontology. *AAPG Annual Convention* May 11–14, 2003, Salt lake City, Utah.

PŘIKRYL, R., LOKAJÍČEK, T., SVOBODOVÁ, J. & WEISHAUPTOVÁ, Z. 2003. Experimental weathering of marlstone from Přední Kopanina (Czech Republic) – historical building stone of Prague. *Building and Environment*, **38**(9–10), 1163–1171.

PŘIKRYL, R. & PŘIKRYLOVÁ, J. 2004. "Leithakalk" limestones in the Lednice-Valtice area (southeast Moravia, Czech Republic): their occurrences and properties. *In*: PŘIKRYL, R. & SIEGL, P. (eds) *Architectural and Sculptural Stone in Cultural Landscape*. The Karolinum Press, Prague, 149–156.

SMITH, B. J. & PŘIKRYL, R. 2007. Diagnosing decay: the value of medical analogy in understanding the weathering of building stones. *In*: PŘIKRYL, R. & SMITH, B. J. (eds) *Building Stone Decay: From Diagnosis to Conservation*. Geological Society, London, Special Publications, **271**, 1–8.

SMITH, B. J., TÖRÖK, Á., MC ALISTER, J. J. & MEGARRY, Y. 2003. Observations on the factors influencing stability of building stones following contour scaling: a case study of oolitic limestones from Budapest, Hungary. *Building and Environment*, **38**(9–10), 1173–1183.

TÖRÖK, Á. 2002. Oolitic limestone in polluted atmospheric environment in Budapest: weathering phenomena and alterations in physical properties. *In*: SIEGESMUND, S., WEISS, T. S. & VOLLBRECHT, A (eds) *Natural Stones, Weathering Phenomena, Conservation Strategies and Case Studies*. Geological Society, London, Special Publications, **205**, 363–379.

TÖRÖK, Á. 2004. Leithakalk-type limestones in Hungary: an overview of lithologies and weathering features. *In*: PŘIKRYL, R. & SIEGL, P. (eds) *Architectural and Sculptural Stone in Cultural Landscape*. The Karolinum Press, Prague, 157–172.

TÖRÖK, Á., ROZGONYI, N., PŘIKRYL, R. & PŘIKRYLOVÁ, J. 2004. Leithakalk: the ornamental and building stone of Central Europe, an overview. *In*: PŘIKRYL, R. & SIEGL, P. (eds) *Architectural and Sculptural Stone in Cultural Landscape*. The Karolinum Press, Prague, 89–93.

WINKLER, E. M. 1994. *Stone in Architecture, Properties, Durability*. Springer-Verlag, Berlin.

Raman spectra of reduced carbonaceous matter as a tool for determining the provenance of marbles: examples of 'graphitic' marbles from Czech quarries

ANETA ŠŤASTNÁ*, JAN JEHLIČKA & RICHARD PŘIKRYL

Institute of Geochemistry, Mineralogy and Mineral Resources, Charles University in Prague, Faculty of Science, Albertov 6, 128 43 Praha 2, Czech Republic

**Corresponding author (e-mail: astastna@gmail.com)*

Abstract: This study reports the Raman data for reduced carbonaceous matter (CM) in crystalline marbles, related to a determination of their provenance. Raman microspectrometry was tested on 'graphitic' marbles formed under different metamorphic conditions (regional v. contact) with various geological ages, and with distinct types of CM from the Bohemian Massif (Czech Republic). First, various modes of occurrence of CM were examined by optical microscopy (OM). The Raman results exhibited spectral variations of CM with different metamorphic grades and types. Several marble groups could be distinguished: (1) well-ordered CM/graphite of higher grade regional metamorphosed marbles; less graphitized metamorphosed organic matter ('disordered' CM); (2) low-grade regional metamorphosed rocks, and (3) contact metamorphosed marbles.

Provenance determination of marbles represents a classical task for marble-producing regions all over the world. Many different observational and analytical approaches have been developed according to the mineralogical-petrographic, geochemical and physical properties of the materials to be studied. One methodology, based on a variable combination of optical microscopy, cathodoluminescence, carbon and oxygen stable isotopes, electron spin and paramagnetic resonance (or inductively coupled plasma-mass spectrometry) has been successfully employed to determine the major source regions of white Mediterranean marbles (Barbin *et al.* 1992; Baïetto *et al.* 1999; Attanasio *et al.* 2000; Lapuente *et al.* 2000, 2002; Gorgoni *et al.* 2002; Green *et al.* 2002). However, some distinct territories such as the Bohemian Massif (Czech Republic) are rich in numerous varieties of impure crystalline marbles (calcite or dolomite), commonly including admixtures of non-carbonate minerals (silicates) and/or organic matter transformed through various degrees of metamorphism. These marbles differ from white marbles not only in their composition but also in their fabric, type, and grade of dominant metamorphism. This fact has an impact on the proper selection of analytical methods for sourcing these marbles. Geochemical methods alone cannot always be successfully employed for these types of marbles due to variations in the results.

The application of Raman microspectrometry to marbles which include carbonaceous matter (CM) represents a very useful additional methodology for these problem marbles (see also Šťastná *et al.* 2009). The characterization of CM from graphitic rocks determined by this technique has been employed relatively widely in the context of metamorphism studies (Jehlička & Beny 1999; Beyssac *et al.* 2002; Jehlička *et al.* 2003) and in art (Coupry 2000). However, this technique has never been applied to the sourcing and provenance determination of marbles. Raman microspectrometry allows us to estimate the structural order and detect the presence of structural defects in CM. Modifications of the Raman characteristics of CM are strongly related to: (1) the intensity of metamorphism, (2) the mode of occurrence of the CM (accumulated versus disseminated) and (3) the orientation of the laser beam to the polished thin section (in relation to the foliation) (Wopenka & Pasteris 1993).

Silesian marbles (from the northeast part of the Bohemian Massif, Czech Republic) are the most outstanding in terms of their exploitation as dimension stones. For a long time, the Silesicum area had the highest concentration of marble quarries in the Czech Republic. Marbles mined here have been used to decorate many famous Czech architectural monuments and have been also exported to Poland, Germany, Austria and Serbia (Rybařík 1994). For

From: PŘIKRYL, R. & TÖRÖK, Á. (eds) *Natural Stone Resources for Historical Monuments*.
Geological Society, London, Special Publications, **333**, 175–183.
DOI: 10.1144/SP333.17 0305-8719/10/$15.00

example, Prague Castle, Prague's Vitkov Memorial, Prague's underground stations, the Janáček Theater in Brno and the Thermal Spring Colonnade in Karlovy Vary were all decorated using Lipová marbles. Tišnov marbles were also exploited locally, and can be found in churches in Předklášteří, Brno and other Moravian sites.

This study is concerned with distinguishing different types of 'graphitic' marbles on the basis of the Raman spectra of the carbonaceous material disseminated in these rocks. The results have implications for provenance determination of CM containing marbles.

Analytical procedure

Optical microscopy (OM) of the CM was performed using a Leica DMLP optical microscope (Laboratory of Optical Microscopy, Institute of Geochemistry, Mineralogy and Mineral Resources, Charles University in Prague). The modes of CM occurrence were photographed with an Olympus C-2000 camera.

The Raman microspectrometry (RM) was applied to 'graphitic' marbles at the Institute of Geochemistry, Mineralogy and Mineral Resources, Charles University in Prague. Micro-Raman analyses of CM were performed on a multichannel Renishaw *InVia* Reflex spectrometer, coupled with a Peltier-cooled changed-coupled device (CCD) detector. Excitation was provided by the 514.5 nm line of a continuous-wave 10 mW Ar-ion laser. Polished thin sections were used for analysis. The samples were scanned from 1100–2000 cm^{-1} at a spectral resolution of 2 cm^{-1}. The scanning parameter for each Raman spectrum was taken as 10 seconds. Ten scans were accumulated for each experimental run, in order to provide a better signal-to-noise ratio. Raw spectra were used without curve fitting or band deconvolution. The WIRE program v. 2.0 was employed to determine the peak position, the peak width (i.e. full width at half maximum, FWHM) and the relative intensity.

In general, multiple spot analyses on different areas of the same sample yielded similar spectra, thus confirming the spectral reproducibility. Samples No. 243 and 290 B contained CM, characterized by several types of Raman spectra which corresponded to different degrees of structural arrangement (see the results and discussion).

Sampling areas and geological background

'Graphitic' marbles which had evolved under different metamorphic conditions (regional v. contact) and with different geological ages and distinct modes of CM occurrence were chosen from Czech quarries. These grey or white marbles are located within three parts of the Bohemian Massif (Czech Republic): the Lugicum (Krkonoše-Jizera Terrane) in the north, the Silesicum (Branná Group and Mantle of Žulová Granite Pluton) in the northeast and the Moravicum (Olešnice Unit and Devonian cover) in the southeast (Fig. 1).

The carbonate formations of the areas studied form marble lenses and intercalations (mainly in metapelites), unlike the marbles in the Mantle of the Žulová Granite Pluton. These 'graphitic' marble deposits are predominantly of Silurian-Devonian age (Jitrava, Křižany, Pilínkov, Horní Hanychov, Rašovka, Bohdíkov, Branná, Žulová, Staré Hradisko, Horní Lipová, Tišnov-Květnice and Tišnov-Dřínová), with the exception of the Lysice marbles which range in age (according to various studies) from Neoproterozoic to the lower Palaeozoic (Houzar & Novák 2001; Chlupáč *et al.* 2002). Most of the 'graphitic' marbles were affected by regional metamorphism to various degrees. Low-grade greenschist facies are typical of the grey marbles from the Krkonoše-Jizera Terrane or Tišnov marbles (Chlupáč *et al.* 2002; Hladil *et al.* 2003). On the other hand, grey calcite marbles from Lysice (Moravicum) were metamorphosed under amphibolite-facies conditions (Houzar & Novák 2002). Staré Hradisko and Žulová marbles, close to the Granite Pluton Žulová (Silesicum), dominantly involved contact metamorphism (Cháb & Žáček 1994). The metamorphic situation of the grey marbles is very complex; pressure-temperature (PT) conditions have not been examined and dating has not been conducted in all the relevant areas.

In the present work, 17 samples were tested from 13 different marble quarries. The locations and main characteristics of the studied marbles are listed in Table 1. The samples examined in this study mostly come from abandoned historical quarries, except for the Horní Lipová and Tišnov-Dřínová quarries which are still in operation.

Results and discussion

Fabric of various types of carbonaceous matter

Results of microscopic studies of polished thin sections indicate that 'disordered' CM and graphite form various types of crystallites in the studied samples (Fig. 2). The majority of low-metamorphosed grey marbles display a linear fabric along the graphitic strips and veins (Fig. 2a, b). The 'disordered' CM is mainly fine grained and forms an accumulation with other accessory minerals such as muscovite, chlorite, quartz or haematite in distorted veins. The foliation is visible macroscopically in

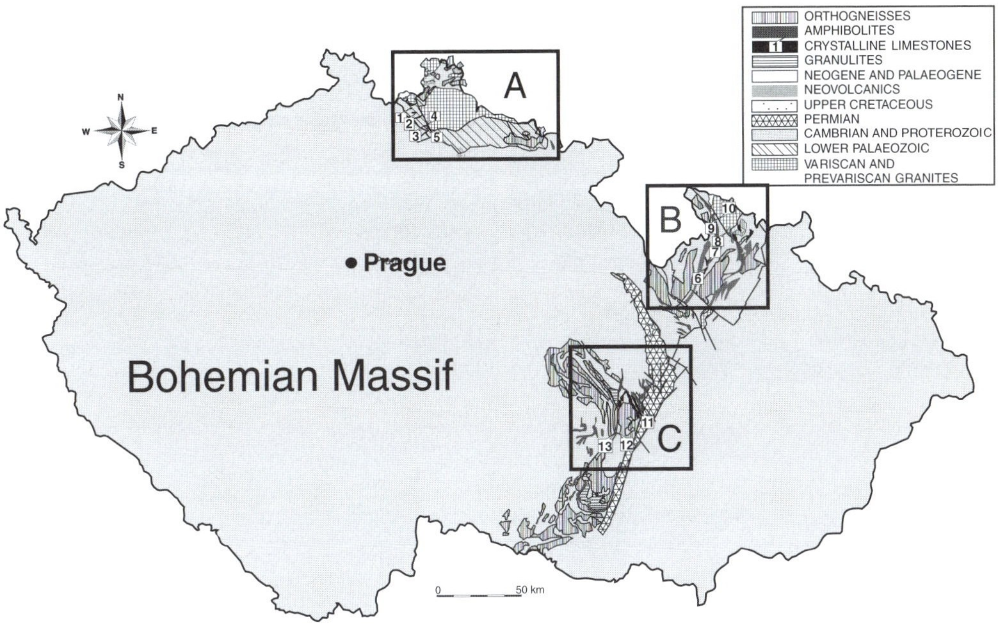

Fig. 1. A simplified geological map of the three studied areas of the Bohemian Massif containing the sampled quarries. The legend of the geological map is only valid for areas A, B and C. A, Lugicum (Krkonoše-Jizera Terrane: 1, Jitrava, 2, Křižany, 3, Pilínkov , 4, Horní Hanychov, 5, Rašovka). B, Silesicum (Branná Group: 6, Bohdíkov, 7, Branná, 8, Horní Lipová and Mantle of the Žulová Granite Pluton: 9, Žulová, 10, Staré Hradisko). C, Moravicum (Olešnice Unit: 11, Lysice and Devonian cover: 12, Tišnov-Květnice, 13, Tišnov-Dřínová).

these samples. Isolated large grains of CM are typical of white, coarse-grained marbles from Staré Hradisko and Žulová, where contact metamorphism is dominant. These crystallites are often elongated and exhibit embayed grain boundaries (Fig. 2c, d). CM does not affect the grey colour here because of the low content and different mode of occurrence of the CM. On the other hand,

Table 1. *Location of all studied marbles from the Czech Republic with a description of the carbonaceous matter (CM)*

Geological setting		Quarry (sample No.)	Dominant metamorphism	Mode of occurrence of graphite/'disordered' CM
Svratka Dome, Moravicum	Olešnice Unit	Lysice (294)	Medium regional	Rims of carbonate grains
	Devonian cover	Tišnov-Dřínová (287)	Low regional	Foliated strips and veins
		Tišnov-Květnice (290)	Low regional	
Silesicum	Branná Group	Horní Lipová (212)	Medium regional	Randomly located clusters
		Branná (229, 208)	Low-medium regional	Clusters forming parallel strips
		Bohdíkov (207)	Low regional	
	Mantle of Žulová Granite Pluton	Staré Hradisko (224)	Contact	Isolated elongated grains with embayed rims
		Žulová (225)	Contact	
Krkonoše-Jizera Terrane, Lugicum	Ještěd Crystalline Unit	Jitrava (241B)	Low regional	Foliated strips and veins
		Křižany (132)	Low regional	
		Rašovka (204)	Low regional	Rims of carbonate grains
		Pilínkov (243)	Low regional	Foliated strips and veins
		Horní Hanychov (244)	Low regional	

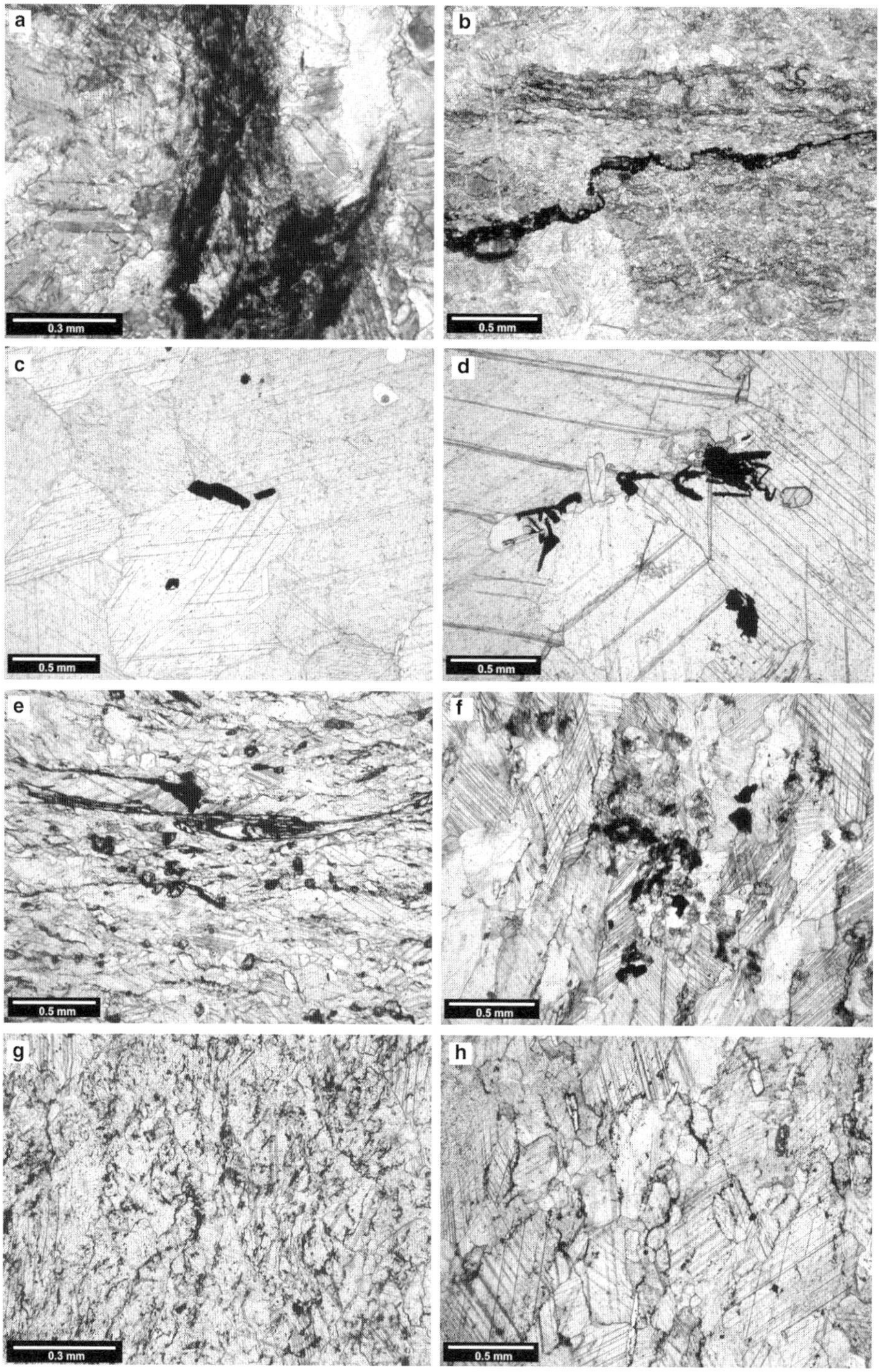

Fig. 2. Microscopic images of the fabric of carbonaceous matter in the studied marbles (plain polarized light). (**a**) Pilínkov (sample No. 243, Lugicum). (**b**) Tišnov-Dřínová (sample No. 287 B, Moravicum). (**c**) Staré Hradisko (sample No. 224 A, Silesicum). (**d**) Žulová (sample No. 225 B, Silesicum). (**e**) Branná (sample No. 229 B, Silesicum). (**f**) Horní Lipová (sample No. 212 B, Silesicum). (**g**) Rašovka (sample No. 204, Lugicum). (**h**) Lysice (sample No. 294 A, Moravicum).

graphite crystallites and well-ordered CM are very fine-grained. It is disseminated randomly or in clusters in Horní Lipová marbles (Fig. 2f), or it can be accumulated in the rims of carbonate grains in Lysice marbles (Fig. 2h). Under microscopic observation, the grey marbles often exhibited similar characteristics; morphology of the metamorphosed organic material and therefore OM alone cannot identify and distinguish CM reliably for the purposes of provenance determination. Rašovka marble is an example of this; it exhibits a fabric similar to Lysice marbles but belongs among the very low-metamorphosed rocks (Fig. 2g v. h).

Raman microspectrometry of metamorphosed carbonaceous matter

Raman spectroscopy enables one to distinguish between CM in various crystalline states. Generally, the Raman spectrum of graphite is classified into either the first-order spectrum (about 1200–1700 cm^{-1}) or the second-order spectrum (about 2350–3350 cm^{-1}) (Wopenka & Pasteris 1993). Well-ordered graphite yields only a single Raman peak around 1582 cm^{-1} (the 'ordered' or O-peak) in the first-order Raman spectrum. For less well-crystallized carbonaceous matter, the first-order Raman spectrum contains an additional 'disordered' or D-peak at around 1355 cm^{-1} and also a peak occurring near the O-peak at about 1620 cm^{-1} (Nemanich & Solin 1979). The second-order Raman spectrum is especially useful for metamorphism research (Yui *et al.* 1996). In this study, only the first-order Raman spectrum was measured. The Raman spectroscopic parameters of all samples from the areas studied are illustrated in Table 2.

The results obtained demonstrate that the Raman spectra of the CM of selected marbles reflect the degree of metamorphism very closely. The relative intensity of the O-peak of the studied samples increases with an increase in the regional metamorphism. Figure 3 displays a very low-grade regional metamorphosed Tišnov marble (a), a low-mid-grade regional metamorphosed Branná marble (b) and a higher grade regional metamorphosed Lysice marble (c). Marbles from the contact zone (Staré Hradisko, Žulová) are characterized by the appearance of the disorder-induced shoulder (D) and peaks with a higher FWHM (Fig. 3d). Several types of CM with different Raman characteristics were found within one sample of marbles from Pilínkov (Krkonoše-Jizera Terrane) and Tišnov-Květnice (Moravicum) (Fig. 4). The differences in the degree of graphitization of the CM in these low-grade metamorphosed

Table 2. *Raman spectroscopic parameters for the samples studied. D-peak ('disordered'), O-peak ('well-ordered')*

Quarry (sample No.)	Peak pos D (Δcm^{-1})	Peak pos O (Δcm^{-1})	Peak width D (cm^{-1})	Peak width O (cm^{-1})	D/O intensity ratio
Tišnov-Květnice (290)	1345.7	1594.9	40.4	73.4	0.85
	1356.1	1601.5	75.6	61.1	1.05
	1350.4	1602.2	72.4	57.8	1.26
Tišnov-Dřínová (287)	1353.2	1606.2	72.5	54.5	1.19
Jitrava (241)	1354.5	1601.6	54.5	61.7	1.04
	1354.2	1602.2	53.8	58.1	1.37
Pilínkov (243)	1355.6	1598.6	36.6	73.0	0.86
	1355.1	1594.1	46.2	70.0	1.11
	1355.0	1591.4	46.9	60.4	1.15
Rašovka (204)	1337.4	1601.1	86.2	54.8	1.42
Křižany (132)	1352.7	1603.3	60.0	53.8	1.33
Horní Hanychov (244)	1354.2	1589.0	37.6	34.5	1.09
Staré Hradisko (224)	1360.7	1585.6	57.5	42.7	0.79
	1359.2	1583.0	58.1	29.2	0.71
Žulová (225)	1359.1	1585.1	58.0	39.5	0.89
	1358.8	1583.9	54.7	34.9	0.83
Bohdíkov (207)	1356.9	1587.2	51.7	57.5	0.88
Branná (229, 208)	1358.3	1584.4	42.5	29.6	0.78
	1359.2	1584.1	35.9	30.5	0.61
Horní Lipová (212)	1351.1	1580.9	42.2	18.0	0.55
Lysice (294)	1356.1	1583.3	40.3	18.8	0.31
	1355.5	1582.4	40.4	18.9	0.29
	1355.9	1582.9	41.0	19.2	0.50
	1356.2	1582.4	41.1	19.4	0.32

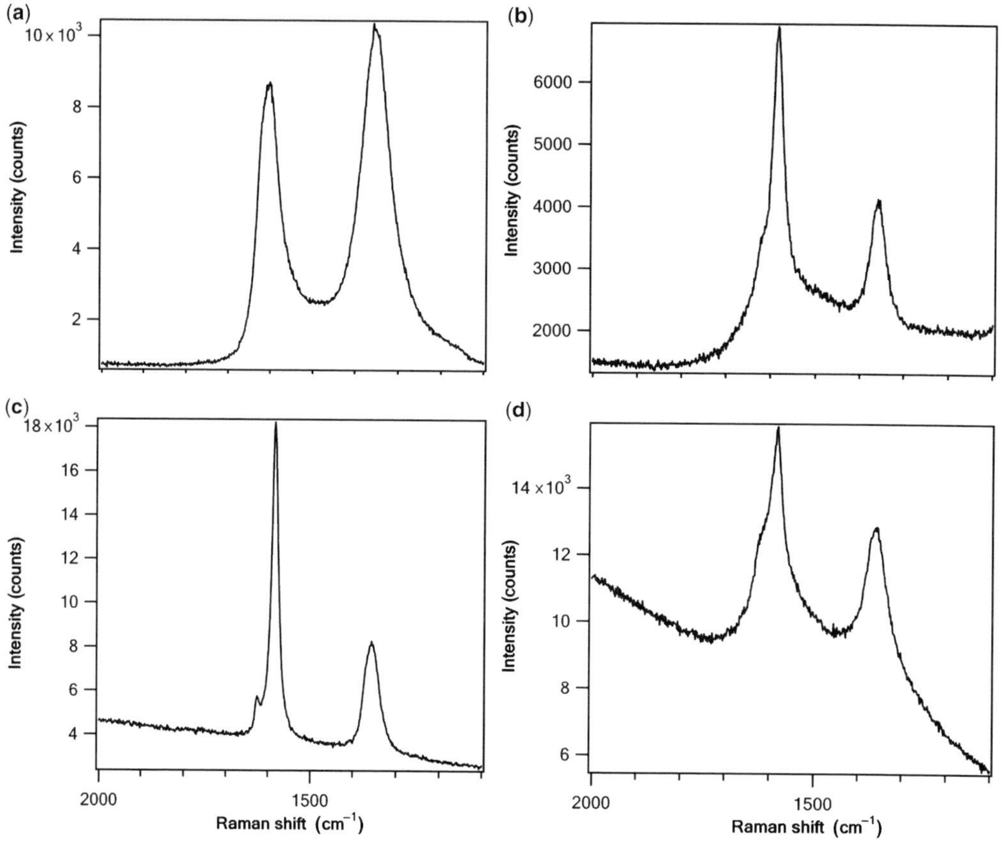

Fig. 3. Representative Raman spectra of carbonaceous matter from the present study of various metamorphic grades and types in order of increasing regional metamorphism: (**a**) Tišnov-Dřínová (Moravicum), (**b**) Branná (Silesicum) and (**c**) Lysice (Moravicum). (**d**) Staré Hradisko (Silesicum), contact metamorphism.

marbles might be related to the initial nature of the sedimentary organic matter and pre-metamorphic history (Kříbek *et al.* 1994). The Raman differences must be interpreted carefully in these cases; semi-quantification of individual grains with different crystallinity yields some information on the dominant phase. Moreover, samples displaying a palette of CM with different crystalline states very rarely occur.

The structural information obtained allows us to discriminate amongst various types of CM-rich marbles with various grades and types of metamorphism. Three simplified groups of Raman spectra, classified according to certain Raman spectroscopic parameters, are depicted in Figure 5: (1) well-ordered CM: graphite of higher-grade regional metamorphosed marbles (Lysice, Horní Lipová) is characterized by the low D/O intensity ratio (0.3–0.6), as well as the low O-peak width (18–19 cm^{-1}); (2) less graphitized organic matter ('disordered' CM) of low-grade regional metamorphosed rocks (Jitrava, Křižany, Rašovka, Pilínkov, Horní Hanychov, Bohdíkov and Tišnov) exhibit high values of the D/O intensity ratio (0.9–1.4) (the O-peak width generally increases above 50 cm^{-1} in this case); and (3) contact metamorphosed marbles (Staré Hradisko, Žulová) and mid-grade regional metamorphosed marbles (Branná) form the middle group, with D/O intensity ratio ranges from 0.6–0.9. An O-peak width of 30 cm^{-1} as the lower limit and 43 cm^{-1} as the upper limit characterize the lattes group. The Raman parameters of Horní Hanychov (Krkonoše-Jizera Terrane) marble deviate towards those of mid-grade and contact metamorphosed samples (O-peak width about 35 cm^{-1}), despite the fact that this belongs to low-grade regional metamorphosed group. Marbles from Branná (Silesicum) also deviate from data typical of low-grade regional metamorphosed rocks. This could be caused by the effects of higher pressures on these 'graphitic' marbles.

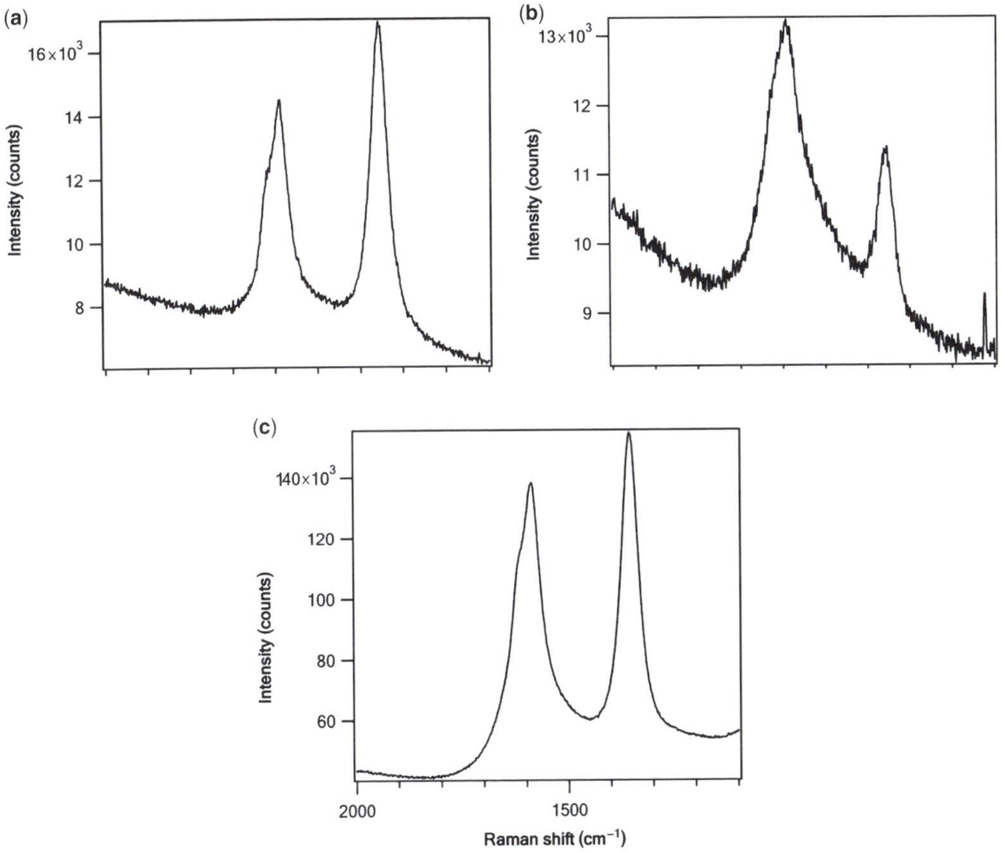

Fig. 4. Various types of carbonaceous matter in a single sample from the Pilínkov quarry (Lugicum, sample No. 243).

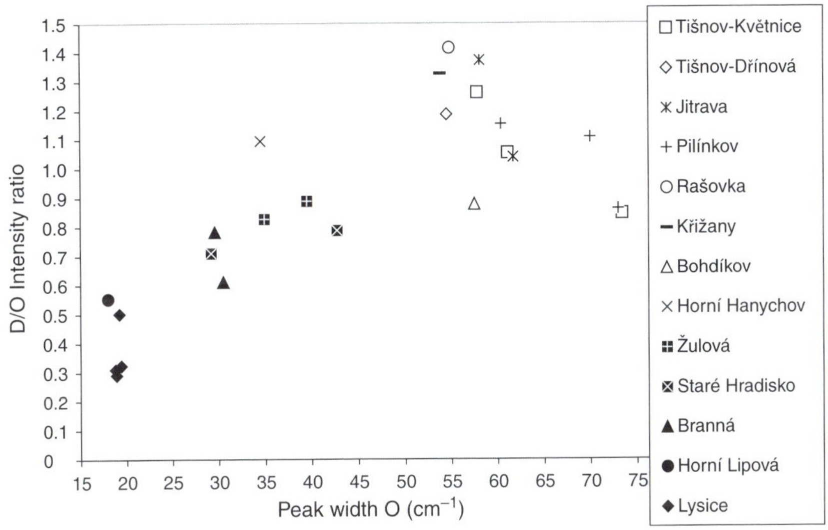

Fig. 5. Plot of the D/O intensity ratio versus O-peak width of carbonaceous matter for all studied samples.

Conclusions

The use of Raman microspectrometry of CM for the determination of the provenance of marbles was tested. The optical microscopic properties and Raman spectra of metamorphosed organic material were investigated on 'graphitic' marbles located within the Bohemian Massif (Czech Republic). The frequencies of major Raman bands and the spectroscopic parameters such as O (D)-peak width and D/O intensity ratio were measured in order to describe various types of CM. Three general types of Raman spectra were distinguished: (1) well-ordered CM/graphite of higher-grade regional metamorphosed marbles, (2) rather amorphous organic compounds as 'disordered' CM of low-grade regional metamorphosed marbles, and (3) contact metamorphosed marbles. Raman microspectrometric analysis also reveals the coexistence of various types of carbonaceous particles, exhibiting different degrees of graphitization within a single sample. These cases have to be qualified individually, based on semi-quantification of the crystallites; however, they do not constitute any limitations for the determination of provenance.

As CM is a very common admixture in various marbles, Raman microspectrometry presents extensive potential as a useful analytical method for fingerprinting dimension stones.

This work was supported by Project MSM 0021620855 'Material flow mechanisms in the upper spheres of the Earth'.

References

ATTANASIO, D., ARMIENTO, G., BRILLI, M., EMANUELE, M. C., PLATANIA, R. & TURI, B. 2000. Multimethod marble provenance determination: the Carrara marbles as a cause study for the combined use of isotopic, electron spin resonance and petrographic data. *Archaeometry*, **42**, 257–272.

BAÏETTO, V., VILLENEUVE, G., SCHVOERER, M., BECHTEL, F. & HERZ, N. 1999. Investigation of electron paramagnetic resonance peaks in some powdered Greek white marbles. *Archaeometry*, **41**, 253–265.

BARBIN, V., RAMSEYER, K., DECROUEZ, D., BURNS, S. J., CHAMAY, J. & MAIER, J. L. 1992. Cathodoluminescence of white marbles: an overview. *Archaeometry*, **34**, 175–183.

BEYSSAC, O., GOFFE, B., CHOPIN, C. & ROUZAUD, J. N. 2002. Raman spectra of carbonaceous material in metasediments: a new geothermometer. *Journal of Metamorphic Geology*, **20**, 859–871.

CHÁB, J. & ŽÁČEK, V. 1994. Geology of the Žulová pluton mantle (Bohemian Massif, Central Europe). *Bulletin of Geoscience*, **69**, 1–12.

CHLUPÁČ, I., BRZOBOHATÝ, R., KOVANDA, J. & STRÁNÍK, Z. 2002. *Geological History of the Czech Republic*. Academia, Prague. (in Czech).

COUPRY, Cl. 2000. Application of Raman microspectrometry to art objects. *Analysis*, **28**, 39–46.

GORGONI, C., LAZZARINI, L., PALLANTE, P. & TURI, B. 2002. An updated and detailed mineropetrographic and C–O stable isotopic reference database for the main Mediterranean marbles used in antiquity. *In*: HERRMANN, J. J. JR., HERZ, N. & NEWMAN, R. (eds) *Asmosia 5. Interdisciplinary Studies on Ancient Stone*. Archetype Publications, London, 115–131.

GREEN, W. A., YOUNG, S. M. M., VAN DER MERWE, N. J. & HERRMANN, J. J. JR. 2002. Source tracing marble: trace element analysis with inductively coupled plasma-mass spectrometry. *In*: HERRMANN, J. J. JR., HERZ, N. & NEWMAN, R. (eds) *Asmosia 5. Interdisciplinary Studies on Ancient Stone*. Archetype Publications, London, 132–142.

HLADIL, J., PATOČKA, F., KACHLÍK, V., MELICHAR, R. & HUBAČÍK, M. 2003. Metamorphosed carbonate sediments of the Krkonoše Mts and Paleozoic evolution of Sudetic terranes (NE Bohemia, Czech Republic). *Geologica Carpathica*, **54**, 281–297.

HOUZAR, S. & NOVÁK, M. 2001. Marbles in the south-eastern margin of the Bohemian Massif (Review) (in Czech). *Vlastivědný sborník Vysočiny, oddílení vìd přírodních*, **15**, 3–33.

HOUZAR, S. & NOVÁK, M. 2002. Marbles with carbonantite-like geochemical signature from variegated units of the Bohemian Massif, Czech Republic, and their geological significance. *Journal of the Czech Geological Society*, **47**(3–4), 103–109.

JEHLIČKA, J. & BENY, C. 1999. First and second order Raman spectra of natural highly carbonified organic compounds from metamorphic rocks. *Journal of Molecular Structure*, **480–481**, 541–545.

JEHLIČKA, J., URBAN, O. & POKORNÝ, J. 2003. Raman spectroscopy of carbon and solid bitumens in sedimentary and metamorphic rocks. *Spectrochimica Acta Part A*, **59**, 2341–2352.

KŘÍBEK, B., HRABAL, J., LANDAIS, P. & HLADÍKOVÁ, J. 1994. The associations of poorly ordered graphite, coke and bitumens in greenschist facies rocks of the Poniklá group, Lugicum, Czech Republic: the results of graphitization of various types of carbonaceous matter. *Journal of Metamorphic Geology*, **12**, 493–503.

LAPUENTE, M. P., MARTINEZ, M. P., TURI, B. & BLANC, P. 2002. Characterization of dolomitic marbles from the Malaga Province (Spain). *In*: HERRMANN, J. J. JR., HERZ, N. & NEWMAN, R. (eds) *Asmosia 5. Interdisciplinary Studies on Ancient Stone*. Archetype Publications, London, 152–162.

LAPUENTE, M. P., TURI, B. & BLANC, P. 2000. Marbles from Roman Hispania: stable isotope and cathodoluminescence characterization. *Applied Geochemistry*, **15**, 1469–1493.

NEMANICH, R. J. & SOLIN, S.A. 1979. First- and second-order Raman scattering from finite-size crystals of graphite. *Physics Review B*, **20**, 392–401.

RYBAŘÍK, V. 1994. *Dimension Stones from the Czech Republic*. Nadace Střední průmyslové školy kamenické a sochařské v Hořicích v Podkrkonoší. (In Czech).

ŠŤASTNÁ, A., PŘIKRYL, R. & JEHLIČKA, J. 2009. Methodology of analytical study for provenance determination of calcitic, calcite-dolomitic and impure marbles from historical quarries in the Czech Republic. *Journal of Cultural Heritage*, **10**, 82–93.

WOPENKA, B. & PASTERIS, J. D. 1993. Structural characterization of kerogens to granulite-facies graphite: Applicability of Raman microprobe spectroscopy. *American Mineralogist*, **78**, 533–557.

YUI, T.-F., HUANG, E. & XU, J. 1996. Raman spectrum of carbonaceous material: a possible metamorphic grade indicator for low-grade metamorphic rocks. *Journal of Metamorphic Geology*, **14**, 115–124.

The 19th century Corsi collection of decorative stones: a resource for the 21st century?

LISA COOKE

4 Lincoln Close, Stoke Mandeville, Bucks, HP22 5YS, UK

(e-mail: lisa.cooke8@btinternet.com.uk)

Abstract: The Corsi collection of decorative stones is arguably the most important and certainly the most diverse and well known of similar collections in Europe. Formed in Rome in the first quarter of the 19th century it consists of 1000 polished sample blocks (*c.* 15 × 7.5 × 4 cm) of natural decorative and semi-precious stone. All the blocks were acquired by Faustino Corsi through other persons, usually a dealer or stonecutter who had them cut to approximately the dimensions of the first model. More than 300 are from stone that had been used in ancient Rome; the others are from stone quarried at a later date. The collection, which is complete, has been in the possession of Oxford University since 1827. The reasoned catalogue by Corsi sheds light on early 19th century ideas about mineralogy and many of the types of stone in use in Rome. Hand specimens are not as important as they used to be for teaching undergraduates, and the decorative arts have little place in modern science. Recent work on provenance, type of stone and nomenclature greatly increase the value of the collection as a resource for identification of ornamental stone used in historical buildings, sculpture and the decorative arts in the 21st century.

The main objective of this paper is to propose that the Corsi collection of decorative stones has the potential to be a useful resource, particularly in the study of antique stones in the 21st century, both because of its underlying quality and heterogeneity and because of recent work to bring the archive up to date. Nowadays opinions of the collection range from 'interesting historical curiosity' to 'of practical value as a reference series'. An outline of the history of the collection and of the diversity and use of ornamental stone during the Roman era and later are discussed in order to provide a context for the topic.

Formation of the collection in Rome

The Corsi collection comprises 1000 sample blocks (Fig. 1) of natural stone, cut to approximately 15 × 7.5 × 4 cm and polished by abrasion on five sides. The 'model' was suggested by the first block purchased for the collection, but sample blocks were cut at different times by different people and the size is only approximately uniform. Some 330 sample blocks are of 'antique' stones (pietre antiche), generally taken to mean that the quarries were unknown in the modern period; the rest were quarried after the Renaissance period. The blocks were considered at the time to be 'very large' (colossale) but their size ensures they are comfortable to hold in the hand and examine on all sides. The Corsi collection was greatly valued in the early 19th century for containing: 'a specimen of almost every known stone (including Granites, Marbles, Porphyries, Jaspers and Serpentines etc.) that has ever been applied to the purposes of ornamental architecture or sculpture …' (note 1). At that time undergraduate teaching made considerable use of hand specimens, pietre antiche were described in books on mineralogy and ornamental stone was in demand for important public and private buildings.

It is difficult to adequately express the full range included in such a comprehensive collection of decorative stones comprising so many types and colours, so a selection will be mentioned. Faustino Corsi realized from ancient writings that any stone that could take a shine when polished used to be known to the ancients as 'marble' from the Greek word 'to shine', and by this definition the collection consists of many different varieties of 'marble'. He points out himself that, at the time of writing, a distinction is drawn by mineralogists between calcareous rocks (known in the early nineteenth century as 'marble') and others of different chemical composition. Approximately half the total number of sample blocks is of calcareous stone including varieties of 'true' marble, limestone, marly limestone, siltstone, mudstone and breccia showing varying degrees of metamorphism, type and shape of clasts, hue and depth of colour. The remainder includes varieties of granite, andesite, gabbro, siliceous breccia and serpentinite as well as agate, amethyst, obsidian and jasper. Together they represent examples of each type of rock

From: PŘIKRYL, R. & TÖRÖK, Á. (eds) *Natural Stone Resources for Historical Monuments.* Geological Society, London, Special Publications, **333**, 185–195.
DOI: 10.1144/SP333.18 0305-8719/10/$15.00

Fig. 1. Examples of sample blocks showing size and quality.

Fig. 2. Etching of Faustino Corsi, frontispiece of catalogue.

formation – igneous, metamorphic and sedimentary. For the most part the different stones were originally sourced from the Roman Empire and Italy, but there are later samples from France, Spain, England the Urals and further afield. The collection acts as an unusual archive of many varieties of decorative stone sourced some time before the 1800s in Italy and Sicily, found in the buildings of Rome. Although the discussion here is largely based on antique stones, many of the so-called 'modern' stones were used in architecture, furniture and decorative items in Italy and elsewhere in Europe, especially during the 18th and early 19th century. It also includes examples of those used for other purposes, such as hard stones that were being used by the early 19th century for inlaid furniture, cameos and other jewellery, and even lithographic stone from Bavaria not long after lithography was invented. Work is still in progress to confirm or refute as far as possible the provenance given by Corsi of individual sample blocks of modern stone, though some have a provenance that it will not be possible to trace.

The collection was formed between 1800 and 1827 by Faustino Corsi (1771–1845) (Fig. 2). The son of a lawyer, his own professional legal career was in the Vatican. He was born and lived all his life in Rome surrounded by ruins of the ancient city where Roman temples and basilicas had been converted to churches and where architectural detail, works of art and other artefacts were ornamented with carved and polished stone. Dealers and stonecutters were selling examples of polished stone cut as blocks or tablets. The opportunity for a rational 'enlightened' study was there to be seized.

Corsi was pioneering in his approach to research and the systemization of decorative stone. His original aim was to make a record of the stones used by the ancient Romans. These were generally known as '*antique*': '*on appelle "marbres antiques", ceux qui ont été employé par les anciens, et ceux dont les carriè sont épuisées ou perdués pour nous*' (Brard 1821 p. 274). Corsi was convinced that examples of the stones from the many quarries in the Roman Empire could be recovered from the ruins of Rome, cut to size and polished by local stonecutters. Sample blocks of antique stone in the collection were for the most part obtained from dealers and craftsmen in Rome. Much of the decorative stone to be seen in the buildings of Rome had, however, been quarried during and since the Renaissance. Corsi recognized this, and made up the majority of sample blocks with so-called 'modern' stone, which he called D'Italia whether originally sourced in the Italian peninsula or elsewhere. Agents in other cities of the Italian peninsula obtained stone from what Corsi believed were working quarries. In Rome there were many dealers, studios and workshops where objects could be obtained that incorporated reused antique stone with stone recently imported from distant continents along relatively new trade routes. These included examples of what Corsi called 'the most famous and most beautiful stones used in Italy' such as rock crystal from the island of Madagascar, although it was known to occur in the Alps, labradorite, amazonite and lapis lazuli. All the sample blocks in the collection, however exotic, were cut approximately to size and polished to his model. Apart from the sense of uniformity, this had the consequence of making them much easier to compare with examples in inlaid work than crystals or unpolished blocks would have been. Because he was so familiar with the architecture and sculpture of Rome, Corsi was well aware that although his sample blocks may be comparatively large for a collection, in reality the stones are not shown to as much advantage for comparison or appreciation as they would be in use. He therefore drew attention to many examples of the various stones in churches, museums and monuments of Rome (see Cooke 2000).

Corsi as a pioneer seeking provenance and universal nomenclature

Published work and acclaim

Corsi was, as far as we know, the first person in the modern era to form a wide-ranging collection and catalogue of decorative stone as a reference for posterity and a greater public rather than for his own delight, and he was one of the first to use the larger sample blocks which were then becoming fashionable with collectors (Mariottini 2004). More importantly, his was the earliest attempt to create a systematic classification for a collection of antique ornamental stone (in contrast to a mention in a publication) and to develop a 'universal' nomenclature based on provenance with the potential to be common to all mineralogists. Corsi was aware of the great work of taxonomy of Carl Linné (1707–1778), also known as Linnaeus.

It appears that Corsi, who had a wife and family for whom to provide, had no opportunity to travel abroad to seek ancient quarries unlike the English collector Edward Dodwell, traveller and author living in Rome. Quarry samples from ancient sites were not available as comparators as they are today. Like many of his contemporaries in Italy, England and other European countries, Corsi's early education was based largely on ancient history and classical literature. Not unnaturally, his method of determining provenance for antique stones was to find references to them in Greek and Latin writings, especially Pliny the Elder's Natural History. These tended to identify a stone by its source, such as Parian or Thasian, but to offer few descriptions. He developed a binomial Latin nomenclature from his observations of stone type and deductions of provenance from these passing references, a surprising number of which turned out to be correct and are still used today. With such inadequate data available to him, it was almost inevitable that Corsi's high endeavour should result in mistakes; these usually occurred because of the misinterpretation of an ancient text (Cooke 2002). An example of this relates to the pink, white and brownish fine-grained marble or marble breccia, heavily crushed and sheared, now known as fior di pesco, *Marmor Chalcidicum*. This stone originates from the Greek island of Euboea (modern Evvoia) near Eretria, some 20 km south of Chalkis, named '*Marmor Molossium*' by Corsi. Corsi infers that its provenance is 'in Epiros, the land of the Molossians' most probably because he misunderstood a text by Paulus Silentiarius. Silentiarius had written with considerable poetic licence describing the many varieties of decorative stone of Haghia Sophia in Constantinople (modern Istanbul) when it was re-dedicated in AD 537 after a disastrous fire.

The stonecutters of antique decorative stone in Italy had already invented vernacular names that had been transmitted orally through generations of artisans. For the most part these were descriptive, such as *giallo antico*, *serpentino* (like a serpent's skin) or *granito rosso*, but were widely known and translated into other languages. Many Italian names can be traced at least to the late 16th century; they were written down with descriptions and details of where they could be seen in Florence by a Dominican monk, Agostino del Riccio (Gnoli & Sironi 1996). Corsi collated the Italian names for antique stones still used by the stonecutters with the Latin names, thus making them more accessible to classical scholars. Examples of Corsi's system of nomenclature includes (with modern descriptions given here): '*Marmor Phrygium, pavonazzetto*' from Phrygia in modern Turkey, a white calcitic marble breccia in hematite-rich matrix; '*Marmor Syenite, granito antico*' from Aswan, Egypt, probably the most famous granite in the world (its use spanning five millennia), which can be coarse or fine grained with potassium feldspar, quartz and biotite; '*Marmor Porphyrite, porfido rosso antico*' from Mons Porphorites, now Gebel Dokan, Egypt, a porphyritic andesite-dacite which at certain times has been the most prestigious stone in use.

Corsi's Latin names alongside the Italian names were available to a wide readership through the three incremental editions of his book on decorative, building and semiprecious stones used in the Roman period, *Delle Pietre Antiche* (1828, 1833, 1845). In the first edition, the chapter on decorative stone is clearly based on catalogue entries referring to antique stone in this collection (Corsi 1825, 1827). The later editions of the book contain substantial additions and a few corrections, while the number of citations is increased to 450 in the third edition (Corsi 1845; see Napoleone 2001) compared with 121 in the catalogue. For many years Corsi's work was considered the best, if not the only, comprehensive source of Latin nomenclature, descriptions and bibliography of antique decorative stone.

Corsi, showing clearly that his work was intended to enlighten a wide readership, noted contemporaneous ideas of the formation of the different types of rock as he understood them, partly based on published work of French mineralogists and mining engineers such as Brongniart (1807), Brochant (1807), who incidentally pays tribute to the work of Professor Abraham Werner of Freiberg with whom he had studied, and Brard (1821). These and other contemporaneous authors wrote about antique marbles, but did not appear to be interested in their provenance.

Corsi was not educated in natural philosophy or mineralogy but in the classics and in jurisprudence. Like many others during the so-called 'Age of Enlightenment' he approached research of his collection, and later his book, logically and systematically as a gifted amateur with a real love of the subject. As a token of the esteem in which he was held by contemporaries in Rome, Corsi was elected a member of the Accademia dell' Arcadia in the company of famous professional contributors to the arts, such as the sculptor Antonio Canova (1757–1822) and the micro-mosaicist Giacomo Raffaelli (1753–1836). He has again been recognized with acclaim more recently as the first specialist in the study of archaeological marble of the modern epoch (Dolci 1992 p. 31) and the most important scholar-collector of the first quarter of the 19th century, the first half of which was the golden age for the study and collection of antique marbles (Gnoli 1971).

Historical background to the use of similar stones

The Roman Empire

In forming this collection and researching and writing about antique stones (pietre antiche), Corsi had the opportunity to indulge his passion for the amazing repertory of worked rocks and minerals that was so much part of his life in Rome. He learnt from classical writings that in ancient Roman times many types and varieties of decorative stone had been obtained and traded widely throughout their Empire where they had been used evermore extravagantly, bringing prestige to donors of public buildings and to owners of private ones. Ancient quarries of marble and other decorative stones have been re-discovered in modern Turkey, Greece, Egypt, Tunisia, Italy, Spain and France. The wide range of use of coloured decorative stone of the Roman period was clearly apparent in the winter of 2002–03 in a comprehensive exhibition held in Trajan's markets in Rome. The displays showed many examples that had been obtained from excavation and are now in museums or private ownership in, for example, Naples, Rome, Paris, Copenhagen and Israel. They included columns, reliefs, flooring, sculpture, large shallow water basins (*labri*), furniture, extraction tools, evidence of the means of transport and selections from other collections of sample blocks of antique stones. The very informative, well-illustrated catalogue, generally known as 'marmi colorati ...' (De Nuccio & Ungaro 2002) is still in print by public demand, and is obtainable in Rome from good bookshops and major museums such as the Capitolini (personal communication, 2009 from M. Mariottini).

Varieties of semi-precious stones and gems were used in the luxury arts during the Roman period, a selection of sample blocks of which are included in the collection although we have not ascertained provenance. Martin Henig, an acknowledged expert in this field who has written several commentaries on important collections of Roman gems, writes in more general terms of the use by men (since the 3rd century BC) of signet rings of semi-precious stone 'mainly cornelian and onyx', and of jewellery (more usually worn by women) which might incorporate rock crystal or lapis lazuli with other materials. Less frequently cups, sometimes carved in relief, and small freestanding portraits and figurines were also made of these and other hard stones (Henig 1983).

Both during the Roman period and later, coloured ornamental stone and white marble, which was usually painted brightly in antiquity, not only bestowed decorative richness but suggested wealth, gravitas, culture and learning whether used in public buildings such as temples or churches, law courts, bath-houses, government offices, papal and secular palaces or in grand villas and mansions.

Since the Roman Empire

During the Late Antique and Medieval periods, reuse of stone in refurbished or new buildings was not unusual. One striking and colourful example from the Medieval period is the use of characteristic non-figurative mosaics developed by the Cosmati family from the work of Byzantine mosaic workshops, which were already becoming greatly appreciated in Italy. Cosmati work, noted for intricate geometric patterns usually of small pieces of antique stone but sometimes including glass (inlaid into white or light coloured marble), could be found in paving, altars, screens, candlesticks and other ecclesiastical furniture as far from Rome as Westminster Abbey in London.

After the Renaissance, ever greater quantities of coloured stone were recovered from the ruins of ancient cities by excavation as town-planning schemes, archaeological, antiquarian and treasure-hunting interests began to increase. Fragments lay scattered on or near the surface of rough roadways and grazing land, while half-buried columns, capitals and friezes were abundant in the cityscape, which to us would appear astonishingly rural. In Rome and the cities of the Italian peninsula and islands, and to a lesser extent in other towns and cities that had been part of the Roman Empire, antique stone continued to be reused, sometimes with modern stone, usually of fairly local origin. Massinelli (2000) suggests that by 1550 in Rome, mosaic and inlaid work using soft stones (*pietre tenere*) were undertaken. From the 1550s, Milan was the most important centre in Europe for carved vessels and panels of hard stones (*pietre dure*). The founding of the Opificio delle Pietre Dure in Florence in 1588 (Giusti 2006) increased the market for inlaid decorative stone in fine furniture and panels, a few examples of which are still made in traditional family workshops in Florence but to a lesser extent. The Opificio today is concerned with conservation and research and houses a museum as well as holding long-term stocks of some stone, for example so-called 'soft' jaspers of Sicily and antique Aswan granite. An important collection of *pietre dure* furniture, as well as displays of micromosaics and unusual decorative objects made of stone, can be seen in the Gilbert Collection at the Victoria and Albert Museum, London. A particularly ornate example of the latter is the setting of a clock given by Pope Pius VII to Napoleon Bonaparte commemorating his coronation in Paris that was made in Rome in1804 by Giacomo Raffaelli (Massinelli 2000).

The Grand Tour and dispersal of stone and names from Italy

For at least a hundred years the stonecutters and dealers of many cities such as Rome, Milan, Florence and Naples had been selling small, polished samples fashioned from off-cuts and fragments of coloured stone. These were in the form of approximately square tablets 3–7 cm and approximately 1 cm deep, with the depth tending to become less through time. The tablets were popular among participants in the Grand Tour who found they made attractive souvenirs that were easy to handle and transport. Tablets were often sold in sets (studii), approximating 100 in number with Italian names supplied in an inventory. Sometimes tablets were inlaid in table tops made in workshops in Italy or elsewhere, for example Paris or England. Corsi himself formed two collections of tablets (approximately 6 cm square): one set of antique and one of modern decorative stones. Inset in tabletops they are now in the Natural History Museum, London, with associated manuscript inventories of names that have been electronically recorded by scanning. Sets (studii) of the smaller tablets, sometimes boxed and often set in wooden trays, may now be found in museums and country houses as far apart as Boston (Mass), Edinburgh (Scotland) and Derbyshire (England) while collections of larger sample blocks are plentiful in Rome and can also be found in Brussels and other cities.

The Grand Tour, at its height in the 18th century, ensured dispersal of antique stone and the Italian vernacular names throughout Europe and, to a far lesser extent, North America. Souvenir sets of tablets as well as columns, reliefs, sculpture, furniture and fragments for 'Italian' flooring were shipped home, often in large quantities. These were incorporated into palatial houses by wealthy individuals, for example the Sixth Duke of Devonshire (Fig. 3). Much of the coloured antique stone, with the more fashionably desirable sculpture and coins, was excavated under licence granted by the Vatican who took first refusal of discoveries but often granted further licences for sale or export of the remainder (Bignamini 2004). Modern stone, especially from Italy, France and Spain, also began to be included in the architectural grand designs of Europe.

Eminent English admirers of Corsi: acquisition of the collection

Spencer Joshua Alwyne, Lord Compton (1790–1851) and later second Marquess of Northampton, whose family was related by marriage to that of the Duke of Devonshire, was elected Fellow of the Royal Society (FRS) 1830 and President of the Royal Society 1838–1848. Lord Compton was President of the Geological Society London during the period 1820–1822. He lived in Rome for ten years at about this time with his wife (before she died in childbirth) and began to take an active part in archaeology as well as geology, both interests staying with him throughout his life. He excavated part of Hadrian's Villa at Tivoli, giving two samples of stone from the site to be included in the collection (one through Stephen Jarrett; see below). His influence may possibly have been indirectly responsible for the donation of the collection to Oxford; Corsi knew him personally and asked his advice when negotiating the sale of the collection:

> '. . . you ask me if I would be happy with £500 without the English stones . . . Lord Compton has told me they are not easy to get over there unless from the Duke of Devonshire . . . such stones separated from the collection are of little interest therefore I intend them to stay with the collection . . .' (note 2)

William Spencer Compton Cavendish (1790–1858) Sixth Duke of Devonshire, who was one of the wealthiest men in Britain, shared with Corsi a passion for decorative stone (Cooke 2004). A few

Fig. 3. Giallo antico reused for decorative display, Sculpture Gallery, Chatsworth.

years after inheriting the title and considerable family estates, the Duke began his first adult journey across Europe; he remarked later that at Rome the love of marble possesses most people like a new sense. He stayed in Rome on at least six occasions for several months at a time, visiting archaeological sites and monuments, museums, sculptors' studios, workshops and dealers in marble objects (note 3). He had an estate in Ireland, where he was likely to have met the landscape painter Gaspare Gabrielli (1770–1828) who had spent more than ten years there, becoming Vice President of the Royal Dublin Society of Artists. Returning to Rome in 1815 Gabrielli became friendly with Corsi; within four years he was acting as personal guide (cicerone) and agent to the Sixth Duke. The Duke met Corsi and saw the collection, later donating seventeen samples of Derbyshire stones and minerals each with cursory documentation as to their locality within the county. They were from working quarries: some were very small scale worked in quiet seasons by tenant farmers mostly for lime-burning; others were relatively large-scale commercial mines of decorative stones popular at the time. There are sample blocks of the bituminous limestone known as Derbyshire Black Marble with a wide range of decorative uses from inlay and etchings to columns (Tomlinson 1996); an early sample of the very rare hematitic limestone known as Duke's Red, of which the small reserve is kept under a staircase in the service area at Chatsworth House; examples of several varieties of fossiliferous limestone, including Hopton Wood stone that was popular for columns and flooring (Thomas 2000); and fluorites, including Blue John which was used for jewellery, ornaments, inlays and large objects for which special techniques were developed in the workshops (Ford 2000).

Sydney Smirke (1798–1877) gravitated towards Rome as a young man, as had his older brother Robert Smirke (later Sir Robert) and many notable English and Scottish architects on the Grand Tour. Later he designed several important buildings in London, including the famous domed Reading Room at the British Museum (now used as an exhibition space and replaced by the British Library). He saw and admired the Corsi collection in Rome. Possibly thinking that acquisition of such a fine collection by the British Museum would enhance his reputation, which at that time was very much that of a younger brother, he brought it to the attention of Charles Konig (1774–1851), Keeper of the Department of Natural History at the British Museum since 1813 on his return to London. Konig duly arranged for the Corsi collection to be on the agenda at the next meeting of Trustees of the British Museum on 11 December 1826. He requested a recommendation from the well-known geologist, William Buckland, who must have made it known he had seen it in Rome. The Trustees, however, declined on this occasion to purchase the collection (see Cooke & Price 2002).

The Reverend William Buckland (1784–1856) was of considerable intellect, widely respected, and a great enthusiast of geological research, collecting and lecturing. He was elected FRS in 1818, President of the Geological Society London 1826–1827 and again 1839–1841, winning the society's

Wollaston Medal in 1848. He was appointed Canon of Christ Church Oxford 1825–36 and later Dean of Westminster (1845–56). While he was holding concurrently the two posts of Reader in Mineralogy and the first Reader in Geology at Oxford University, he made a 'geological Grand Tour' in 1826 during sabbatical leave taken to celebrate his marriage. After a considerable amount of fieldwork and collection of specimens in the Alps and elsewhere in Europe, he and his bride travelled on to Rome. When Buckland replied to Konig's request for a 'testimonial' for the attention of the Trustees of the British Museum, his undeniable enthusiasm is very clear even from a small excerpt:

> 'I have no hesitation in saying that it [the Corsi Collection] is quite unique in its kind, and such as is never likely to be made by any other individual, that it is in the highest degree interesting not only in an Geological and Mineralogical point as affording splendid specimens of the most beautiful rocks existing, but also as illustrating the History of the Arts both of Ancient and Modern Times . . .' (note 1).

Stephen Jarrett is important to the history of the Corsi collection although not in the public eye. As a wealthy undergraduate student of Magdalen College Oxford in his third (final) year, he travelled to Rome in the winter of 1826–1827. There he met Corsi and was much impressed by the collection, then consisting of 900 sample blocks, which he arranged to buy to donate to the university with approximately 250 copies of the catalogue. He also requested that a further 100 sample blocks be added, and that a supplementary catalogue be printed at his expense (Corsi 1827). After his return he was honoured by being granted an Honorary MA, Buckland giving the laudatory oration at the degree ceremony. Jarrett's generosity was further rewarded by being granted special dispensation to study at the university for a further year.

Henry Miers (1858–1942), one-time Principal of London University and later Vice-Chancellor of Manchester University, was a energetic man of great talent, especially for scientific research, teaching and administration. He won many awards during his lifetime, including the Wollaston Medal of the Geological Society London in 1934. His varied interests ranged from mineralogy, crystallography and gemstones to classical Greek sculpture, museums and libraries, football, and hot-air ballooning. In December 1895, on being appointed Wayneflete Professor of Mineralogy, he became responsible for the Corsi collection in the Oxford (University) Museum, (Cooke & Price 2002). Miers, some 70 years after Buckland, similarly regarded the collection as containing 'unrivalled examples (and some unique) of antique marbles' (note 4). He was 'anxious to make it one of real use as a reference series' (note 5). To this end he corresponded with William Brindley who, in the course of exploratory journeys for the marble sculptors and importers Messrs. Farmer & Brindley & Co, had rediscovered several ancient quarries, for example on the Greek island of Chios and near Thessaly on the Greek mainland (Brindley 1886). He was considered 'to know more about marble quarries than any man living' in his time (note 6). Brindley examined the sample blocks of antique marble and gave his opinion as to provenance, pointing out some of Corsi's misapprehensions. Brindley's visits and remarks were recorded by Mary Porter, a teenager fresh from living in Rome and already with an interest in the Corsi collection (Porter 1973). Miers, noticing her obvious interest, asked her to translate the catalogue for in-house use, and to rearrange the sample blocks guided by a classificatory system recently published by the Curator of Mineralogy at the United States National Museum in Washington DC (Merrill 1903). This was probably to reflect the fact that the Corsi collection was definitively in a scientific setting and was now in the care of a mineralogist. It had been somewhat neglected in the Oxford Museum, the centre for science in the university, and while in the overall care of the Radcliffe Librarian who was the Professor of Medicine with rather different priorities. Porter was encouraged by Miers to work for a degree in crystallography, a field in which she was particularly gifted. Long before crystal structure was defined by X-rays, she carried out crystal calculations for 31 years for the Barker Index (Porter & Spiller 1951) and was also granted a lifetime Honorary Research Fellowship at Somerville College, Oxford. On retirement in the early 1960s she rearranged and re-labelled the sample blocks in the order of Corsi's catalogue, suggesting that it was better than the order inspired by Merrill.

Corsi's enduring reputation—mentioned in publications since 1970

Corsi was not only greatly thought of during his lifetime but is still admired well over a century and a half later. His work is given some prominence by the doyen of late 20th century studies in antique marble, who also actively sought to rediscover ancient quarries – Raniero Gnoli (1971) – and in many illustrated Italian publications (e.g. Borghini 1989; Dolci & Nista 1992; Lazzarini 2004). Another collection made sometime before 1870, found in Berlin, was published in German as an illustrated handbook and is regarded as a useful reference (Mielsch 1985). The two books published in English on decorative stone both mention Corsi and illustrate a selection of sample blocks from the collection, while reflecting the main interests of their authors. The first is the Collected Papers of John Ward-Perkins (1912–1981), Director of the British School at Rome for 29 years, who was a fine archaeologist with a love of classical architecture who extensively researched the Roman marble trade. In the informative book there are two colour plates each showing seven sample blocks from the Corsi collection, and notes and maps showing the location of the best-known quarries of the Roman Empire. The extensive bibliography is divided into 11 sections such as 'General marble' and 'Quarries, quarrying and trade' (Dodge & Ward-Perkins 1992). The second is an illustrated source book by a mineralogist that includes a guide to minerals and an overview of many varieties of decorative stone (Price 2007). It provides useful information and references, especially on elementary mineralogy and ancient and working quarries. A proportion of individual samples of natural stone illustrated are of antique stone from the Corsi and other collections held in Oxford University Museum of Natural History (OUMNH). Other stones are from the early modern period (some in the Corsi collection) and yet others are examples from every continent of the world from active quarries.

Since Corsi's work, taking advantage of modern technology (especially in science, communication and transport), are the interdisciplinary papers presented and published in English (with a small proportion of papers in French) at international conferences of the Association for the Study of Marble and Other Stones in Antiquity (ASMOSIA). Meetings are held in different countries, including Greece, Italy, France, Spain and the USA, at approximately two and a half yearly intervals. These meetings serve as a forum for discussion between geologists, museum curators, art historians, archaeologists and conservators; the proceedings report on the latest analyses, discoveries and other topics relating to antique stone. The Corsi collection archive records important relevant new findings (see Herz & Waelkens 1988; Waelkens *et al.* 1992; Maniatis *et al.* 1995; Schvoerer 1999; Herrmann *et al.* 2002; Lazzarini 2002).

Confidence in Corsi as source of information in 21st century

The last decades of the 20th century saw increasing and more widespread interest by academics in different aspects of stone used in antiquity, which continues to this day. Many sources of information are now available to bring the archive of the collection up to date. Decorative stones from the Roman period are, with experience, usually recognizable macroscopically by characteristic colouring and formation. The number of quarries is finite, and those of importance have been rediscovered. Identification of the majority of sample blocks can be afforded a high degree of confidence, although they are from secondary sources. Scientific analyses of samples of similar stones in laboratories have also played a part. White, grey and black marble, and indeterminate limestone in the collection cannot be assigned localities with accuracy without invasive tests, which are not considered appropriate.

Individual sample blocks have been examined and opinions given as to original provenance by geologists experienced both in handling antique decorative stone and in hunting for ancient quarries, notably Lorenzo Lazzarini and James Harrell. Lazzarini, active member of the Scientific Committee of ASMOSIA and one time President (now of Istituto di Architettura di Venezia and previously of 'La Sapienza' university in Rome) is well known for teaching stone recognition and conservation, for the re-discovery of ancient Greek and other quarries and for analysis of antique stone. As well as many papers, he has written a detailed illustrated monograph on antique Greek polychrome stones, of which there are examples in the collection. Lazzarini (2007) shows many illustrations and aspects of the stones ranging from quarries, usage and sample blocks to thin sections, tables of scientific analyses, and a comprehensive general bibliography. Harrell, one-time member of the Scientific Committee of ASMOSIA, kindly provided a detailed report for the Oxford University Museum of Natural History (OUMNH) on the Egyptian granites and porphyries in the collection. He teaches in the Department of Earth, Ecological and Environmental Sciences, University of Toledo University (Ohio, USA), and has conducted similar work on ancient Egyptian stones and quarries, especially of igneous rocks. Trevor Ford (OBE, FGS) who has a special interest in the geology and minerals of Derbyshire (Ford *et al.* 1996) has retired as Senior Lecturer and Associate Dean of Science at the University of Leicester. Ford, and Ian Thomas who is based at the National Stone Centre in Derbyshire, have independently localized the provenance of samples of the Derbyshire stones in the collection. Various other people, such as Patrick Rogers who has studied and written about the marble of Westminster Cathedral and Ian MacDonald, a marble merchant based in Wiltshire (both with considerable interest in and experience of antique stone and Italian quarries), have also discussed individual samples with Monica Price for the archive.

These combined opinions have been synthesized with published work to make the collection and catalogue reliable as a reference for antique stones, and for the modern Derbyshire stones. Assistant curator, Monica Price is researching the modern Italian stones in the collection.

The relevance of Corsi's work in the 21st century

Different viewpoints

This historic, groundbreaking collection is still complete after some 180 years. The quality and wide variety of the sample blocks of the collection ensure their value for comparison for identification. Quarry variation is shown by the several different sample blocks of certain stones including, for example, more than ten each of giallo antico, yellow limestone from Numidia (modern Tunisia) and of *africano* a meta-breccia from Teos on the western coast of modern Turkey.

Delle Pietre Antiche had immeasurable influence on a wide readership. For nearly a century its pronouncements were accepted unquestioningly in many countries by a great number of people. Corsi's Latin names are still of practical as well as historical interest, even when suggesting what is now known to be an incorrect provenance. For about 80 years or more since the first edition, geologists and museum curators labelled items accordingly. In some cases there has not been an opportunity for correction, which can be confusing today unless 'briefed'. The original catalogue should also be read with caution or mental footnotes.

As a teaching aid in a general undergraduate syllabus in Earth Sciences, or indeed for museum display, the collection has lost much of its glamour. This is largely because hand specimens now play a minor role in teaching and ornamental stone is neither as admired nor considered as relevant as it used to be. Tastes in architecture have changed; antique marble is not generally appreciated for its beauty or history as it was in the 19th century. The commercial market for decorative stone is now run differently, with new sources and markets being exploited world wide.

Access to the collection and catalogue

Personal communication is virtually the only way to access the collection and catalogue today, even although it may be of value to those with a special interest in, or responsibility for, conservation of antique flooring, mosaics and inlaid furniture. Other actual and potential users include architectural historians, historians of the decorative arts, museum and country house curators and archaeologists and specialist advisors (often geologists) who may provide reports of stones found during excavation. Enquiries about the collection and requests to view should be addressed to: The Assistant Curator of Mineral Collections, Oxford University Museum of Natural History, Parks Road, Oxford OX1 3PW, England, UK, e-mail: mineralogy@oum.ox.ac.uk.

Future possibilities

The opportunity of setting up an internet database of antique decorative stone using the Corsi collection as a basis should not be overlooked. The work of the Czech Geological Survey (Schulmannová & Skarková 2004) provides a model for data which are finite. In the first instance, possibly with a view to adding other 19th century collections, the potential database would comprise digital images and information synthesized during study of the Corsi collection. It should be seen as an aid to identification and would not attempt to include physical, mechanical or technical parameters, which are either relatively obvious or irrelevant for the kind of interior use to which the stone would be most likely be put. However, a salutary example of some pitfalls of a type of stone, used in antiquity in one climate and used in the 20th century in a very different climate with little understanding or care for its properties, is shown in a study of regional weathering of Roman travertine (Sidraba *et al.* 2004).

Conclusions

Recent work on provenance, type of stone and nomenclature greatly increases the value of the collection as a resource for identification of ornamental stone used in historical buildings, sculpture and the decorative arts. In my opinion, the Corsi collection is of historical and practical significance, providing a useful archival resource for the 21st century. This echoes that of Porter: 'a collection made before 1825 in Rome becomes increasing valuable with each fifty years, both historically and archaeologically' (note 7).

I am indebted to the Trustees of the Chatsworth Estates and thank Diane Naylor in particular, for making sources of information available including the 6th Duke's Journal and Sculpture Accounts.

References

BIGNAMINI, I. 2004. British excavations in the Papal states during the eighteenth century: written and visual sources. *In*: BIGNAMINI, I. (ed.) *Archives & Excavations*. Archaeological Monographs of the British School at Rome 14. BSR. London, 91–108.

BORGHINI, G. (ed) 1989. *Marmi Antichi*. Ministero per l Beni Culturali e Ambientali – Istituto Centrale per il Catalogo e la Documentazione, Rome. Reprinted 1997. Editioni De Luca Rome.

BRARD, C. P. 1821. *Minéralogie Appliquée aux Arts*. 2. F. G. Levrault, Paris.

BRINDLEY, W. 1886. Marble: its uses as suggested by the past. *Journal of the Royal Institute of British Architects*, **3**, 89–97

BROCHANT, A. J. 1807. *Elementi di Mineralogia*. (Italian translation from French) 1. Cairo; Milan.

BRONGNIART, A. 1807. *Traité Élémentaire de Minéralogie avec des Applications aux Arts*, **1**. Crapelet, Paris.

COOKE, L. 2000. The catalogue of Faustino Corsi and the monuments of Rome. *In*: CALVI, G. & ZEZZA, U. (eds) *Quarry, Laboratory, Monument 1*. Proceedings of International Congress in Pavia 2000, September 6–30. La Goliardica Pavese, Pavia, 239–243.

COOKE, L. 2002. The Corsi collection: review of some original provenances. *In*: LAZZARINI, L. (ed.) *ASMOSIA VI. Interdisciplinary Studies on Ancient Stone*. Proceedings of the Sixth International Conference of the Association for the Study of Marble and Other Stones In Antiquity, Venice, June 15–18, 2000. Ausilio, Padua, 537–544.

COOKE, L. 2004. The shared passion for stone of William Spencer Cavendish, 6th Duke of Devonshire, and Faustino Corsi, a lawyer in Rome. *In*: PŘIKRYL, R. & SIEGL, P. (eds) *Architectural and Sculptural Stone in Cultural Landscape*. Karolinum Press, Prague, 41–51.

COOKE, L. & PRICE, M. T. 2002. The Corsi Collection in Oxford. *In*: HERMANN, J. J., HERZ, N. & NEWMAN, R. (eds) *ASMOSIA 5 Interdisciplinary Studies on Ancient Stone*. Proceedings of the Fifth International Conference of the Association for the Study of Marble and Other Stones In Antiquity, Museum of Fine Arts, Boston, 1998. Archetype Press, London, 415–420.

CORSI, F. 1825. *Catalogo Ragionato d'una Collezione di Pietre di Decorazione Formata e Posseduta in Roma*. Da'Torchi del Salviucci, Rome.

CORSI, F. 1827. *Supplemento al Catalogo Ragionato d'una Collezione di Pietre di Decorazione Formata in Roma ... e posseduta dall' Univesità di Oxford*. Da'Torchi del Salviucci, Rome.

CORSI, F. 1828. *Delle Pietre Antiche*. Da'Torchi del Salviucci, Rome.

CORSI, F. 1833. *Delle Pietre Antiche*. (2nd ed.). Tipografia Salviucci, Rome.

CORSI, F. 1845. *Delle Pietre Antiche*. (3rd ed.). Tipografia di G Puccinelli, Rome.

DE NUCCIO, M. & UNGARO, L. (eds). 2002. *I Marmi Colorati della Roma Imperiale*. Marsilio, Venice.

DODGE, H. & WARD-PERKINS, B. (eds). 1992. *Marble in Antiquity: Collected Papers of J. B. Ward-Perkins Archaeological Monographs of the British School at Rome 6*. BSR, London.

DOLCI, E. 1992. Cultura del marmo e collezioni erudite. *In*: DOLCI, E. & NISTA, L. (eds) *Marmi Antichi da Collezione La Raccolta Grassi del Museo Nazionale Romano*. Museo Civico del Marmo, Carrara, 19–39.

DOLCI, E. & NISTA, L. (eds) 1992. *Marmi Antichi da Collezione La Raccolta Grassi del Museo Nazionale Romano*. Museo Civico del Marmo, Carrara.

FORD, T. D. 2000. *Derbyshire Blue John*. Ashbourne Editions, Ashbourne, Derbyshire.

FORD, T. D., SARJEANT, W. A. S. & SMITH, M. E. 1996. Minerals of the Peak District. *Peak District Mines Historical Society Bulletin*, **12**(1), 16–56.

GNOLI, R. 1971. *Marmora Romana*. Dell'Elefante, Rome. (2nd edition 1988)

GNOLI, R. & SIRONI, A. (eds). 1996. *Agostino del Riccio Istoria delle Pietre*. Umberto Allemandi & C., Turin.

GIUSTI, A. 2006. *Pietre Dure and the Art of Florentine Inlay*. Thames & Hudson, London.

HENIG, M. 1983. The luxury arts. *In*: HENIG, M. (ed.) *A Handbook of Roman Art*. Phaidon, Oxford, 152–165.

HERMANN, J. J., HERZ, N. & NEWMAN, R. (eds). 2002. *ASMOSIA 5: Interdisciplinary studies on ancient stone*. Proceedings of the Fifth International Conference of the Association for the Study of Marble and Other Stones in Antiquity. Museum of Fine Arts, Boston, USA. June 1998. Archetype Publications, London, 11–15.

HERZ, N. & WAELKENS, M. (eds). 1988. *1st ASMOSIA meeting*. Transactions of NATO Advanced Research Workshop May 9–13, 1988 (NATO ASI Series Vol. 153) Classical Marble: Geochemistry, Technology and Trade. Il Ciocco, nr. Lucca, Italy. Kluwer Academic Publishers, Dordrecht.

LAZZARINI, L. (ed). 2002. *ASMOSIA VI. Interdisciplinary Studies on Ancient Stone*. Proceedings of the Sixth International Conference. Dipartimento di Storia dell' Architettura, Istituto Universitario di Architettura di Venezia, Venice, Italy. 15–18 June 2000. Ausilio, (AAGEP), Padua.

LAZZARINI, L. ed. 2004. *Pietre e Marmi Antichi*. Cedam, Padua.

LAZZARINI, L. 2007. *Poikiloi Lithoi ... I Marmi Colorati della Grecia Antica*. Accademia Editoriale, Pisa.

MANIATIS, Y., HERZ, N. & BASIAKOS, Y. (eds). 1995. *ASMOSIA III. The Study of Marble and Other Stones Used in Antiquity*. Transactions of the 3rd International Symposium. Laboratory of Archaeometry Greek National Centre for Scientific Research, 'Demokritos', Athens, Greece. May 17–19, 1993. Archetype Publications, London.

MARIOTTINI, M. 2004. Per una storia del collezionismo dei marmi antichi. *In*: LAZZARINI, L. (ed.) *Pietre e Marmi Antichi*. Cedam, Padua, 135–189.

MASSINELLI, A. M. 2000. *The Gilbert Collection Hardstones*. Philip Wilson, London.

MERRILL, G. P. 1903. *Stones for Building and Decoration*. 3rd ed., John Wiley & Sons, New York.

MIELSCH, H. 1985. *Buntmarmore aus Rom im Antikenmuseum Berlin*. Staatlich Museen Preussischer Kulturbesitz, Berlin.

NAPOLEONE, C. (ed). 2001. *Delle Pietre Antiche Il trattato sui marmi romani di Faustino Corsi*. Franco Maria Ricci, Milan.

PORTER, M. W. 1973. *Diary of Henry Alexander Miers 1858–1942*. Privately printed at The Oxford University Press, Oxford.

PORTER, M. W. & SPILLER, R. C. 1951. *Barker Index of Crystals A Method for the Identification of Crystalline Substances*. Cambridge University Press, Cambridge.

PRICE, M. T. 2007. *Decorative Stone the Complete Sourcebook*. Thames and Hudson, London.

SCHVOERER, M. (ed). 1999. *ASMOSIA IV. Archéomatériaux Marbres et autres roches*. Actes de la IVe Conférence Internationale. Centre de Recherche en Physique Appliqué à l'Archéologie Université de Bordeaux. Bordeaux 3, France. 9–13 Octobre 1995. CRPAA-Presses Universitaires de Bordeaux, Bordeaux.

SCHULMANNOVÁ, B. & SKARKOVÁ, H. 2004. Internet database of decorative and building stone. *In*: PŘIKRYL, R. (ed.) *Dimension Stone New Perspectives for a Traditional Building Material*. Balkema, Leiden, London, 153–155.

SIDRABA, I., NORMANDIN, K. C., CULTRONE, G. & SCHEFFLER, M. J. 2004. Climatological and regional weathering of Roman travertine. *In*: PŘIKRYL, R. & SIEGL, P. (eds) *Architectural and Sculptural Stone in Cultural Landscape*. Karolinum Press, Prague, 211–228.

THOMAS, I. A. 2000. Tarmac's Derbyshire heritage. Part 2. *In*: FENN, R. W. D., MACLEOD, G. S. & BROMWICH, N. P. (eds) *Tarmac Papers 4*. Tarmac, Wolverhampton, 267–336.

TOMLINSON, J. M. 1996. *Derbyshire black marble*. Peak District Mines Historical Society Special Publication 4. Peak District Historical Society, Matlock Bath, Derbyshire.

WAELKENS, M., HERZ, N. & MOENS, L. (eds). 1992. *2nd ASMOSIA meeting: Ancient Stones: Quarrying, Trade and Provenance*. Catholic University of Leuven, Belgium. October 16–20. 1990. Acta Archaeologica Lovaniensia, Monographiae 4. Leuven University Press, Leuven.

Notes – unpublished sources

(1) MS: letter from W. Buckland to C. Konig 2 December 1826. Mineralogical Department archive file 10/41-5, Natural History Museum, London (formerly British Museum, Natural History).

(2) MS: letter from F. Corsi to S. Smirke 18 November 1826. Min. Dept. archive file 10/41-1, Natural History Museum, London (translation from Italian).

(3) MS: 6th Duke's Note book and Sculpture Accounts. Archive of the Chatsworth Estates.

(4) Miers, H. A. 1899 Twelfth Annual Report of the Delegates of the University Museum.

(5) MS: letter from H. A. Miers to W. Brindley, 29 November 1899 (with annotations by Brindley). Corsi file, Mineralogical Collections archive, OUMNH.

(6) MS: letter from T. G. Jackson (architect) to Miss Tatham (niece of Henry Miers) 23 October 1904. Corsi file, Min. Colls. archive, OUMNH.

(7) Typescript, anon., 2 June 1964, attributed to M. W. Porter by Price. Corsi file, Min. Colls. archive, OUMNH.

Working for an electronic database of historical stone resources in Friuli-Venezia Giulia (Italy)

ANNA FRANGIPANE

Udine University, Faculty of Engineering, Department of Civil Engineering, via delle Scienze 208, Udine, Italy (e-mail: anna.frangipane@uniud.it)

Abstract: Friuli-Venezia Giulia is situated in northeast Italy, on the border with Austria and Slovenia. The availability of stone from many different sites across the geographical area of the Carnian and Julian Alps, the foothills and the eastern Karst has meant that, historically, a great variety of stone materials has been employed in the region. The sources of these materials are well documented as are the locations of earlier quarries, providing evidence that small-scale local quarrying operated throughout the region for more than two thousand years. In the modern period, from the beginning of the 20th century, these quarries were progressively abandoned. It is only recently that the area of historical building resources has started to attract new attention and detailed studies have been carried out into the use of natural stone materials, both from the geological and architectural points of view. The results of these studies are of great importance for the history of construction, and they also provide useful support for restoration. However, there is a great need for a comprehensive system to organize the considerable amount of data from historical documents and recent research. At present, much of the information available is accessible only to specialists, a situation which hinders the sharing of knowledge and the development of the field. In order to meet this need, a research project is in progress which aims to create an electronic database of the historical stone resources of Friuli-Venezia Giulia. Based on the historical listings of quarries included in two 19th century surveys of Friuli-Venezia Giulia, the database will describe the sites with existing buildings and other constructions in which the quarried materials were employed supplementing the data, where possible, with illustrations and specific references. The aim of the work is not only to demonstrate the great variety of stone resources in Friuli-Venezia Giulia and the influence of this on the history of building, but also to allow the cross-disciplinary correlation between architectural, material and geographical data. Once the database has been implemented, it is hoped that it will provide a continuously upgraded resource for future conservation and restoration projects.

Despite being a small region with an area of less than 8000 km^2, Friuli-Venezia Giulia in the northeast corner of Italy has a very varied geology. This is reflected in the diversity of the landscape between the northern mountains, the western and eastern foothills, the central moraines, the south alluvial plain and the Karst area bordering Slovenia (Fig. 1). The significant variety of stone building material available is a direct consequence of this geological variety (Fig. 2), the complexity of which is described by a stratigraphic series (Fig. 3) ranging with continuity from Ordovician to Pliocene (Taramelli 1877; Selli 1963; Martinis 1977; Abramo & Michelutti 1988; Venturini 2006). The Carnian and Julian Alps are a source of marble, carious dolomite, schist and limestone. The foothills provide sandstone and limestone, while the moraines are the source of conglomerate, clay and gravel. The alluvial plain is rich in gravel and clay. The Karst area is the source of fossiliferous limestone. Starting from this background, the on-going research presented here aims at organizing a detailed database which will highlight these geographical origins, taking advantage of the availability of a great number of detailed studies regarding the local use of stone material in architecture and other constructions. At present this information is dispersed between archives, local publications, international contributions and archaeological and historical notes.

As noted above, Friuli-Venezia Giulia presents a broad variety of stone building materials. There are detailed studies on the geology of the sites of origin, the consequence of the almost continuous stratigraphic series linking the Ordovician to the Quaternary periods, and detailed studies of specific buildings. However, there is no updated method of linking stones to buildings in a systematic and comprehensive way; the task is challenging. The project is in its early stages which are reported here, consisting of the initial collection of readily available material and its analysis. The planned database is intended to link all the material collected, correlating information on sites and buildings and providing basic mineralogical and petrographic data for each of the stone types described. A useful point of

From: PŘIKRYL, R. & TÖRÖK, Á. (eds) *Natural Stone Resources for Historical Monuments.*
Geological Society, London, Special Publications, **333**, 197–209.
DOI: 10.1144/SP333.19 0305-8719/10/$15.00

Fig. 1. Geographical map of the study area, the Friuli-Venezia Giulia region (Map of Friuli V.G., originally 1 : 150 000, Tabacco Publisher, Tavagnacco, Udine, I, 2002, modified).

reference for the project is similar work recently carried out on the stone resources of the Trentino Alto Adige region and published under the title '*Atlante della Pietra Trentina* - Atlas of Trento Area Stone' (Cattani & Fedrizi 2005).

Most of Friuli's historical quarrying sites have been abandoned and knowledge of their locations has been lost, due to the interruption in the use of traditional building materials which was a consequence of the urbanization and industrialization of the late 1950s. This lacuna was clearly illustrated by the reconstruction carried out after the 1976 earthquake, which seriously damaged the vernacular architecture of the region, as well as most of its historical town centres. The construction of new buildings saw stone walls abandoned in favour of reinforced concrete structures (Fumo *et al.* 2004). When employed, stone materials were confined to decorative elements and, for economical and logistical reasons, the choice was limited to a few local stones or to the cheapest stones of other Italian regions. The same lack of awareness characterized, and unfortunately continues to characterize, most of the restoration work in the region.

However, good examples clearly show the importance of the choice of the correct building material in restoration. A case in point is the restoration of the Royal Palace of Venaria Reale, near Turin, where the choice of original stone prompted the reopening of previously abandoned quarries.

This type of research reflects, in a certain sense, a new attitude toward local identity growing today in Italy at different levels. Thanks to increased cultural awareness, regional urban planning is showing a

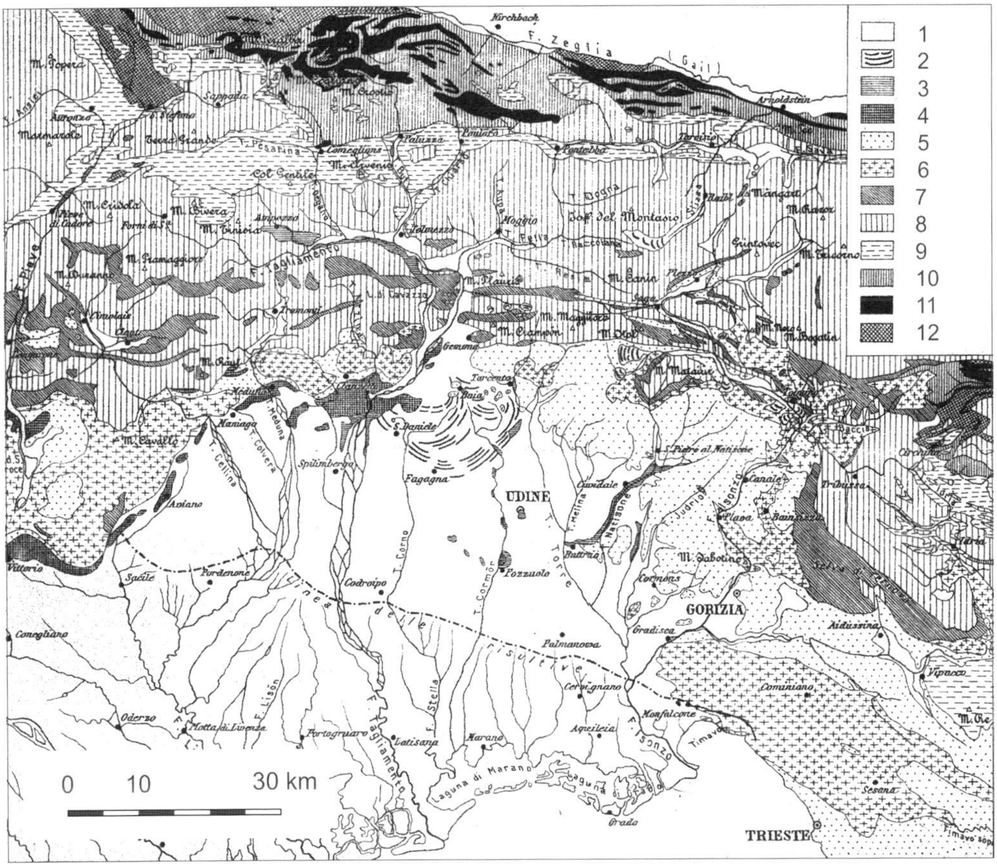

Fig. 2. Geological map of the study area. 1, Quaternary period; 2, moraines; 3, conglomerates; 4, Miocene; 5, Eocene; 6, Cretaceous; 7, Jurassic; 8, Upper and Middle Trias; 9, Lower Trias, Permian; 10, Carboniferous; 11, Devonian, Silurian; 12, phyllites (from Gortani 1926, modified).

timid return in the direction of using local resources, even if legislation and economic considerations still constitute huge obstacles.

Although the current work is limited to the response to a need for shared knowledge, both practical and historical, it also has a wider relevance at a local level. The correlation of different features of the 'spirit of place', defined as the relationship between stone buildings and their original materials, represents a new sensitivity also emerging at an international level towards the need for cultural identity. The specific traits of this identity are local, but their meaning has wider recognition.

Historical notes

Human settlement since the prehistoric period has been demonstrated throughout the region and traces of this presence of man have been revealed more and more frequently by recent studies (Pessina & Carbonetto 1998). The civilization of 'castellieri' is documented from 1600–200 BC, both in the plain (Quarina 1943; Cassola Guida & Corazza 2002, 2003) and in the Karst area (Loseri Ruaro *et al.* 1984). Different environmental conditions brought about the use of different building materials in the construction of these elementary walled camps which were built on a natural or an artificial earth elevation and gave their name to this civilization. Earth, wood and pebbles were used in the hill and plain areas, while fossiliferous limestone was adopted in the Karst area. The link between the building and the area of provision was direct, regardless of distance. Walls were part of the landscape, a feature that necessarily characterizes poor economies.

With the arrival of the Romans (1st century BC) the situation changed drastically. The beginning of the building culture in the region could be fixed at this date. For the Roman Empire, Friuli-Venezia

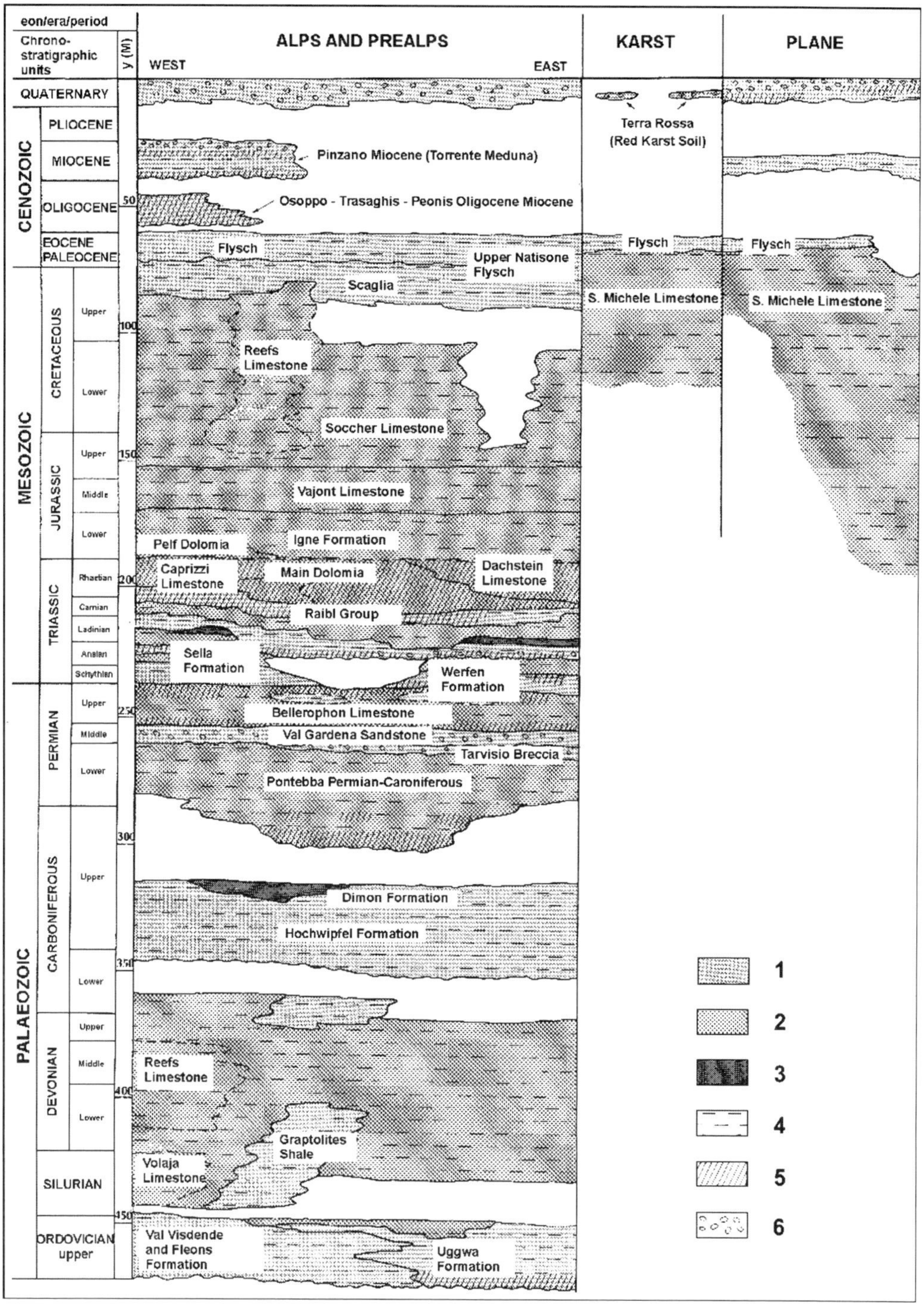

Fig. 3. Stratigraphic scheme of Friuli-Venezia Giulia region. 1, mainly terrigenous rocks; 2, mainly carbonate rocks; 3, igneous rocks; 4, marine environment; 5, transition environment; 6, continental environment (from Martinis 1977, modified).

Giulia represented the border with the east, an open door to the Pannonian plains as the frequent invasions from the 4th to the 16th century later demonstrated. The Romans established their base at the port of Aquileia, one of the most important of the empire, on the banks of the Torre river. Other important Roman settlements in the region were Zuglio – *Julium Carnicum* – a municipality in the mountain area today named Carnia and Cividale – *Forum Julii* – which lies on the banks of Natisone river at the meeting point of the eastern valleys at the border with the Slav regions.

With few exceptions in the field of sculpture, *Julium Carnicum* and *Forum Julii* present the use of local building materials, the carious dolomite of the mountains and the local fine limestone breccia of the hills (Visentini 1983). At Aquileia, on the other hand, thanks to the presence of important trade links, marbles from further away are present (Pensabene 1987) together with the Karst fossiliferous limestone and the white limestone of Istria. The presence of this Istria Stone is a feature of building construction in the region, both monumental and vernacular, throughout its history (Frangipane 2004). Its use was interrupted only in very recent years, when the Adriatic was divided by new national borders.

Aquileia was destroyed by the army of Attila the king of the Huns in 452. Since then, numerous populations crossed the ancient border of the *X Regio*, the *Venetia et Histria*, sometimes occupying the region. The history of those early centuries is still poorly known. In contrast, there is ample documentation of the reign of the Longobards who established their capital at Cividale, where they left work of great importance for the history of Italian art. Despite the considerable interest in formal features and stylistic influences, the use of construction materials in these centuries has only rarely been investigated. Little can therefore be asserted in this regard beyond suggesting that local material was probably employed as well as that coming from the dismantling of Roman buildings, as reported in archive documents (Perusini 1954).

Permanent government of the region was established under the German Empire, and ruled from the 10th century by the Patriarch of Aquileia in a continuously unstable relationship with the local nobility. It was again a period of walls: the walls of the customs barriers to German countries; the walls of the hill castles system overlooking the plain west to east; and the walls of the first markets at the junctions of the roads crossing the region which roughly followed the old Roman consular routes. The traces of these defensive structures are still present in the unrefined stone screens surrounding orchards as well as major estates. The building material employed was that found close at hand and ready to use. The variety of pebbles from the rivers and the alluvial plains is evidence of the connection between the history of local settlements and the environment. It was also the period of churches; the visible sign of a government system based on the Patriarchal power established first at Aquileia then moving to Cividale and finally establishing itself in Udine is evident. Churches of great importance, such as the Duomos of Aquileia, Udine, Cividale, Gemona, Venzone, and Trieste and churches of minor importance sprung up all around the region.

With the arrival of Venetian domination in Friuli via the province of the town of Udine in the early 15th century (1420), the scale of intervention changed. Representative buildings such as the Town Halls of Udine and Gemona, which attempt to imitate the formal presence of Istria and Verona stones with local materials (Spadea *et al.* 2000), first exemplify the Venetian taste in architecture. This was the period of urban renovation and inland investments. From that point onwards, the use of a calcareous breccia, the Piasentina Stone of the eastern hills, dominated in building construction and finishing in Udine. Istria stone, on the other hand (sometimes even sourced from quite distant quarrying areas), became the stone of major artistic work carried out by well-known artists.

An important building event which stretched over the entire 17th century was the construction of the Palmanova fortress, in the middle of the plain south of Udine. Designed to prevent Turkish invasion, its supplies of stone came from the coast via an artificial canal which was its link to the sea (Pavan 1993).

In the decades which followed, despite the passing of time, the building history of the main towns changed only in its formal aspects. In the town of Trieste, the beginning of the expansion of which can be traced to the declaration of the free market by the Austrian Emperor Charles VI in 1719 (Caputo & Masiero 1987), there was massive use of fossiliferous limestone from the coast and of local sandstone. In Udine, Piasentina Stone coupled with pebbles prevailed. Meanwhile, the development of a countless number of small quarries to satisfy immediate needs was a constant feature at local levels.

As elsewhere, the use of natural stone was interrupted by the advent of concrete. An important supply of marl was found in the region, especially in the eastern hills close to Cividale (Marinoni 1881).

Historical stone resources in the region

The availability of stone resources in Friuli-Venezia Giulia was demonstrated since the 16th century by the local historian Jacopo Valvason di Maniago (1565) who, describing the mountain area, referred to its marbles which adorned the palaces of Venice. They represented an important feature of the area as shown by a second, later reference by the local writer Nicolò Grassi (1782), describing again the mountain area. Since then, many studies have investigated the use of stone resources in the region.

However, with the exception of the two 19th century inventories described below, they addressed the topic in either quite a broad way or in a very specific one, which does not permit the definition of a complete state of the art.

Five zones can be identified in the region, each of which is characterized by the presence of one or more specific stones:

(i) the mountains to the north;
(ii) the western hills;
(iii) the eastern hills;
(iv) the moraines in the centre and the alluvial plain in the south of the region; and
(v) the Karst area in the east.

The provenance and characteristics of the main stone types are listed in Table 1.

The mountain area

The mountain area, corresponding to the *Canal del Ferro* (Iron Valley) and to the Carnian area, groups four different source areas (Carulli & Onofri 1966).

Since the earliest times, the compact Devonian limestone reefs of the Timau area have supplied a dark or pale grey marble, veined by white calcite paths, known by the names 'Timau Black', 'Timau Flowery Black', 'Carnian Dark Gray' and 'Carnian Pale Grey'. This material was widely used in the area and in the main town of Udine to create important architectural elements such as the fountain by Giovanni da Udine, a pupil of Raphael, in the New Market Square (1542).

Devonian marbles of weak metamorphism, pale grey with white and pale pink veins, come from the Forni Avoltri area. They are known by the name of 'Carnian Peach Flower'. Due to their restricted availability, they were employed mainly for decorative purposes in limited quantity, to build altars, stairs, holy water stoups and portals or window bands.

In the Lower and Middle Devonian limestone reefs of the Paularo area the 'Alpine Red', the 'Onyx Red' and the 'France Red' marbles were quarried. They are red marbles, veined by white and black intrusions. Their use was mostly limited to the area of quarrying, due to the difficulties in road transport. A type of 'Carnian Gray' was also quarried there.

The Upper Jurassic limestone reefs of Verzegnis (Tolmezzo area) are the quarrying site of the so-called 'Almond Nut', 'Brier-Root', 'Porphyritic', 'Flowery Porphyritic' and 'Vermillon Brown' marbles. Characterized by their intense red colour, they are mostly of recent use.

In the area of Socchieve (Marinelli 1898), a less valuable building material can be found: the Carnia carious dolomite locally named *tof*. This was typical not only of the local vernacular architecture but also, as previously mentioned, of the Roman remains in Zuglio.

The western hills

The stones of the western hills come from two different sources (Carulli & Onofri 1966), located in the areas of Clauzetto and Aviano. The Cretaceous reefs of the Clauzetto area supplied the stone to the masters of the Lombard school of sculpture active in Friuli during the 16th century as recently shown by archive studies referring to the portal of the duomo of Tricesimo, north of Udine, dated 1505 (Bergamini & Goi 1982). The 'Clauzetto Stone' is a fossiliferous Cretaceous limestone in which visible pale brown fossil elements are distributed in a homogeneous beige matrix. Quarried in the Aviano area, this homogeneous stone is still employed all over the region due to the location of quarries close to important road connections.

The western hills are also the source of a yellowish, fine-grained, dolomitic sandstone which was probably employed for the realization of the medieval portal of the Duomo of Udine (Spadea *et al.* 1996) and in the vernacular architecture of the area.

Also quarried there was a yellowish alluvial conglomerate which characterizes building in the town of San Daniele del Friuli, site of the Andrea Palladio's Doric Town Gate (1579).

The eastern hills

The eastern hills, situated between the towns of Buia and Cividale, are the site of quarrying of the Piasentina Stone. This is an Eocene calcareous breccia of variable component grain size. Dark to pale grey and very compact, it has a micritic matrix containing abundant quartz elements (Carulli *et al.* 1968). Its use, documented from the late Middle Ages to today, is so widespread in the historical architecture of Friuli that it makes more sense to describe what is not built with Piasentina Stone rather than what is. It is such a versatile and resistant stone that it can be adopted for almost every use. Its principle feature is the very rough, rustic-style finishing, widely employed for its effectiveness by Andrea Palladio in Udine for the Bollani Arch (1556) and Antonini Palace (1570) and in Cividale for the Provveditori Palace (1565). The size of the component grains gives the name to three different types: 'Coarse

Table 1. *Main stone types used as building and decorative material in Friuli-Venezia Giulia region*

Commercial name (*Italian*/English)		Quarrying area	Petrographic designation	Stratigraphic position
nero Timau	Timau Black	Carnian Alps (Timau district)	Compact crystalline limestone	Upper Devonian[1]
nero Timau fiorito	Timau Flowery Black			
grigio carnico scuro	Carnian Dark Gray			
grigio carnico chiaro	Carnian Pale Grey			
Fior di pesco carnico	Carnian Peach Flower	Carnian Alps (Forni Avoltri district)	Compact breccia limestone	Devonian[1]
dolomia cariata	Carious Dolomite	Carnian Alps	Dolomitic breccia	Trias[2]
rosso alpino	Alpine Red	Carnian Alps (Paularo district)	Compact breccia limestone, weakly crystallized	Middle and Lower Devonian[1]
rosso oniciato	Onyx Red			
rosso tipo Francia	France Red			
grigio carnico	Carnian Grey			
mandorlato	Almond Nut	Carnian Alps (Tolmezzo district)	Compact crystalline limestone	Upper Jurassic[1]
noce radica	Brier-Root			
porfirico	Porphyritic			
porfirico fiorito	Flowery Porphyritic			
bruno vermiglio	Vermillon Brown			
pietra di Clauzetto	Clauzetto Stone	Western Hills (Clauzetto district)	Compact limestone	Cretaceous[1]
pietra di Aviano	Aviano Stone	Western Hills (Aviano district)	Dolomitic limestone	Cretaceous[1]
'*pietra gialla*'	'Yellow Stone'	Western Hills (Aviano district)	Dolomitic sandstone	Miocene[3]
pietra piasentina	Piasentina Stone	Eastern Hills	Calcareous breccia	Eocene[4]
Vernadia	Vernadia Stone	Eastern Hills	Compact sandstone	Lower and Middle Eocene[5]
marmo di Vallemontana	Vallemontana Marble	Eastern Hills (Faedis district)	Limestone	Cretaceous (?)[5]
Chiocciolato	Snailed	Karst (Gorizia district)	Limestone	Cretaceous[6,7]
Gabria	Gabria			
nero Vallone	Vallone Black			
Repen Vallone	Vallone Repen			
Aurisina chiara	Pale Aurisina	Karst (Trieste district)	Limestone	Cretaceous[6,7]
Roman Stone	Roman Stone Aurisina			
Aurisina granitello	Granite Aurisina			
Aurisina fiorita	Flowered Aurisina			
Repen chiaro	Pale Repen			
Repen Zolla	Zolla Repen			
breccia italiana	Karst breccia	Karst (Trieste district)	Calcareous breccia	Eocene[6]
stalattite	Stalactite Stone	Karst (Trieste district)	Calcareous alabaster	Recent formation[7]
masegno	Karst Sandstone	Karst (Trieste district)	Sandstone	Eocene[6,8]
pietra d'Istria	Istria Stone	Istria	Limestone	Cretaceous[9]

Source: [1]Carulli & Onofri (1966); [2]Taramelli (1877); [3]Spadea *et al.* (1996); [4]Carulli & Onofri (1968); [5]Marinoni (1881); [6]Carulli & Onofri (1969); [7]Cucchi & Gerdol (1985); [8]Callegaris *et al.* (1999); [9]dalla Costa & Feifer (1981).

Piasentina', 'Medium Piasentina' and 'Fine Piasentina'.

Vernadia Stone is a compact sandstone originating from the Lower and Middle Eocene periods (Marinoni 1881). Historically, it was quarried at the same sites as Piasentina Stone and employed in vernacular architecture and in the simplest town buildings. A white Cretaceous limestone named 'Vallemontana marble' was quarried in a narrow valley in the area up to the 1950s and employed in small quantities in the town of Udine. It is one of many examples of exploitation of a stone resource located in a unique site, whose historical background is often difficult to define due to its limited use in time and space. Such resources, however, have an important role in the precise definition of building material employed in the region. It is also the case for St Agnese stone, which comes from the mountains overlooking the town of Gemona. It is a pale red ammonite limestone, definitely used for the columns of the Duomo and for the dressing of the local Town Hall and possibly brought to Udine for the dressing of the earliest façade of the Town Hall. In this case, as in others, the lack in archive documents makes it impossible to specify the origin of the stone. This has prevented, up to now, interesting considerations of the trade in building materials.

The moraines and the alluvial plain

Pebbles of various sizes, abundant both in the moraine hills, in the southern alluvial plain and in gravel river beds, have always been employed in these areas for the construction of bearing walls in conjunction with heavy corner stones.

In different geological periods, stones from mountain desegregation have been transported to all parts of the region by its most important rivers: the Tagliamento and the Torre. The diversity of mountain stone materials is interestingly reflected in the variety of pebbles constituting the river beds, the moraines and their alluvial plains. These pebbles are a feature of both vernacular architecture (Bertagnin & Frangipane 2005) and important buildings; however, they often constitute a point of weakness in bearing structures as the destructive effects of the 1976 earthquake demonstrated, due to the absence of cohesion between the elements and the presence of mortars poor in lime.

Huge erratic boulders of various origin, present both within the glacial and alluvial cores, were also employed for particular architectural components such as corners and portal base blocks.

The Karst area

The formation of the calcareous plateau which constitutes the Karst area ranges from the Upper Cretaceous to the Eocene period (Carulli & Onofri 1969). On the basis of the different materials quarried there, the Karst can be divided into two areas located in the provinces of Gorizia and Trieste.

The Gorizia Karst was the site of quarrying of a limited number of limestones which were highly appreciated for their aesthetic value: the dark limestones known as 'Snailed' and 'Gabria' which have brighter visible fossil inclusions; the 'Vallone Black', a dark bituminous limestone of organic origin with no visible fossils; and the 'Vallone Repen', a compact pale brown limestone with abundant darker visible fossils, both long and narrow.

The Trieste Karst, on the other hand, is an area of very extensive quarrying (Valussi 1957) as documented by the remains in Aquileia or by the traces of carving tools visible in the 'Roman quarry' in Aurisina, overlooking the Trieste coast (D'Ambrosi & Sonzogno 1962). Recorded by Scamozzi in his treatise (1615), but almost abandoned until the 18th century, the Trieste Karst quarries subsequently became the principle source of supply of building material for the development of the harbour and the town of Trieste. Following the construction of the Trieste–Vienna railway, completed in 1857, the Trieste Karst limestone was exported all over the Austro-Hungarian Empire and used in important buildings such as the Opera Houses in Budapest and Graz and the New Empire Palace and the Parliament House in Vienna (Cucchi & Gerdol 1985). The limestones quarried here are the 'Pale Aurisina', the 'Roman Stone Aurisina' and the 'Granite Aurisina', all pale brown limestones of different shades with darker minced fossil fragments of variable dimensions. 'Flowered Aurisina' is a compact pale brown limestone with visible darker fossil elements. 'Pale Repen' and 'Zolla Repen' are grey limestones of different shades with darker, visible, long and narrow fossils. 'Karst breccia' is a compact breccia limestone, basically pale brown in colour with white, pale pink and brown elements. In the Trieste Karst yellow and red Stalactite Stone was also quarried. This is a rare and precious limestone of chemical origin, formed by the deposition of carbonates in underground caves.

The local sandstone, quarried in the extended Eocene formations, is commonly used across this region (Callegaris *et al.* 1999) and mainly employed in wall construction.

Stone from more distant sources

A summary of the historical stone resources of Friuli-Venezia Giulia cannot omit a fundamental reference to the use of Istria Stone, the white fossiliferous Cretaceous limestone widely employed in Venetian architecture since the Middle Ages (dalla

Costa & Feifer 1981). In Friuli-Venezia Giulia, Istria Stone had a very important role in the typical architecture of the areas of Venetian influence in the province of Udine. Istria Stone was chosen for the dressing of church façades such as those of the Church of St James (early 16th century), the Manin Chapel (early 18th century) by the Venetian architect Domenico Rossi, the Church of St Antony (1733) by the Venetian architect Giorgio Massari in Udine and the Duomo (early 17th century) at Palmanova. It also features in the portals and window frames of important palaces such as the Udine Pawn Palace, dating from the early 17th century.

The quarry inventories of the 19th and 20th centuries

The diversity and quantitative importance of stone resources was considered in detail for the first time at the end of the 19th century, above all in the studies by Camillo Marinoni and Luigi Pitacco. They compiled two important inventories which today constitute a fundamental source of data for any kind of research on the topic.

In the work '*Itinerario mineralogico del Friuli* - Friuli Mineralogical Tour' (1881) the natural science scholar Camillo Marinoni presents an accurate enquiry into the presence of minerals, thermal waters and stone resources in the area. 194 sites of building materials are listed following a watershed order, with a brief indication of their location. The inventory specifies the type of stone using the fairly simple categories of limestone, sandstone, conglomerate, marble, carious dolomite, schist and other materials, for example gypsum, *marmorino*, mill stone, fire stone and clay. The places where the rough material was worked are also provided. A separate table organizes marble varieties according to colour and sites. A detailed description of the uses precedes the inventory, citing stone dimensions, decorative stones, materials for mortar and cement and materials for brick production.

The Marinoni inventory is part of a wider work, the 'Statistical year book' of the Friuli region. It consists of a basic geological description of material without any specific reference to building use. In order to organize the available information within the on-going research project, the Marinoni survey was summarized in a working database. This specified for each location the watershed, valley, municipality, location and site. The inventory lists the name of the material, a note of any particular features and, where relevant, the presence of stone cutting and other information.

The locating of sites on a map illustrates the distribution of quarries in the area (Fig. 4). These mainly border the plain, but they are also present in some cases in the major mountain valleys. As is well known, this circumstance is a constant in the location of early quarries. The difficulties of transport lead everywhere to the exploitation of sites close to water or main road systems.

The work by Luigi Pitacco, '*Descrizione delle pietre e dei marmi che si impiegano nelle costruzioni in provincia di Udine* - Description of stones and marbles used in buildings in the province of Udine', written in 1884, was merely commercial in intent. Its aim, described in the introduction, was to help develop the exploitation of stone resources in Friuli. The work lists 126 quarrying sites, organized by municipality. A detailed description of the area of quarrying is provided, together with a precise definition of the material and of its uses.

The entry for each quarry specifies the name of the site, the relevant municipality, its altitude, the distance from and the name of the nearest public road. Prices of the stone material are provided with reference to the quarry site, to the village and to the nearest railway station. The description of the nature, quality and uses of the stones includes their common and lithological names, density, colour, common use and qualities. Extra notes complete some of the records.

As part of the on-going research, as in the case of the Marinoni inventory, a database of the Pitacco inventory was created. The locations of the quarrying sites were indicated on a map, grouping stone resources into the categories of limestone, sandstone, conglomerate, marble and others, in order to allow direct comparison with the Marinoni data. These data match almost perfectly. The minor differences which exist can be explained in terms of the different purposes of the two works. Marinoni seeks to provide a complete description of stone resources in the area, focusing on different materials irrespective of quantities. The Pitacco inventory, however, seeks to promote the quarry industry, neglecting sites of minor importance from a commercial point of view.

Thanks to the detailed references to the location of quarrying sites, these two inventories constitute a continuous reference for any kind of research regarding the use of stone in historical buildings. As a consequence of the different purposes they were compiled for, they complement each other. They provide a substantial starting point for the projected setting up of a complete database of stone resources of Friuli. They both lack, however, data concerning the eastern coast and Karst quarries, for which other sources are required.

The 20th century is characterized by the progressive abandoning of small-scale quarries and by the concentration of the exploitation of a few

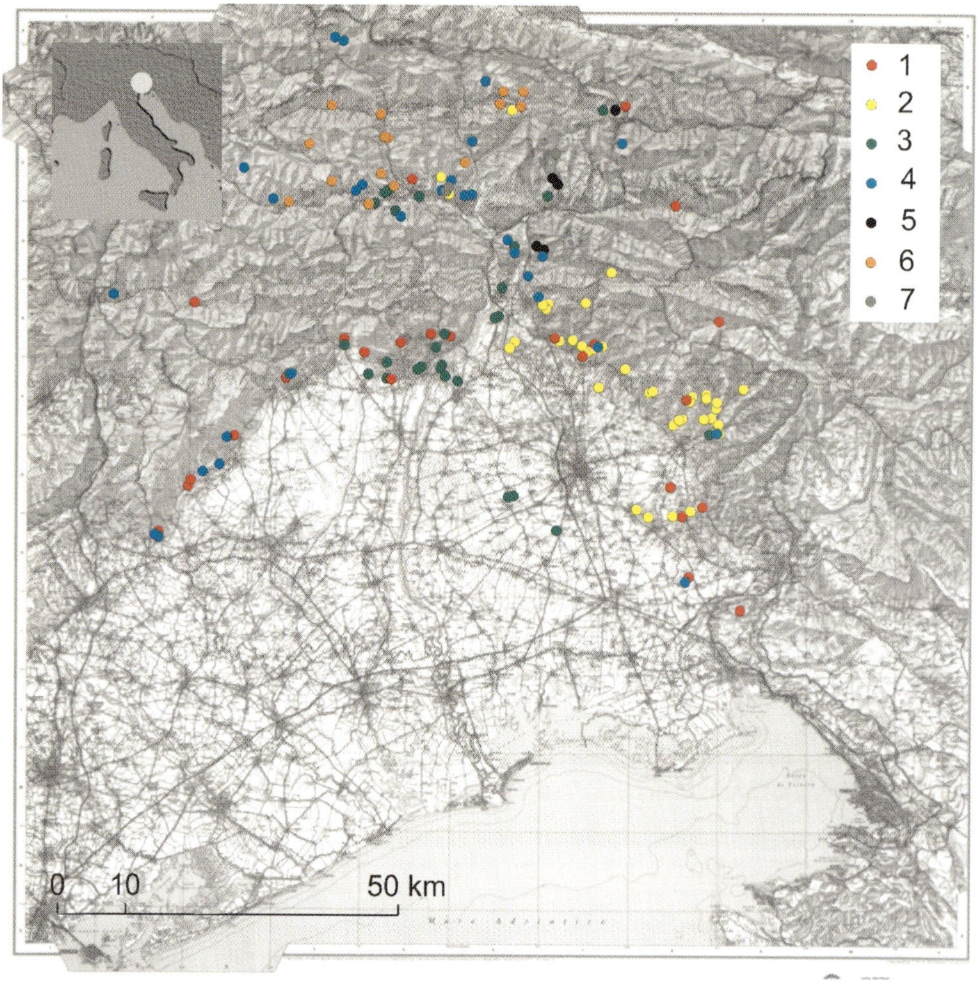

Fig. 4. Location of Friuli area quarrying sites, as reported in Marinoni Inventory (1881). 1, limestone; 2, sandstone; 3, conglomerate; 4, marble; 5, other; 6, carious dolomite; 7, schists (Map of Friuli V.G., originally 1 : 150 000, Tabacco Publisher, Tavagnacco, Udine, I, 2002, modified).

important sites which dominate the sector. This is due to the size of the deposit quarried or for logistical reasons, such as location close to the railway or to the motorway system. In addition, in the course of the century, the sector experienced a progressive reduction in the local use of building stone. This was coupled with an increasing development of export, albeit in limited quantities.

Advertising activity aimed at promoting the product was carried out in the 1960s, leading to the realization of a number of studies of regional stone resources. Those by Carulli and Onofri: '*Il Friuli. I marmi* - Friuli. Marbles' (1966) and '*Marmi del Carso* - Karst marbles' (1969) in particular allow us to draw a picture of the active quarrying sites in that period and highlight the drastic reduction in the number of quarries (Fig. 5), a trend that has continued in the last 30 years as recently shown by a study by Trevisan *et al.* (1996).

Conclusions

As described in the introduction, the research currently in progress aims to compile a comprehensive database of historical stone resources in Friuli-Venezia Giulia.

Starting from the collection of published references (of which only a limited number are presented here) and of archive documents over the whole period of the history of construction in the region, the research has focused up to now on the two

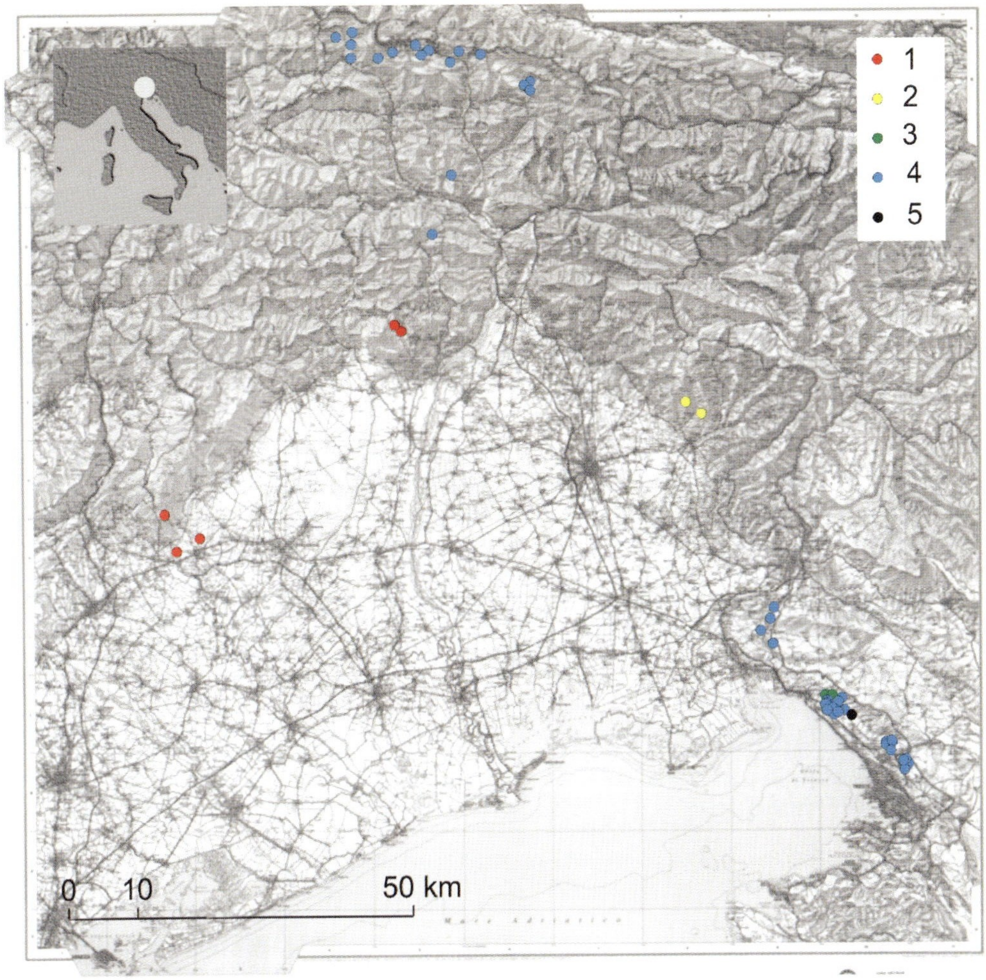

Fig. 5. Location of active quarrying sites in the 1960s, as reported by Carulli & Onofri (1966, 1969). 1, limestone; 2, sandstone; 3, conglomerate; 4, marble; 5, other (Map of Friuli V.G., originally 1 : 150 000, Tabacco Publisher, Tavagnacco, Udine, I, 2002, modified).

19th century surveys described above, organizing them in two working databases.

The next steps in the project consist of (i) developing the reference material collection by investigating further local studies, town archive documents and previous researches and (ii) surveying the quarries listed in the inventories in order to properly locate them and to collect stone samples.

Once the data collection phase has been completed the research will continue by creating, for each building with documented provenance stone data, the basic card for the planned database of materials.

The need to organize a huge amount of data suggests the idea of a starting card which will supply general information and will contain at least a picture of the building, a map locating it and a map indicating the source quarry, together with a detailed picture of the stone appearance and finishing. Basic data, such as the period or the year of realization, the author (where known) and an information box containing all the references concerning the building will also be included. They will be integrated by basic mineralogical and petrographic data.

In order to focus on the identification of stone resources, the research will concentrate on the buildings of the region for which reliable data are available. Although this will initially provide only a broad sampling, the aim is to create a working tool to be developed in the future through the addition of subsequent research results. It is hoped that the project will support professionals in future conservation and restoration planning, and promote a renewed culture of materials.

This is a challenging project, which is currently in its initial stages. The complete absence of work of this kind in the field in the region until now encourages the hope that the sharing of knowledge offered by the publishing of the cards on the web will provide a valuable resource.

References

ABRAMO, E. & MICHELUTTI, G. 1988. *Guida ai suoli forestali nella regione Friuli-Venezia Giulia.* Regione Autonoma Friuli-Venezia Giulia, Udine.

BERGAMINI, G. & GOI, P. 1982. Bernardino da Bissone a Tricesimo. *In: Proceedings of the 59th Congress of the Società Filologica Friulana, Friuli Philological Society.* Tricesimo (Udine), 26 September 1982, 351–362.

BERTAGNIN, M. & FRANGIPANE, A. 2005. From the river to the house: wall patterns in traditional buildings in the Friuli plain (NE Italy). *In: Proceedings of the International Seminar 'Theory and practice of construction: knowledge, means, models'.* Ravenna, 27–29 October 2005, 1357–1366.

CALLEGARIS, R., DOLCE, S. & BRESSI, N. 1999. *Flysch. Trieste tra marna e arenaria.* Comune di Trieste, Assessorato alla Cultura, Museo Civico di Storia Naturale, Trieste.

CAPUTO, F. & MASIERO, R. 1987. *Trieste e l'impero: la formazione di una città europea.* Marsilio, Venice.

CARULLI, G. B. & ONOFRI, R. 1966. *Il Friuli. I marmi.* Del Bianco, Udine.

CARULLI, G. B., NIMIS, G. P. & ONOFRI, R. 1968. *La Pietra Piasentina.* Del Bianco, Udine.

CARULLI, G. B. & ONOFRI, R. 1969. *I Marmi del Carso.* Del Bianco, Udine.

CASSOLA GUIDA, P. & CORAZZA, S. 2002. *Il Tumolo di Sant'Osvaldo.* Università degli Studi di Udine, Udine.

CASSOLA GUIDA, P. & CORAZZA, S. (eds) 2003. *Il Castelliere di Variano.* Comune di Basiliano, Basiliano (Udine).

CATTANI, E. & FEDRIZI, F. (eds) 2005. *Atlante della Pietra Trentina: Antichi e Nuovi Percorsi.* Guida Pratica all'Utilizzo. Nicolodi, Trento.

CUCCHI, F. & GERDOL, S. 1985. *I Marmi del Carso Triestino.* Camera di Commercio, Industria, Artigianato e Agricoltura, Trieste.

DALLA COSTA, M. & FEIFFER, C. 1981. *Le Pietre nell' Architettura Veneta e di Venezia.* La stamperia di Venezia editrice, Venice.

D'AMBROSI, C. & SONZOGNO, G. 1962. *La Cava Romana. Marmi e Pietre del Carso e dell'Istria.* Cava Romana, Aurisina, Trieste.

FRANGIPANE, A. 2004. Dimensioned Istria stone walls in the Friuli Plain (north-eastern Italy). *In: Proceedings of the 8th International Scientific Conference 'Heritage Values of Historic Structures - the History of Structures'.* Cluj- Napoca, Romania, 28–30 October 2004, 94–111 and CD-Rom.

FUMO, M., FRANGIPANE, A., DI GANGI, A. & CALVANESE, V. 2004. Thirty years of earthquakes in Italy: How intervention culture changed. *In: Proceedings of the 8th International Scientific Conference 'Heritage Values of Historic Structures - the History of Structures'.* Cluj- Napoca, Romania, 28–30 October 2004, CD-Rom.

GORTANI, M. 1926. *Guida geologica del Friuli* (including map). Stabilimento Tipografico Carnia, Tolmezzo (Udine).

GRASSI, N. 1782. *Notizie Storiche della Provincia della Carnia.* Fratelli Gallici alla Fontana, Udine.

LOSERI RUARO, L., CASSOLA GUIDA, P. & MONTAGNARI KOKELJ, E. (eds) 1984. *Caput Adiae.* La Protostoria. Comune di Trieste, Musei Civici di Storia ed Arte, Trieste.

MARINELLI, G. (ed.) 1898. *Guida della Carnia.* Friuli Alps Society, Udine.

MARINONI, C. 1881. *Sui Minerali del Friuli.* Statistical Year Book of Udine Province, Seitz, Udine, 21–84.

MARTINIS, B. (ed.) 1977. *Studio geologico dell'area maggiormente colpita dal terremoto friulano del 1976 - Geology of the Friuli Area primarily involved in the earthquake.* National Research Council, Geological and Mining Science Committee, Italian Geodynamis Project, Rome.

PAVAN, G. (ed.) 1993. *Palmanova fortezza d'Europa 1593–1993, Catalogue of the exposition.* Palmanova – Codroipo, 6 June - 15 November 1993, Marsilio, Venice.

PENSABENE, M. 1987. *L'importazione dei manufatti marmorei ad Aquileia.* Antichità Altoadriatiche, II, Arti Grafiche Friulane, Udine, 365–399.

PERUSINI, G. 1954. Aquileia ... cava di pietre. *Aquileia Nostra. Bulletin of the National Association for Aquileia*, **XXIV–XXV**, 141–142.

PESSINA, A. & CARBONETTO, G., 1998. *Il Friuli prima del Friuli. Preistoria friulana: uomini e siti.* Vittorelli, Gorizia.

PITACCO, L. 1884. *Descrizione delle pietre e dei marmi naturali che si impiegano nelle costruzioni in provincia di Udine.* Tipografia di G.B. Doretti e soci, Udine.

QUARINA, L. 1943. Castellieri e Tombe a Tumolo in Provincia di Udine. Ce Fastu? *Bulletin of the Friuli Philological Society*, **XIX(1)**, 1–2, 54–86.

SCAMOZZI, V. 1615. *L'Idea dell'Architettura Universale.* International Centre of Studies Andrea Palladio - Centro Internazionale di Studi Andrea Palladio, Vicenza, 1997.

SELLI, R., 1963. Schema geologico delle Alpi Carniche e Giulie. *Annali del Museo Geologico di Bologna "Giovanni Cappellini"*, Bologna, Series 2, 1–136.

SPADEA, P., PERUSINI, T. & FRANGIPANE, A. 1996. Dolostones used in middle age in Friuli (Ne Italy). *In*: RIEDERER, J. (ed.) *Proceedings of the 8th International Congress on Deterioration and Conservation of Stone.* Berlin, Germany, 30 September – 4 October 1996, 155–157.

SPADEA, P., PERUSINI, T., FRANGIPANE, A. & MADDALENI, P. 2000. The Loggia del Lionello of Udine (15th century): weathering of the stone facing. *In:* GALAN, E. & ZEZZA, F. (eds) *Protection and Conservation of the Cultural Heritage in the Mediterranean Cities:*

Proceedings of the 5th International Symposium, Sevilla, 5–8 April 2000, 146–147.

TABACCO. 2002. *Map of Friuli V.G.* 1 : 150 000. Authorization n. 1093 by Tabacco Publisher, Tavagnacco (Udine).

TARAMELLI, T. 1877. *Catalogo Ragionato delle Rocce del Friuli.* Salviucci, Rome.

TREVISAN, M., MENCHINI, G., ASTORI, A. & FERUGLIO DAL DAN, C. 1996. *Studio sul settore del marmo e della pietra ornamentale nella provincia di Udine.* Camera di Commercio, Industria, Artigianato e Agricoltura, Udine.

VALUSSI, G. 1957. *La pietra calcarea in Italia e nel Carso triestino.* Università degli Studi di Trieste, Facoltà di Economia e Commercio, Istituto di Geografia, Report N. 1, Smolars, Trieste.

VALVASON DI MANIAGO, J. 1565. *Descrizione della Cargna.* Jacob e Colmegna, Udine, 1866.

VENTURINI, C. 2006. *Evoluzione geologica delle Alpi Carniche.* Pubblicazioni del Museo Friulano di Storia Naturale, N. 48, Comune di Udine, Edizioni del Museo Friulano di Storia Naturale, Udine.

VISENTINI, M. 1983. Anche i Romani (forse) usavano la pietra Piasentina. *Quaderni della F.A.C.E,* **63**, 51–61.

Electronic database of historical natural stones of the Czech Republic: structuring field and laboratory data

HANA KAMPFOVÁ[1] & RICHARD PŘIKRYL[2]*

[1]*Institute of Rock Structure and Mechanics, Academy of Sciences of the Czech Republic, Prague, Czech Republic*

[2]*Institute of Geochemistry, Mineralogy and Mineral Resources, Faculty of Science, Charles University in Prague, Albertov 6, 128 43 Prague 2, Czech Republic*

**Corresponding author (e-mail: prikryl@natur.cuni.cz)*

Abstract: The structure of the electronic database of natural stones used in the Czech Republic (named *Deka*) has been completed. The system is based on MySQL, and is comprised of the following crucial items: general data, data regarding the geographic location and status of the quarry; petrographic description, associated mineralogical data and chemical analyses; technical data covering physical and mechanical properties as well as sculptural workability; parameters about the deposit (quarry); and data describing historical exploitation and utilization. Along with the structure of the database, other practical problems such as access to the database, insertions, corrections, erasing a record or reference, etc. are discussed. The database is designed for geologists, historians and restorers resolving problems of a stone's provenance from monuments, architects involved in the selection of promising stone types for new projects and for educational purposes to geologists, restorers, etc.

Appendix 1: A description of the eight types of data used in the dimension store database is provided at http://www.geolsoc.org.uk/sup18394

The use of natural stone for private, public and religious structures is a constant distinguishing feature throughout human history (Shadmon 1996). A full understanding of the cultural heritage built from natural stone requires knowledge of previously exploited resources, the typical visual features plus physical properties of the varieties of quarried stones as well as appropriate laboratory techniques that can facilitate the location of sites from where a stone type (e.g. sampled from a restored monument) was originally extracted. Regional or national potentials of natural stone are often documented in various types of collections for example, those stored in museums such as Corsi's collection of dimension stones used in antiquity, deposited at the Oxford Museum of Natural History (Cooke 2004) or stored within the archives of national geological surveys (Ehling 2004). Extensive information can be derived from directories of dimension stone deposits for certain regions and/or a specific epoch (e.g. Hanisch & Schmid 1901; Gnoli 1988; Borghini 1989; Grimm 1990; Lazzarini 2004, 2007). The increasing interest in natural stone, due to the restoration of monuments (Curran *et al.* 2006) and the development of new projects (Pivko 2004) requires a solid knowledge of the available resources as well as of the historical stone types used.

The information stored in those stone collections, deposited in archives or in previously published books (often not available) can be easily distributed to the present scientific community by their conversion into electronic databases accessible through internet web pages. Some recent papers refer to these electronic databases focused on quarried stones (Schulmannová & Skarková 2004) as well as to historical resources of dimension stone in particular regions (Giampaolo *et al.* 1998, 1999; Uhlir *et al.* 2004). These, and many others sources, highlight the fact that electronic databases present one of the most appropriate means for data storage and presentation.

This paper focuses on the possible presentation of a broad range of data collected during both field and laboratory work on the historical dimension stone resources of the Czech Republic (Přikryl *et al.* 2001, 2002, 2004; Přikryl 2007). This paper endeavours to provide more detailed insight into the type of data included in the *Deka* database, based on the structure and benefits for potential users. After integrating heterogeneous data (e.g. mineralogical, petrographical, geochemical, rock physical and mechanical properties and technological and exploitation), the database should serve as a primary source for researchers involved in the exploration of natural stone deposits, testing of

From: PŘIKRYL, R. & TÖRÖK, Á. (eds) *Natural Stone Resources for Historical Monuments*. Geological Society, London, Special Publications, **333**, 211–217.
DOI: 10.1144/SP333.20 0305-8719/10/$15.00

their properties or searching for source localities during the pre-restoration research of historical monuments (Přikryl 2007). The information stored in this database might also be of interest to teachers, restorers, sculptors and architects (who might explore the data on stone workability and appearance for the selection of different stone varieties for their current projects). Although it is currently planned that some of the basic data will be published in the form of an atlas, the analytical data is still expanding and can more easily be presented in electronic form. The lithotheque is expected to comprise 800–1000 different types of natural stones quarried within the Czech Republic in the past upon completion.

However, the database will be capable of comprising about 65 000 records without any need to change its structure. The database contains information on the geographic location and state of the quarry, petrographic descriptions including mineralogical data and chemical analyses, technical data covering physical properties and sculptural workability, parameters of the deposit (quarry) and data related to the stone's historical exploitation and utilization.

Elements of electronic databases

The use of information systems (IS) in the form of electronic databases for data storage and administration provides numerous advantages to possible users. Database software is optimized for searching within various types of data, providing referential integrity of the data and protecting against unauthorized use.

The most effective IS are constructed on the principle of a powerful server and simple client computers, without any loss of user-friendliness. First, these systems should work on the internet interface. The web browser comprises part of each modern operating system. Databases are managed by Database Management System (DBMS) programs. Along with expensive commercial software such as Oracle, MS SQL Server, Sybase or InterBase, there are also freeware alternatives (mSQL, MySQL or PostgreSQL) that can be used.

Indirect access to the database is managed through scripts which are part of HTTP protocol (together with HTML instructions). The most extensive script language is PHP, in contrast to the similar ASP, is distributed as freeware and is independent of platform; both Unix (Linux) and MS Windows PHP versions are available.

PHP is not bound to a specific server, as it can be run on any of them. A script is connected to the database. According to its functions and the user settings, it sends a requirement such as an SQL command to the database server. The server responds with the required results to the PHP script. After compilation of the data, the script displays them to the user in a nicely organized form. DBMS has no possibility of solving all tasks, because each database server contains some of its own protocol over which PHP scripts (clients) are used to communicate with it. Therefore, a client must support the protocol of each database server. This obstacle is resolved by the universal database interface, which allows the transfer of requests to the database server in a standardized form. It represents the possibility of transferring scripts between individual database servers without any likelihood of script code modification.

Database of dimension stones

System used

The electronic database of Czech natural stones is designed to be run using the MySQL freeware system. This system, like most of the DBMS, is based on a relational database model. This consists of one or more tables (collecting certain types of data) composed of columns (items, attributes) and rows (records). In the case of the use of additional tables, each must have a unique name and a foreign key which represents for their proper correlation to other tables in a database. Each record is uniquely identified by a primary key.

Structure of dimension stone database

The electronic version is based on the previously published structural details of the Czech dimension stone database (Přikryl 2007; Přikryl *et al.* 2001, 2002). This database consists of several types of data divided into eight basic groups (see Appendix 1 in supplementary publication), upon which the individual tables of the electronic database are based.

The *rocks* table comprises the most important part of the database. It includes 77 items providing data on the stone's properties such as mineralogical composition, physical and technological parameters, location (including geographical coordinates), exploitation data and historical utilization. The attribute *id* represents the primary key for this table. It is determined by the order of the rock entry. This number is not changeable, and is assigned to the rock automatically (it is of the *auto increment* type). For items that must be presented in graphic form, the database table called *rocks* includes the items *graph1* and *graph2*, allowing for the insertion of graphical files (e.g. X-ray diffraction patterns, fracture orientation diagrams, etc.).

This graphic documentation for each record can be accompanied with additional figures. This data is gathered in the *graphs* table. The HTML code enables these items to be displayed in both *.jpg and *.gif formats.

Independent tables exist for any numerically expressed chemical (and other technical) data. The structure is built to accommodate an arbitrary number of analyses. In this specific database, the number is restricted to 255 by the type *tinyint*, which is the smallest numerical type. For the chemistry of natural stone, major, minor and trace elements are displayed in two tables denominated *an_hlpr* (major elements) and *an_stpr* (minor and trace elements, if available). When dealing with specific rock types such as sedimentary and crystalline limestones and/or carbonate-rich rocks (e.g. marlstones), the database offers the possibility of displaying results of carbonate isotopic composition (*an_stis* database table) and carbon-sulphur analyses (database table *an_csan*). Specifically, isotopic data are of great importance for the correct analysis of stone provenance (Craig & Craig 1972). For impure carbonate-rich rocks, the data on non-carbonate phases are displayed within the *an_nezb* database table, giving the phase composition of the insoluble residuum.

The practical use of the stone, its performance in use as well as resistance to weathering agents (durability) are highly dependent on its physical properties (e.g. Currier 1960). This broad family of physical parameters is covered by the *an_tevl* database table that allows the input of minimal, average and maximal values for each property.

A list of references is also attached to every rock's description. Each rock can have a different number of citations which may be repeated. Two independent tables are therefore necessary: *references* and *cc*. The *references* table includes citations with their primary ID key, which is cross-referenced with the ID key of the rock in the *cc* table.

User access to the database

Although the database can be controlled directly by MySQL commands, this approach is quite complicated. Permission to access the database is provided through the *security.php* script implemented on all PHP files. This script starts by the command *require* which guarantees, contrary to command *include*, that processing will be halted in the case of the absence of the *security.php* file script. The accidental or wilful removal of the security file or any unauthorized access to the database is therefore impossible.

Access to the database by entering one's name and password presents the first basic level of protection. The second more important level of security is hidden from the user. It represents direct entry into MySQL database. The *function.php* file manages access for three different levels of users: 'readers', 'writers' and administrator. Due to the fact that running trials proceed on a local server, it was not necessary to set user names and passwords. In ordinary usage, it will be necessary to protect the contents of this function from improper use.

After the successful run of the security subroutine, the PHP script continues with the functional processing. The commonly used functions and constants are summarized in one file *function.php*, which enables rapid changes of the project parameters.

The *maxprovypis* parameter determines how many items will be printed on the screen. If such a limitation was not set, then the searching would proceed through more records and the system could eventually collapse. The *directory* parameter sets the implicit path to the directory in which graphs and other image documentation will be saved. The *maxfilesize* restricts the maximum size of the *.jpg and *.gif files. Potential deliberate or inadvertent overloading of a disc with stored data is treated by this command.

The parameters *max_length_search* and *length_reference* indicate the maximum length of printed records. In the case where real length surpasses the maximum value, the text on the monitor is shortened and only displayed in the full version when fistnote is set to be an asterisk (by clicking on it). The *number_of_graphs* indicates the maximal count of other image documentation, which it is possible to assign in the rock's settings. It is possible to add other image documentation, which exceeds this limit in the optional *correct record*. Parameters *number_of_samples[xxxx]* were defined to simplify and clarify the cycles which process data from the tables of chemical and technological properties. In the case when some new data on technical properties is added, it will not be necessary to repair all sequences which process these data.

Working with the natural stone database

General

After a successful login, the processing of scripts follows. It will automatically start the processing of the scripts of which *Main.php* is the first. This script, analogous to *up.php* (top bar) and *down.php* (bottom bar), is the guidepost for the following tasks which can be accessed: add record; search record; repair record; repair citation; and logout.

Inserting a new record (new stone type)

A primary script, which manages the insertion of a new stone type, is *insert.php*. This page enables the user to insert all 77 items in the *rocks*, *graphs*, *analytical results* and *references* tables. Due to the huge amount of data, some categories (chemical composition of main elements, accessory and trace elements, composition of isotopes, results of carbon-sulphur analyses, values of insoluble residua, technological properties and the insertion of already-used citations) must be separated into independent pages in new windows. The opening-up of new pages and the transferring of data between them are allowed by JavaScript. The JavaScript instructions are directly integrated into the HTML code. These are directly interpreted in the user's browser (in contrast to PHP script). Therefore, the set of rocks could be partly restricted if the user has prohibited execution of JavaScript in their browser.

The sheet 'Inserting rock' is subdivided into the following units: trade name; general specification; petrographic description; mineralogical data; petrographic classification; chemical composition; technological properties; architectural and sculptural workability; geology of the deposit; exploitation and documented use; notes; other figures; and list of references.

These units contain particular items (for details see Appendix 1, sup18394). Most of these items are inserted as a text string, maximal length 255 symbols (*varchar*). Items, which cannot be squeezed into this 255 symbols limit can be inserted in the text area (*textarea*) with a maximum size of 65 535 symbols. These items can be left empty if such data are missing.

The main 'send' button delivers all required data to the database. If the record is loaded into the database correctly and no failure message appears, the window automatically returns to a blank field for insertion of a new rock's data entry.

Searching in the database

Searching for a specific rock (dimension stone) type will probably be the most common application by the end user. Quick searching can be easily performed by using the primary key of rock in the table *rocks* (see above). Other search criteria can also be used.

Using limestones and other carbonate-rich rocks as an example, the main window allows searching according to: name of the rock; locality; type of the rock; X-ray diffraction of insoluble residuum; petrographic classification; geological position; exploitation period; notes; and references.

The next categories are displayed in the new window after clicking on the relevant link. After inputting the data within a secondary window and clicking the 'Save' button, the data are moved into the main window by JavaScript. They are then placed into hidden variables types.

It is not necessary to know the entire or exact content of the category. The missing portions of a string can be replaced by the '%' symbol. In the *search.ph* program, the '%' symbol is automatically supplemented after all text strings with the exception of the name of the rock which is also in front of all text strings. In the case when a rock name is unknown, the symbol '%' must be entered manually. This adaptation was selected because of the possibility to print-out a rock whose name begins with a specific letter or string (without specifying the full name or string). A MySQL query is not case sensitive.

In categories containing numeric data, there is no requirement to know the exact values. The search proceeds by setting lower and upper boundaries. When one of the limit values is not stated, all values which lie above (or below) the entered number will be chosen from the database. If the exact value is known, then the parameters 'from' and 'to' equal to this value will be set. The 'workability closely' category presents the only exception. The range of the search here is set by the insertion of a dash.

Correcting and erasing a record

Correction of an incorrectly inserted record or completion of an incomplete record can be performed through '*correct record*'. The executive supervisory program for this action is *correct.php*. First, the rock must be selected from all of the records recorded. This print-out is divided by x items (x is defined in *function.php* file through the parameter *maxprovypis*) to increase clarity. It is possible to filter the printout by the name of rocks, for more facile orientation.

After clicking on the 'correct' button, the window will open with the selected rock. This window is similar to the form of the original of data input. It is possible to correct and complete all data, including the addition or deletion of all image documentation and citations.

If the whole record is unsuitable for any reason (e.g. it is a double entry), clicking on the 'delete' button will erase it from all tables in the database. When deleting the last record, any subsequent entry will then receive the (newly) unblocking ID number. In any other case, the ID number will not be affected.

Correcting and erasing a reference

This function is similar in the organization of its structure to that of the correction and deletion of a record. The executive supervisory program is *reference.php*. It is also possible, as in the previous case, to filter the list for citations and consequently to correct or delete them.

Discussion

General

There are many stone 'libraries' and databases in the world which are being digitalized step-by-step. The collection of the Federal Institute for Geology and Natural Resources in Germany (Ehling 2004) can be cited as an example. The main disadvantage of these is their non-accessibility, which is limited to the possibility of personal arrangements in Berlin. Other interesting projects are the database established by Austrian researchers (Uhlir *et al.* 2004) focusing on stone varieties used in the Alpine region during Roman times or the Italian database of decorative stones Italithos, which is focused on rock types quarried in Italy (Giampaolo *et al.* 1998, 1999). In the Czech Republic, the 'Internet database of decorative and dimension stones' project, which is run under Oracle, is mostly focused on recently quarried varieties; it was initiated by the Czech Geological Survey (Schulmannová & Skarková 2004).

Most of the dimension stone databases are presently still under development. Evidently, there is need for coordinated actions among these various groups. This is not a simple issue, however, Each region and country has its unique geological situation and mining history. It is therefore clear that each database aims to fit 'its' own conditions; it will not be straightforward to find a common structure within the immediate future. For instance, Stein (2004) touched upon the idea of an International Register of rocks used as dimension stone. According to this author, the system must offer a neutral and invariable mode of communication which will guarantee the possibility of cooperation over all different levels of the dimension stone industry.

Material v. virtual dimension stone databases

Material collections of products of nature including dimension stone (e.g. Cooke 2004) comprise important parts of museum collections. The museum collections are generally not easily accessible for broad research tasks, especially when 'destructive' tests (e.g. chemical analyses, physical rock testing) are required. The major advantage of a material stone archive (a lithotheque) is apparent when contact with real stone is required. This is important for example, for restorers, stonemasons and also for geologists. The major problem of such a collection, on the other hand, resides in its physical deposition in a unique place and its inaccessibility to most other potential worldwide users. These collections cannot be multiplied, and these depositions require enormous space requirements.

Written and published data (a book or atlas on dimension stones) presents another possibility for data presentation. These can be read without the necessity of any additional equipment, but older editions are often no longer available (Grimm 1990) or do not contain colour plates indicating the rock's appearance (Hanisch & Schmid 1901).

The uniqueness of electronic databases, accessible through the internet, lies in their global availability. This advantage has been recognized during the last decades and is broadly used in many branches of the Earth sciences such as geology (Baxter & Horder 1981), mineralogy and crystallography (Wang *et al.* 1994; Chicagov *et al.* 2001; Hoppe & Ruck 2004), hydrology (Henriksen *et al.* 2003) and astrogeology studies (Arvidson *et al.* 2003).

Linguistic issues

An important point of any globally available database is the language(s) used within it. Although English has became the dominant and universal language in the sciences during the past decades, a stone database designed for a certain region (country) where other national language(s) are spoken must also be available in this language. However, the English version would surely increase its world-wide attractiveness/usefulness. For this reason, the Czech natural stones database is to be launched in both Czech and English.

The use of different languages can bring certain difficulties that arise from the unique linguistic features of the individual languages. For example, the Czech characters č or š, when sorting the records alphabetically in English, must occur just after c or s and not after z. In the database described in this paper, such problems can be resolved by changes in the set-up of the configuration file. In the Windows system it is possible to find this file in the main directory *C:* as a *my.cnf.*; after opening in some text editor it is necessary to replace the *latin1* encoding by *win1250* at line 'default-character-set = latin1'. For this change to be accepted it is necessary to reboot MySQL. However, the current system is still not able to resolve the problem with the unitary Czech letter 'ch', which the system considers as the combination of the English letters 'c' and 'h'.

Comparison to other stone electronic databases

The database presented in this paper is designed for a maximum of 65 535 records of stones. If such a range is not sufficient, it is possible to extend it up to 2^{64} records without problem. The number of Czech stone types is not expected to exceed 1000 records according to the previous archival, literature and filed studies of historical stone resources (Přikryl *et al.* 2001, 2002, 2004).

It is clear that a search must suit this huge amount of data. As mentioned above, the user has the possibility of searching in every category simultaneously, in contrast to the Italian database (www.italithos.uniroma3.it). The search efficiency has therefore been remarkably enhanced. As the search is probably the most complex part of the program, it also presents the most likely source of possible errors. The effort involved in providing this complex searching tool led to the complicated structure of the current program. The search is performed at several levels, which must then be connected together. These interconnections were the weakest link in the program during testing, and deserve further and careful assessment.

In the category of chemical composition, technological properties and architectural and sculptural workability, entered in numerical values, the searching is carried out by means of intervals. The user enters the lower or/and upper boundaries of these intervals. The Italian database also enables searching by means of numerical ranges of values (Giampaolo *et al.* 1998, 1999).

In the *Deka* database, similar to other decorative stones databases (for example the database of Italian decorative stones at www.italithos.uniroma3.it), it is possible to search by different categories. However, there are some differences in searching methods. In the Italian database, one has to determine the area of those categories over which the search will proceed. The great disadvantage of this approach is the impossibility of combining separate sub-areas of the main categories. In the *Deka* database, searching proceeds across all categories. In the case that all categories are employed in one search form, it would go unnoticed and become complicated. Hence, JavaScript was used to divide particular categories into 'sub-windows'.

Errors can appear directly in the source program but can also be caused by the programming environments of MySQL and PHP. At present, version 5 is available for PHP (Release candidate 1 from 18th March 2004). For the database of the decorative stones of the Czech Republic, we have used the PHP Triad 2.2.1 for Windows package version 4.1.1, version 3.23.47 of MySQL and version 1.3.23 Apache (HTML environment). If the database is run in different versions or on different systems (Windows v. Unix) there could possibly be some compatibility problems.

Conclusions

The major advantage of an electronic version of a natural stone database lies in the integration of various data types and for case of accessibility for the various experts working in the field of natural stone research. The data integrated in such a system are otherwise split through many different sources (books, unpublished research and technical reports, producer data leaflets, etc.) that are generally unavailable. On the other hand, the electronic databases provide easy user access and the possibility to expand, renew or change the data presented by the database user.

The electronic database of natural stones would be of major interest to three specific groups of users. In no special order of importance, the first encompasses those geologists that work professionally with natural stone in searching for new or previously used resources, looking for source localities when carrying out materials research on samples taken from restored monuments or when dealing with weathering/durability studies. Restorers and technologists of cultural heritage structures represent the second group of potential users. For this group, the database might be beneficial when containing information on the durability and general stone properties (composition to correctly select the cleaning and conservation systems, physical properties to evaluate degradation progress and/or effectiveness of various conservation treatments). Thirdly, the database might also be of interest to architects and historians working in the field of monument care and cultural heritage. For this group of potential users, the information on historical sources, periods of their usage and the stone's application in specific monuments would be of importance and interest.

Although the structure of the Czech natural stone database has been successfully tested, any software application is in some sense 'a living organism' and nobody can expect its finishing point ever to be reached. In particular, databases are sensitive to various combinations of input parameters, entered by several users.

This study has been financially supported by the Ministry of Education, Youth and Sport of the Czech Republic (project No. MSM 0021620855 'Material flow mechanisms in the upper spheres of the Earth' and project No. MSM 520000001 'Research and development project in restoration of sculptural artistic monuments based on historical and current knowledge'). The comments from anonymous reviewers helped to improve the text.

References

ARVIDSON, R. E., ANDERSON, R. C. *ET AL.* 2003. Physical properties and localization investigations associated with the 2003 Mars Exploration rovers. *Journal of Geophysical Research – Planets*, **108**(E12), 8070.

BAXTER, G. G. & HORDER, M. F. 1981. Databanks and databases in geology – introduction. *Journal of the Geological Society*, **138**, 573–574.

BORGHINI, G. (ed.) 1989. *Marmi Antichi.* De Luca ed., Roma.

CHICHAGOV, A. V., VARLAMOV, D. A., DILANYAN, R. A., DOKINA, T. N., DROZHZHINA, N. A., SAMOKHVALOVA, O. L. & USHAKOVSKAYA, T. V. 2001. MINCRYST: a crystallographic database for minerals, local and network (WWW) versions. *Crystallography Reports*, **46**(5), 876–879.

COOKE, L. 2004. The shared passion for stone of William Spencer Cavendish, 6th Duke of Devonshire, and Faustion Corsi, a lawyer in Rome. *In*: PŘIKRYL, R. & SIEGL, P. (eds) *Architectural and Sculptural Stone in Cultural Landscape*. Charles University in Prague, The Karolinum Press, Prague, 41–51.

CRAIG, H. & CRAIG, V. 1972. Greek marbles: determination of provenance by isotopic analysis. Science, **176**, 401–403.

CURRAN, J., SMITH, B., STELFOX, D., SAVAGE, J. & WARKE, P. 2006. Natural stone database for northern ireland: a research-industry partnership funded by EU building sustainable prosperity programme. *Geophysical Research Abstracts*, **8**, paper No. EGU06-A-04782.

CURRIER, L. W. 1960. Geological appraisal of dimension-stone deposits. US Geological Survey, Bulletin 1109, United States Government Printing Office, Washington.

EHLING, A. 2004. Dimension stone collection at the Federal Institute for Geosciences and Natural Stone Resources (BGR). *In*: PŘIKRYL, R. (ed.) *Dimension Stone 2004*. Taylor & Francis Group, London, 123–125.

GIAMPAOLO, C., DE RITA, D., CAPICOTTO, B. M., GODANO, R. F. & DI PACE, A. 1998. Un data base sui litotipi italiani. *L'Informatore del Marmista*, **443**, 31–39.

GIAMPAOLO, C., ROMANI, G., PAZZELLI, L. & DI PACE, A. 1999. Un "data base" sui litotipi italiani per i beni culturali. esempio: il "marmo" di cottanello. *Plinius*, **22**, 202–203.

GNOLI, R. 1988. *Marmora Romana*, 2nd edn. Dell'Elefante, Rome.

GRIMM, W. D. 1990. *Bildatlas wichtiger Denkmalgesteine der Bundesrepublik Deutschland*. Bayerisches Landesamt für Denkmalpflege, Arbeitsheft 50, München.

HANISCH, A. & SCHMID, H. 1901. *Österreichs Steinbrüche*. Verlag von Carl Graeser & Co., Wien.

HENRIKSEN, H. J., TROLDBORG, L., NYEGAARD, P., SONNENBORG, T. O., REFSGAARD, J. C. & MADSEN, B. 2003. Methodology for construction, calibration and validation of a national hydrological model for Denmark. *Journal of Hydrology*, **280**(1–4), 52–71.

HOPPE, D. & RUCK, M. 2004. A crystallographic database for minerals. *Angewandte Chemie – International Edition*, **43**(45), 6024–6024.

LAZZARINI, L. (ed.) 2004. *Pietre e Marmi Antichi*. Cedam, Padua.

LAZZARINI, L. 2007. *Poikiloi Lithoi, Versiculores Maculae: i Marmi Colorati della Grecia Antica*. Accademia Editoriale, Pisa.

PIVKO, D. 2004. World's quarries of commercial granites – localization and geology. *In*: PŘIKRYL, R. (ed.) *Dimension Stone 2004*. Taylor & Francis Group, London, 147–152.

PŘIKRYL, R., SVOBODOVÁ, J. & SIEGL, P. 2001. Search for historical resources of dimension stone in the Czech Republic. *In*: SANDRONE, R. (ed.) *Proceedings of the International Workshop "Dimension Stones of the European Mountains"*. June 10–12, 2001, Luserna san Giovanni – Torre Pellice (Italy), 307–309.

PŘIKRYL, R., SVOBODOVÁ, J. & SIEGL, P. 2002. Provenancing of dimension stones and "Atlas of monumental stones of the Czech Republic". *In*: PŘIKRYL, R. & PERTOLD, Z. (eds) *Česká Ložisková Geologie na Počátku 3. Tisíciletí*. Charles University in Prague, Faculty of Science, Prague, May 20, 2002, 105–110.

PŘIKRYL, R., SVOBODOVÁ, J. & SIEGL, P. 2004. Historical dimension stone resources in the Czech Republic. *Roc Maquina*, **53**, 28–31.

PŘIKRYL, R. 2007. Understanding the earth scientist's role in the pre-restoration research of monuments: an overview. *In*: PŘIKRYL, R. & SMITH, B. J. (eds) *Building Stone Decay: From Diagnosis to Conservation*. Geological Society, London, Special Publications, **271**, 9–21.

SCHULMANNOVÁ, B. & SKARKOVÁ, H. 2004. Internet database of decorative and building stone. *In*: PŘIKRYL, R. (ed.) *Dimension Stone 2004*. Taylor & Francis Group, London, 153–155.

SHADMON, A. 1996. *Stone: An Introduction*. Intermediate Technology Publications, London.

STEIN, J. 2004. Introducing an international register of natural stone. *In*: PŘIKRYL, R. (ed.) *Dimension Stone 2004*. Taylor & Francis Group, London, 157–162.

UHLIR, C. F., SARTORI, A., MÜLLER, H. W., HEMMERS, C. & TRAXLER, S. 2004. SRI – A comprehensive web-database for Roman stone monuments. *In*: PŘIKRYL, R. (ed.) *Dimension Stone 2004*. Taylor & Francis Group, London, 163–167.

WANG, A., HAN, J. Y., GUO, L. H., YU, J. Y. & ZENG, P. 1994. Database of standard Raman-spectra of minerals and related inorganic crystals. *Applied Spectroscopy*, **48**, 959–968.

Ornamental stones in the cultural heritage of Campania region (southern Italy): the Vitulano marbles

FRANCESCO ALLOCCA[1], DOMENICO CALCATERRA[2]*, GABRIELLA CALICCHIO[1], PIERGIULIO CAPPELLETTI[1], ABNER COLELLA[1], ALESSIO LANGELLA[3] & MAURIZIO DE' GENNARO[1]

[1]*Dipartimento di Scienze della Terra, Università degli Studi di Napoli Federico II, Naples, Italy*

[2]*Dipartimento di Ingegneria Idraulica, Geotecnica ed Ambientale, Università degli Studi di Napoli Federico II, Naples, Italy*

[3]*Dipartimento di Studi Geologici e Ambientali, Università del Sannio, Naples, Italy*

**Corresponding author (e-mail: domenico.calcaterra@unina.it)*

Abstract: The marbles exploited between Vitulano and Cautano (Benevento province, Campania region, Italy), have been widely utilized (at least from the 18th to the beginning of the 20th century) in several important monumental buildings of the region, due to their good properties and peculiar aesthetic qualities. They are limestones deriving from the filling of palaeo-cavities, carved into an emerged Cretaceous calcareous platform, by calcareous breccias, bauxite and alabastrine deposits. Such cavities are the result of both jointing and karst processes. The studied outcrops of the Vitulano Marbles are affected by a complex joint pattern in terms of attitude and spacing, which results in rock blocks highly variable in volume. Regarding the mineralogical composition, calcite is definitely predominant whereas the insoluble residue is constituted by low amounts of dolomite, bohemite, ematite and kaolinite. The petrophysical characterization put in evidence fairly good geomechanical properties, only partly affected by the geostructural anisotropy of the rock.

Campania region (Southern Italy) is an area characterized by few important deposits of ornamental stones. Historically, a much wider range of local stone types was used for architecture. Both technical and aesthetical characteristics of local marbles contribute to the ornamental and structural quality of historical buildings.

A complete characterization of these materials is still in progress and the present research will take into consideration the so-called 'Vitulano and Cautano Marbles'. In fact, notwithstanding the wide use of this rock in the architectural heritage of Campania, minor information about the mineralogical-petrographical composition, physico-mechanical features and the weathering phenomena are so far available.

This research aims to acquire all this technical information in order to possibly identify the old exploitation sites (in view of a possible commercial 'rediscovery'), and to evaluate the final uses and potential of the deposits. Such a complete set of information will be helpful in the case of restorations and possible uses of this rock in relevant architectural works (Di Girolamo *et al.* 2000; de Gennaro *et al.* 2003; Calcaterra *et al.* 2004, 2007; Carta *et al.* 2005).

Previous use of Vitulano marbles

Since the end of the 19th century, the Vitulano and Cautano marbles formation has been deeply investigated from a geological and palaeontological point of view. However, the several exploited ornamental lithotypes have been differently named leading to confusion (Zaccagna 1890; Salmoiraghi 1892; Penta 1935; Pieri 1950; D'Argenio 1961). All these authors describe the most relevant feature of this formation, that is the peculiar chromatic and textural pattern. This feature promoted uses dating back to the beginning of the 17th century as witnessed by the Neapolitan Baroque architecture (Cantone 1992; Aveta *et al.* 1993); typical examples are the Churches of SS Severino and Sossio, Gesù Nuovo, S. Pietro a Majella and the Certosa di S. Martino (Fig. 1).

Due to these specific features, the use of this stone is also recorded outside the Campania region in areas such as Rome. There, it was used to manufacture the frames of the small doors of the *loggia* close to the main entrance of S. Giovanni in Laterano, some portions of the pavement of Torlonia Chapel in S. Giovanni in Laterano and the major banister of the SS. Apostoli Church (Penta 1935, 1937).

From: PŘIKRYL, R. & TÖRÖK, Á. (eds) *Natural Stone Resources for Historical Monuments.* Geological Society, London, Special Publications, **333**, 219–231.
DOI: 10.1144/SP333.21 0305-8719/10/$15.00

Fig. 1. Examples of uses of Vitulano Marbles in the monumental architecture of Campania Region. (**a**) Caserta Royal Palace – main stairway: banister and coatings. (**b**) Caserta Royal Palace – detail. (**c**) Naples - S. Pietro a Majella Church: altar detail. (**d**) Naples - Gesù Vecchio Church: detail of the inlay banister made with *mischi* (mixed) marbles.

The limited availability of local resources in the 18th century and the consequent increase of costs relegated the use of this stone only to the richest representatives of clergy and nobility in funerary chapels, sacred buildings and noble palaces. Only in the first half of the 18th century, after the arrival of Charles III Bourbon in Naples, were some traditional exploitation sites reactivated to meet the huge demand for natural stones necessary to build the Royal Palaces of Portici, Capodimonte and Caserta (Penta 1935, 1937; Fiengo 1983).

The same 'marbles' were also used in the Archaeological Museum for the restoration of the Duomo, the Gesù Nuovo Church (Penta 1935) and in the Pompei Sanctuary Basilica, S. Carlo Theatre, Stock Exchange Palace (Penta 1935).

In the following centuries, the exploitation activity continued discontinuously as a function of the market trend. There were periods of great unrest, during which the material was also exported abroad to France, England, North America, Australia and Russia where it was used in coating slabs of the Kremlin Palace (D'Argenio 1961).

At the beginning of 1930s, the quarrying continued for few months per year mainly focused on the production of crushed rock and cut-stones (Penta 1933). A renewed interest towards the use of local stone for ornamental purposes in the late 1930s resulted in a significant resumption of the Vitulano marbles. These were used in monumental buildings and representative public spaces such as the New Building Post Office in Naples, the Court of Appeal Hall in Naples and the new buildings of the Provinces of Naples and Caserta (Penta 1937; Federazione Nazionale Fascista degli Esercenti Industrie Estrattive 1939). After this period, the exploitation was definitely abandoned with the only exception being a limited production devoted to the manufacture of small paving elements and furnishing objects.

Geological setting

Former geological studies on the investigated formation improperly define this rock as 'marbles'. Obviously, this term should be viewed as a commercial denomination of a natural stone having the ability to be polished and containing mainly calcium carbonate and not as a geological classification (UNI EN 12670, 2003).

The former geological observations were carried out by Zaccagna (1890) in an engineering report kept in the archives of the Geological Service. The quarries of Vitulano marbles are described in this report, most of them constituted by polychrome calcareous breccias. Zaccagna, as later confirmed by Cassetti (1894), linked these limestones to Cretaceous age on the basis of the occurrence of macrofossils such as rudistae, acteonellae, nerineae, requieniae and trigoniee (D'Argenio 1959, 1961).

The genesis of the Vitulano marbles is strictly related to the palaeogeographic evolution of the Camposauro Massif during the middle-lower Cenozoic. In fact, between the end of the Cretaceous and the Miocene, the calcareous masses of Mt Camposauro were affected as a consequence of uplift by a strong erosive action with widespread karst phenomena. This led to the formation of underground and surface cavities. During this emersion phase, which coincided with the Palaeogene stratigraphic hiatus, the bauxite deposits were formed (D'Argenio 1959, 1961).

Afterwards, during the Langhian transgression, a rill wash of the topographic surface and of the above-mentioned bauxites occurred which enriched the water with pigmented material, prevailingly dark red or yellow. These waters later contributed to cement the brecciated debris inside fractures and fissures previously enlarged by karst phenomena. These fractures along with the polychrome cement, were occasionally filled by floated bauxitic pisoliths (Fig. 2a). Carannante *et al.* (1988)

Fig. 2. (**a**) Floated pisoliths within a bauxitic layer and (**b**) 'Uria quarry' located at *Aia del Palillo*.

considered the whole geological formation including the Vitulano marbles to be of Cretaceous age (lower Aptian – Senonian).

On the basis of different ages and genesis, D'Argenio (1961, 1963, 1967) distinguishes lower and upper Vitulano marbles. In particular, the 'lower' marbles, forming lenticular-shaped deposits of polychrome breccias with thickness and width less than 5 m, are linked to the Cenomanian continental phase period. Their genesis is due to the filling of palaeofractures of the partially karstified emersion substrate with carbonate and subordinate bauxitic fragments of different rounding grades. On the contrary, the so-called 'Upper Vitulano marbles' were attributed to the Senonian emersion phase period (D'Argenio 1967), formed by the filling of palaeocavities originated by either karst dissolution or jointing with calcareous breccias and bauxitic and alabaster deposits.

The 'Vitulano Marbles deposits' falls on the eastern side of Mt Camposauro in Vitulano and Cautano towns within the Benevento province. Eleven exploitation sites are located around Colle della Noce (Fig. 3) over an area of about 2 km^2 (Zaccagna 1890; Penta 1935; D'Argenio 1961). The most interesting quarry (Fig. 2b), historically known as Uria quarry and named *Cave di Marmo* (VT5, Fig. 3) on the topographic map (1:25.000 scale, sheet No. 173 – Vitulano), is a slope quarry located at about 800 m a.s.l. Ornamental stones such as the Uria red, a red cement heterometric breccia, have been exploited for a long time from this site. Due to its intense jointing, the productive horizon is not continuous.

Most of the quarries are currently abandoned as a consequence of the scarce road network and due to the geostructural features of the formation. The occurrence of crystalline calcite veins and of frequent voids between the clasts also makes the rock difficult to shape with a consequent large loss of material.

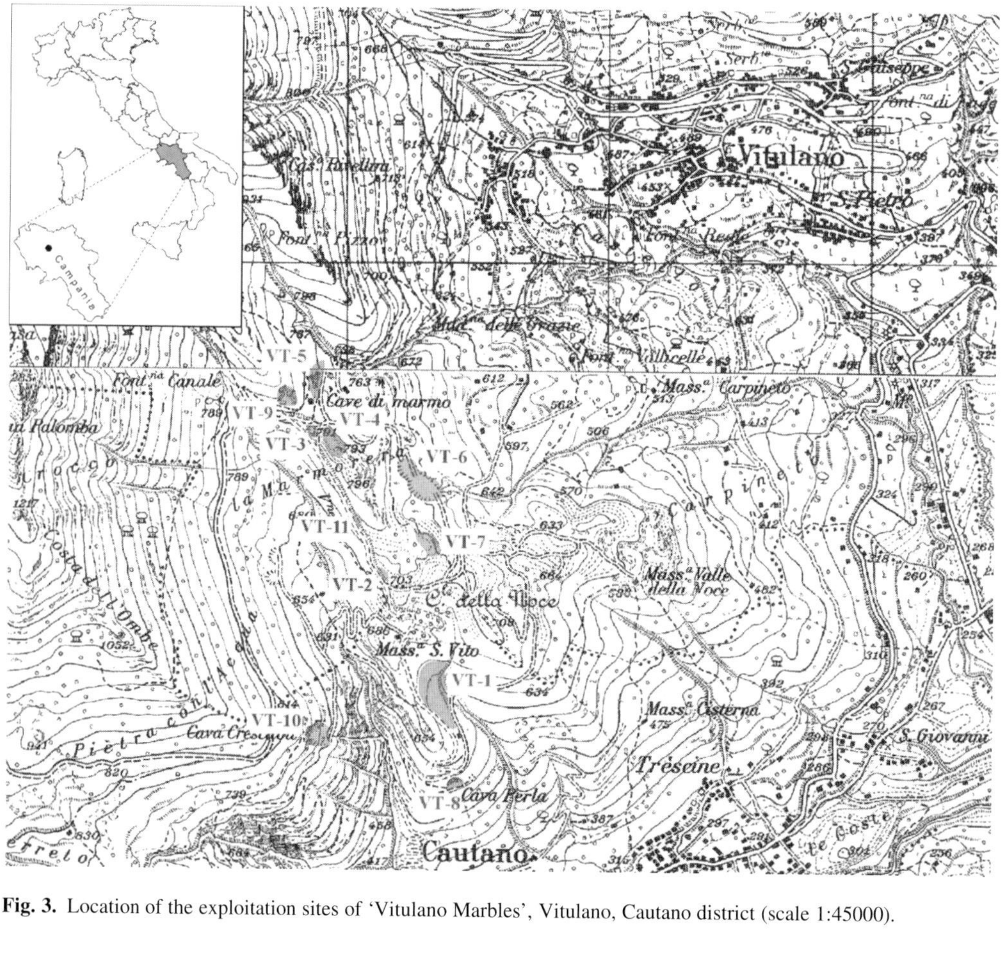

Fig. 3. Location of the exploitation sites of 'Vitulano Marbles', Vitulano, Cautano district (scale 1:45000).

Methods

Structural characterization of exploitation sites

To preliminarily evaluate the potential productivity of each exploitation site, a spatial distribution analysis of the joints finalized to an overall assessment of the rock block unit volumes was carried out. Cross scanlines were performed to measure fractures exposed along the layer or crosscut surfaces, recording data for each discontinuity according to the ISRM suggestions (1978). The distribution of the scanlines, although conditioned by the site accessibility and by the presence of suitable outcrops at easily reachable elevations, can be considered representative of the studied deposit. The unit volume of the potentially exploitable blocks was estimated using the Palmstrøm equation (1996):

$$V_b = \frac{S_1 \times S_2 \times S_3}{\sin\gamma_1 \times \sin\gamma_2 \times \sin\gamma_3},$$

where V_b is the volume of the block, S_1, S_2, S_3 are the spacing of any single discontinuity set (in m) and γ_1, γ_2, γ_3 are the angles between the discontinuity sets.

Mineralogical characterization

Mineralogical characterization was determined by both optical microscope observations (Leitz Laborlux Pol 40) and by X-ray powder diffraction analysis (XRPD) (Philips PW1730/3710) using a CuKα radiation, incident- and diffracted-beam Soller slits, curved graphite crystal monochromator, 2θ range from 3° to 100°, step size 0.02°, and 10 s counting time per step.

Thermal analysis was carried out with a multiple thermoanalyzer Netzsch STA 409.

Specific gravity and bulk density

Specific gravity, expressed in units of kg m^{-3}, was measured with a He-pycnometer (Micromeritics Multivolume Pycnometer 1305) on powdered specimens ($\pm$0.1 to 0.2 % accuracy). Bulk density was obtained by weighing cylindrical specimens (2.5 cm diameter; height $\leq$ 3 cm). Measured apparent and specific volumes allowed the open porosity to be calculated.

Capillarity absorption

The amount of water absorbed as a function of time was measured according to the Italian standard reported in NorMaL 11/85 (1985). Specimens used for this test had a cylindrical shape and a surface/apparent volume (S/V) ratio of 1 cm^{-1} $\leq$ s(cm^2)/v(cm^3) $\leq$ 2 cm^{-1}.

Water absorption by total immersion

This test enables the amount of water absorbed by a natural stone sample after immersion at atmosphere pressure to be determined. Cubic specimens (7.1 cm) have been used and the tests were carried out following the recommendations of NorMaL 7/81 (1981).

Uniaxial compressive strength

Uniaxial compressive strength (UCS) tests were carried out with a Controls C5600 testing device allowing a maximum axial load of 3000 kN. The axial load was increased continuously at a constant rate of 1.0 $\pm$ 0.5 MPa/s. Tests were carried out following the procedure suggested by Italian standard UNI EN 1926 (2000) on cubic shaped specimens (7.1 cm).

Secant modulus of elasticity (Young's modulus)

The Young's modulus and stress-strain curve was determined by means of uniaxial compressive tests on cubic-shaped specimens (7.1 cm) and following the procedures recommended by Italian standard UNI 9724/8 (1992). Two strain gauges were applied to each specimen to continuously record the strain as a function of the applied increasing pressure. The test was carried out with the same device used for the determination of the UCS at a constant load rate of 0.5 MPa s^{-1}; load and strain were continuously recorded by an automatic data logger.

Flexural strength

Flexural strength was measured on five samples (12 $\times$ 3 $\times$ 3 cm in size) of each variety of stone, following the Italian UNI 9724/5 recommendations (1990). The load was constantly increased at a rate below 0.2 MPa s^{-1}.

Abrasion resistance

Abrasion resistance was determined on samples (12 $\times$ 12 $\times$ 3 cm in size) of each variety of stone following the European standard UNI EN 14157 (2005).

Impact resistance

Impact resistance is defined by the following expression:

$$W = Mhg \qquad (2)$$

where W is the break collision energy in joules, M is the mass of a spherical steel ball (1.00 kg), h is the lowest falling height of the steel sphere necessary to break a rock slab (20 × 20 × 3 cm) resting upon a bed of silica sand and *g* is the gravitational acceleration (UNI EN 14158, 2005).

Linear thermal expansion coefficient

After drying to constant mass, the specimen (L = 25 cm; W = 5 cm; H = 2 cm) is subjected to length measurement (L) while maintaining at least two different temperatures. The linear coefficient of thermal expansion between the extreme temperatures is expressed as the unitary change in length for a change of temperature of 1°C (UNI EN 14581, 2005).

Colorimetric measurements

These analyses were carried out by means of a Konica-Minolta CR-400 tristimulus colorimeter, in order to distinguish absolute colour values of the different facies as distinctive features of the lithotypes.

Results and discussions

Site characterization

Structural characterization carried out on the Uria front quarry provided evidence of the presence of many random-oriented fractures due to either tectonic or anthropogenic causes. These fractures strongly influence the type and the shape of the block, making the calculation of the volumes rather approximate.

Five scanlines were carried out for a total length of about 60 m (61 discontinuities surveyed). In each single survey of 5–16 m length, six measures (up to maximum of 25) were performed.

The high jointing grade of the studied facies did not allow discontinuity sets to be observed, as most of them have a random distribution. This is particularly true in sites 2 and 5. In the first case, the jointing is also due to the use of explosives; evidence of this disturbance is provided by the several rose-shaped fractures which radiate from a small central hole where the explosive was placed.

The statistical analysis carried out on the discontinuities of site 1 (Fig. 4a) provided evidence for the occurrence of at least one set of joints, even if their attitude values are not concentrated. Site 4 (Fig. 4b), partly corresponding to a fault slickenside, also confirms the data dispersion with the exception of two peaks (13% and 15%) linked to two joint sets dipping NE and SW, respectively.

Since this statistical approach at each site did not yield enough sets to allow the analytical determination of the potentially exploitable unit volumes, a cumulative analysis of the entire set of measures (61) was carried out. The elaboration (Fig. 4c) therefore allowed at least two discontinuity sets (J1 and J2) to be distinguished, in addition to the system corresponding to the layers (J3). The potential unit volumes were estimated on the basis. However, these values ranging between 0.3 and 15 m^3, as already discussed, should be considered as approximate due to the large number of random fractures.

Petrography

From a petrographical point of view, Vitulano marbles can be essentially classified as two varieties. The red and grey varieties are, characterized by calcareous clasts set in a prevailing red-brownish bauxitic matrix and angular calcareous clasts in a scarce pale grey matrix, respectively. Also worthy of mention are other varieties characterized by textures and structures slightly different from the main types.

Optical microscope analyses revealed quite a complex structure due to the lithological features of the deposit (Fig. 5). Two distinct parts can be distinguished in thin sections, host rock and filling. As far as the carbonatic component is considered, the terminology proposed by Dunham (1962) and

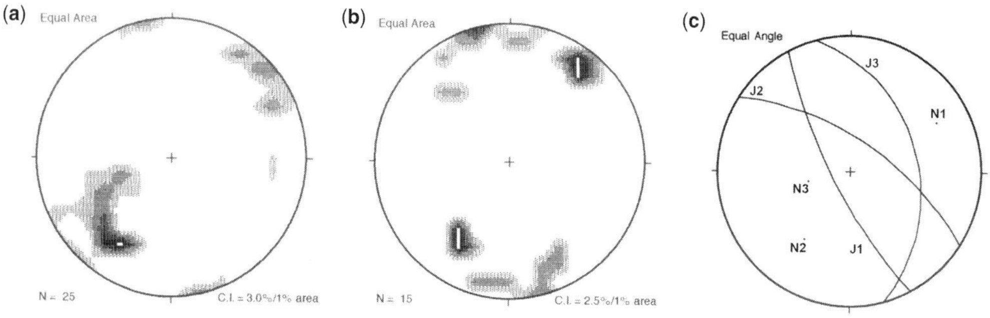

Fig. 4. Stereoplots of the Uria quarry discontinuities: (**a, b**) Contour diagrams of sites 1 and 4 and (**c**), Great circles of the main discontinuity sets in the different sites.

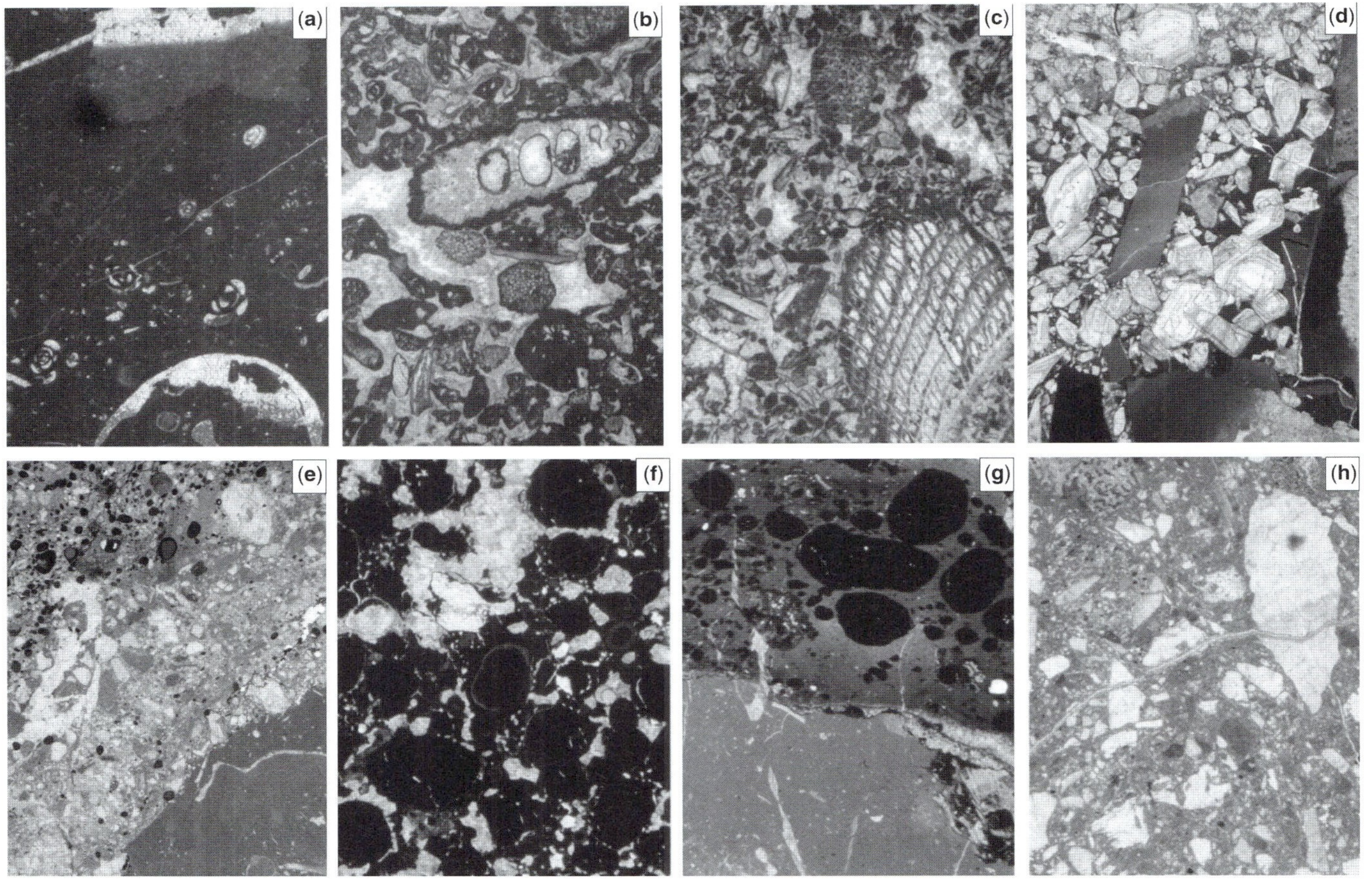

Fig. 5. Optical micrographs of Vitulano Marbles in grey (**a, b, c**) and red (**d, e, f, g**) facies. (**a**) wackestone with bentonic foraminifera and gasteropods (magn. 20×); (**b**) grainstone-packstone with algae, gasteropods, miliolidae and foraminifera (magn. 10×); (**c**) grainstone with abundant bioclasts (green algae dasicladales, radiolitides) and peloids (magn. 10×); (**d**) dolomite crystals filling a cavity (magn. 10×); (**e**) wackestone with gasteropods and ostracodes, with different filling stages clearly evident (magn. 10×); (**f**) bauxitic pisoliths (magn. 20×); (**g**) grainstone micrite and silt filling with red-brown bauxitic pisoliths set in silt sediments of yellow-orange colour due to the presence of Fe and Al oxides (magn. 10×); (**h**) typical filling of karst cavities constituted by lithoclasts of different origin set in a reddish silt matrix (magn. 10×).

integrated with the Folk classification (1962) defines the limestone constituting the host rock as grainstone-packstone with bioclasts in micritic matrix and wackestone-packstone with siltitic matrix. Karst cavities are characterized by different filling deposits: muds and reddish silt, bauxitic pisoliths or alabastrine calcite (Allocca 2003). In particular, limestones of the grey variety host rock are defined as wackestone, packstone and grainstone with benthonic foraminifers (miliolides, textulariides, orbitolinided, discorbides), ostracodes, gasteropoda shells and algae. They demonstrate dissolution structures constituted by karst cavities secondarily paved by silt and filled with spatic cement (biopelmicrite or biosparite; Fig. 5a, b, c). Wackestone also shows the occurrence of abundant peloids. On the contrary, samples of the red variety can be identified as packstone, wackestone and grainstone. Packstone is characterized by dolomite and zoned calcite detrital crystals, set in a micritic matrix (Fig. 5d). Wackestone is characterized by abundant bioclasts (radiolitides fragments, dissolved gasteropoda shells, lamellibranches shells, benthonic foraminifera: miliolides, textulariides), lithoclasts and peloids (Fig. 5e). Finally, grainstone is characterized by green algae dasicladales and radiolitides, abundant lithoclasts and peloids and advanced dissolution.

The filling material is mainly bauxite (residual clays), lithoclasts, red-brown bauxitic pisoliths cemented by spatic calcite of the same colour or included in yellow-orange silt and, nodules and ooids set in a pale brown micritic matrix (Fig. 5f, g). The cement found is of different origin; fractures filled by crystalline micrite (intramicrite) were also recorded (Fig. 5h).

XRD analyses on some samples of the red and grey facies provided evidence, as expected, of calcite as the main component; only rarely was dolomite also detected. Insoluble residue of the red facies, about 1%, is mainly constituted by boehmite, hematite and kaolinite; interstratified I/S and traces of titanium oxides such as anatase and rutile were also observed. Thermal analysis on the same raw samples (Fig. 6) also provided evidence of calcite as the main component, characterized by an endothermic peak at around 900–920 °C. The red facies collected at the Upper Uria quarry also show another endothermic peak at about 560 °C, likely due to the dehydroxilation of kaolinite (Fig. 6, right).

Thermal analysis of the insoluble fraction was only carried out on samples which gave a significant amount of residue (red facies). Two different typologies of residue were identified (Fig. 7). The fraction with prevailing bauxitic content (Fig. 7, left) is characterized by first a weak endothermic reaction at 110 °C (due to dehydration of small amount of clay minerals) and two exothermic reactions at 340° and 470 °C linked to the occurrence of iron oxides and hydroxides. The second residue (Fig. 7, right) is typical of kaolinite group minerals with two well-defined endothermic reactions at 110° and 560 °C.

XRF chemical analyses of both facies account for a high calcium oxide content (54–56 wt%). Na_2O and MnO are close to the lower instrumental detectable limit whereas K_2O is on average around 0.05 wt%; P_2O_5 and TiO_2 are both below 0.1 wt%. MgO only occurs in the red facies ranging around 1 wt%; Al_2O_3 and Fe_2O_3 are higher in the same facies, thus confirming the bauxitic composition of

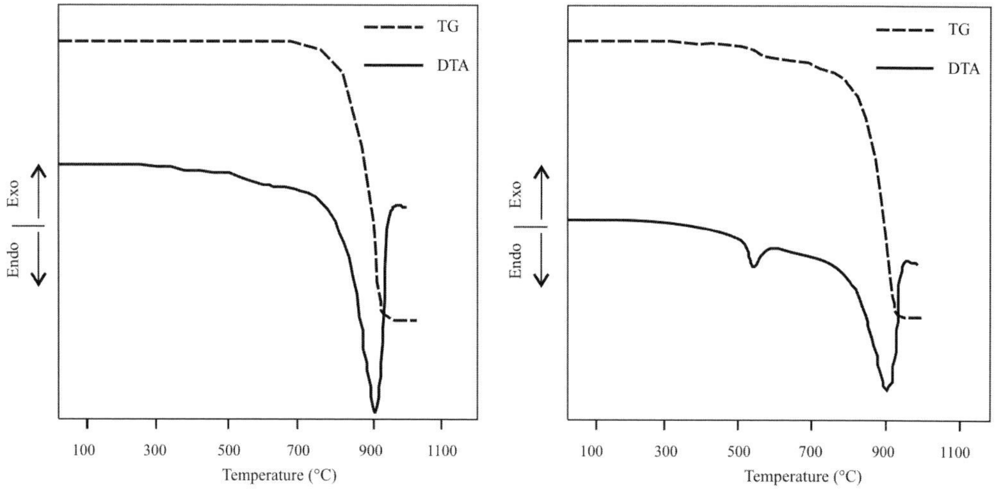

Fig. 6. Thermal analysis of Vitulano samples. Left: grey facies; right: red facies (Allocca 2003).

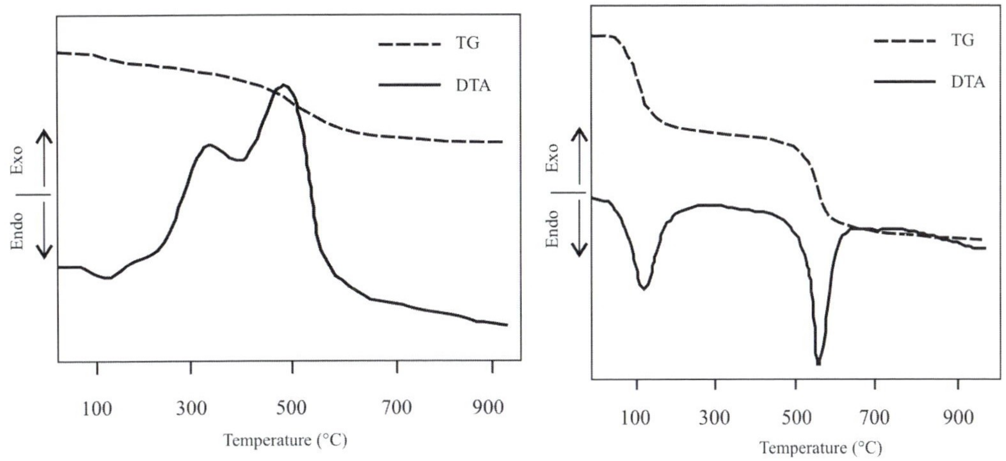

Fig. 7. Thermal analysis of different insoluble residue typologies from samples of the red facies (Allocca 2003).

non-carbonatic fraction characterized by boehmite and diaspore.

Colorimetric measurements were carried out on 10 sampling points. Mean values are 26, 39, 151 (HSV values) for the grey facies, 12, 108, 100 (HSV values) for the dark facies and 12, 82, 112 (HSV values) for the pale type of the red facies.

Physical properties

Very slight differences in the mean values of specific gravity and bulk density between the two facies of Vitulano marble were recorded (Table 1). Data on open porosity (on average 2.71 vol. % for the red facies and 2.02 vol. % for the grey type) and compactness (very close to 1 for both) rank the Vitulano marbles among low porosity materials. Open porosity and compactness may also give an overall evaluation of the resistance and the durability of the rock. Slight differences between porosities of the two facies should be related to the occurrence of bauxitic clasts, which provide the typical red colour to the red facies, and to the presence of microfissures.

Experimental curves concerning the water capillary absorption and water imbibition by total immersion tests also display a homogeneous trend with saturation occurring after about 10–20

Table 1. *Results of the physico-mechanical tests.*

	Vitulano marbles in red facies				Vitulano marbles in grey facies			
	n°samples	mean	max	min	n°samples	mean	max	min
Bulk density (kg m^{-3})	3	**2635**	2649	2629	3	**2645**	2659	2629
Specific gravity (kg m^{-3})	3	**2709**	2729	2699	3	**2699**	2720	2689
Compactness (−)	3	**0.97**	0.98	0.97	3	**0.98**	0.99	0.98
Open porosity (vol. %)	3	**2.71**	2.99	2.29	3	**2.02**	2.35	1.53
Capillarity absorption coefficient (g cm^{-2} s$^{0.5}$)	6	**0.11**	0.12	0.07	6	**0.09**	0.12	0.07
Imbibition capacity (%)	6	**0.19**	0.26	0.15	6	**0.20**	0.24	0.16
Ultrasonic dry velocity (m s^{-1})	6	**6105**	6438	5594	6	**6340**	6631	6214
Ultrasonic wet velocity (m s^{-1})	6	**6256**	6403	6093	6	**6466**	6541	6372
Uniaxial compressive strength (MPa)	5	**97**	114	82	5	**131**	192	74
Elastic tangent (50%) modulus (GPa)	2	**69.5**	–	–	2	**91.0**	–	–
Elastic average modulus (GPa)	2	**57.5**	–	–	2	**71.0**	–	–
Elastic secant modulus (GPa)	2	**61.5**	–	–	2	**77.0**	–	–
Flexural strength (MPa)	3	**14.9**	16.3	12.6	3	**12.5**	13.1	11.4
Impact resistance (kg m^2 s^{-2})	4	**2.45**	2.45	2.45	4	**1.96**	2.45	1.47
Abrasion resistance coefficient (mm)	3	**17.83**	19.00	17.00	3	**17.33**	17.5	17.00
Linear thermal expansion coefficient (mm mm^{-1} °C^{-1})	6	**0.013**	0.038	0.005	6	**0.016**	0.031	0.004

minutes (Fig. 8). Higher values generally recorded in the red facies are in agreement with the slight higher porosity of this lithotype, even although two specimens of the grey facies show highest values comparable to those of the red facies. This confirms the possible occurrence of a random microcrack net.

Uniaxial compressive strength (UCS) tests provided evidence for a mean value of 131 MPa for the grey facies; this was definitely higher than that of the red type (97.4 MPa) which likely suffered from the presence of the terrigenous component. Accordingly, the values of the elastic modules (tangent, mean and secant) are also higher for the grey variety. On this basis, both Vitulano varieties can be defined as strong rocks according to Deere & Miller (1966). Regarding their stiffness, the red marble can be classified as stiff and the grey type as very stiff.

As far as tensile strength is concerned, the high variability of the results (11.4–16.3 MPa) indicates the better performance of the red facies and, once again, confirms the heterogeneity of the materials. On the contrary, the abrasion test determined a similar behaviour of the two varieties with mean values lower than the reference value (20 mm). Moreover, the results of the impact resistance attribute a good attitude to the instant mechanical stress to both facies. Ultrasonic wave velocity, measured on the three main directions of the considered specimens, are quite homogeneous for the grey variety but show a wide variability for the red type. This behaviour should be related to the rock anisotropy due to the occurrence of calcite veins and/or residual pockets visible at macroscopic observation. In saturated samples, this condition fades away likely because the imbibition in other portions of the sample diminishes this effect.

The petrophysical investigation revealed that Vitulano Marbles can be considered geomaterials provided with good physical properties. This evidence is further confirmed by the comparison of their main technical parameters with those of other lithotypes available in Campania (Table 2).

Possible application of Vitulano marbles

On the basis of the currently available data, the most favourable use of Vitulano Marbles is as internal coating surfaces which reduce the exposure to weathering agents. In these conditions, the state of conservation of the stone is always fairly good with a few exceptions due to damage provoked by anthropic actions or surface opacization due to powder or atmospheric particulates. The reaction of the stone to weathering in those exceptions is different. In this case, black crusts by sulphation, surface pitting and discoloration of the most

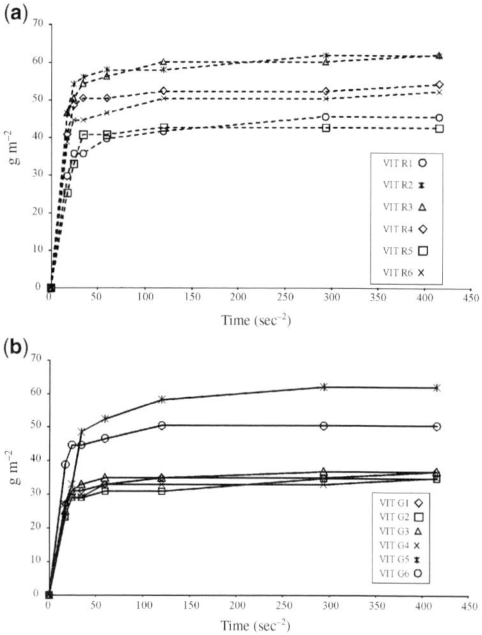

Fig. 8. Capillarity water absorption curves for (**a**) Vitulano grey facies and (**b**) red facies samples.

intensely red coloured portions have been recorded locally.

A problem of stone availability, definitely exists as a consequence of the complex geological-structural setting of the deposit. This makes the qualitative uniformity of this ornamental stone an occasional requisite and affects exploitation.

Conclusions

Vitulano marbles cannot be considered as a resource which can be subjected to a regular intensive exploitation, as regularly occurs in some adjacent regions with other sedimentary lithotypes (namely Trani stone from Puglia region or Perlato Royal from Lazio region). In the case of Vitulano marbles and, more generally, of the Campanian 'historical stones', attention should be paid to a possible 'quality' exploitation. This would be aimed only at restoration of the monumental stone heritage or for the accomplishment of particularly relevant contemporary pieces of work. As far as this latter aspect is concerned, it is suggested that Vitulano marbles, as well as other Campanian historical stones, should be used in one of the future stations of the Naples underground railway. This could try to counterbalance the current tendency which accounts for the use of industrial materials simulating the aesthetic features and the technical

Table 2. *Physico-mechanical properties of the main sedimentary geomaterials from Campania region (modified after Calcaterra* et al. *2007).*

	Vitulano marbles (red facies)	Vitulano marbles (grey facies)	Breccia irpina	Mondragone yellow	Mondragone grey	Cusano	Bellona	Padula
Specific gravity (kg m^{-3})	2709	2699	2749	2700	2684	2699	2799	2779
Open porosity (vol. %)	2.71	2.02	1.65	0.80	0.70	2.77	2.28	3.13
Capillarity absorption coefficient (g cm^{-2} $s^{0.5}$)	0.11	0.09	–	–	–	–	–	–
Imbibition capacity (%)	0.19	0.20	0.27	0.20	0.28	0.57	0.33	0.63
Ultrasonic dry velocity (m s^{-1})	6105	6340	5894	5786	5891	6225	–	5308
Ultrasonic wet velocity (m s^{-1})	6256	6466	6136	6087	6169	6339	–	5689
Uniaxial compressive strength (MPa)	97	131	101	106	75	98	161	75
Elastic tangent (50%) modulus (GPa)	69.5	91.0	82.0	28.5	15.1	96.0	–	80.0
Flexural strength (MPa)	14.9	12.5	–	13.6	5.42	13.3	20.5	11.1
Impact resistance (kg m^2 s^{-2})	2.45	1.96	2.74	–	–	4.80	–	4.41
Abrasion resistance coefficient (mm)	17.83	17.33	20.12	0.43 (°)	0.45 (°)	17.17	–	21.83
Linear thermal expansion coefficient (mm mm^{-1} °C^{-1})	0.013	0.016	–	–	–	–	–	0.014

(°) non-dimensional coefficient (*sensu* Regio Decreto 2232/1939) requisites of the natural stone. Concentrating historical stones of Campanian Region in regularly attended pieces of architecture would be equivalent to creating continuously enjoyable contemporary museums. This could contribute to bringing a resource back into favour. This resource has been exploited for several centuries but is currently disregarded, if not totally abandoned. This would also avoid the loss of a culture which made the stone heritage of the Campania Region worldwide famous, but which is today totally neglected by those involved in the design of great urban works. These projects should take into proper consideration the tradition and availability of regional resources also in the light of the recently approved Regional Plan of Exploitation Activities which, further than providing precise regulation of activities concerning the exploitation of the historical ornamental stones, also recommends actions aimed at their promotion.

This work is dedicated to the memory of two young and smart colleagues who prematurely passed away, Francesco Allocca and Gabriella Calicchio. Their diligent work facilitated an advance in the knowledge of the Campanian geomaterials. Thanks are due to G. Carannante for his support in sedimentological observations on thin sections. The authors are also grateful to the good friend and colleague Sossio Del Prete, for his help in structural and geomechanical surveys. The authors also thank anonymous reviewers and R. Přikryl for their careful revision of the manuscript. The work was carried out within the project *Progetto Dimostratore Campi Flegrei*, the *Centro di Competenza Regionale per lo Sviluppo ed il Trasferimento dell'Innovazione Applicata ai Beni Culturali e Ambientali* 'INNOVA' and under the financial support of PRIN 2003 (MdG) and MARS (*Associazione Pietre Storiche della Campania*).

References

ALLOCCA, F. 2003. *Recupero, salvaguardia e valorizzazione degli antichi siti di estrazione e delle pietre ornamentali utilizzate nell'architettura storica della Campania*. PhD Thesis, Sassari/Napoli Universities.

AVETA, A., AMORE, R. & MEGNA, C. 1993. *Il colore delle città. Note per il restauro delle cortine edilizie napoletane*. L'Arte Tipografica, Napoli.

CALCATERRA, D., CAPPELLETTI, P., LANGELLA, A., COLELLA, A. & DE'GENNARO, M. 2004. The ornamental stones of Caserta province: the Campanian Ignimbrite in the Medieval architecture of Casertavecchia. *Journal of Cultural Heritage*, **2**, 137–148.

CALCATERRA, D., CAPPELLETTI, P., MORRA, V., CALICCHIO, G. & D'ALBORA, M. 2007. I materiali lapidei storici nel quadro dell'attività estrattiva in Campania. *Proceedings of the International Conference CITTAM 2007 'Stone Building between Innovation and Tradition'*, Luciano Editore, Napoli, 133–142.

CANTONE, G. 1992. *Napoli Barocca*. Edizioni Laterza, Napoli.

CARANNANTE, G., D'ARGENIO, B., DELLO IACOVO, B., FERRERI, V., MINDSZENTY, A. & SIMONE, L. 1988. Studi sul carsismo cretacico dell'Appennino campano. *Memorie della Società Geologica Italiana*, **41**, 733–759.

CARTA, L., CALCATERRA, D., CAPPELLETTI, P., LANGELLA, A. & DE'GENNARO, M. 2005. The stone materials in the historical architecture of the ancient center of Sassari: distribution and state of conservation. *Journal of Cultural Heritage*, **6**, 277–286.

CASSETTI, M. 1894. Osservazioni geologiche sul Monte Massico presso Sessa Aurunca in provincia di Caserta. *Bolletino del Regio Comitato Geologico d'Italia*, **25**, 160–166.

D'ARGENIO, B. 1959. Osservazioni geomorfologiche sul gruppo del Taburno. *Bolletino della Società dei Naturalisti in Napoli*, **68**, 151–160.

D'ARGENIO, B. 1961. Osservazioni sulla genesi e l'età dei Marmi di Vitulano e sulla paleogeografia del Monte Camposauro. *Bolletino della Società dei Naturalisti in Napoli*, **70**, 2–12.

D'ARGENIO, B. 1963. I filoni sedimentari del Taburno-Camposauro (Appennino Campano). *Bolletino della Società dei Naturalisti in Napoli*, **72**, 138–143.

D'ARGENIO, B. 1967. Geologia del Taburno-Camposauro (Appennino Campano). *Atti Accademia Scienze Fisiche e Matematiche Napoli, Memorie Geomineralogiche sull'Italia centro-meridionale*, **6**, 35–218.

DEERE, D. U. & MILLER, R. P. 1966. *Engineering classification and index properties for intact rock*. Tech report no AFWL-TR-65-116, Air Force Weapons Lab, Kirtland Air Force Base, New Mexico.

DE GENNARO, R., CALCATERRA, D., DI GIROLAMO, P., LANGELLA, A. & DE'GENNARO, M. 2003. Discovering the stone heritage of southern Italy: technical properties of the Mondragone marble from Campania region. *Environmental Geology*, **44**, 266–276.

DI GIROLAMO, P., SGROSSO, I., DE GENNARO, R. & GIURAZZI, S. 2000. Metamorphic rocks in Campania (Southern Italy): the Mondragone Marbles. *Bolletino della Società Geologica Italiana*, **119**, 761–766.

DUNHAM, R. J. 1962. Classification of carbonate rocks according to depositional texture. *In*: HAM, W. E. (ed.) *Classification of Carbonate Rocks*. AAPG. Memoir, **1**, 108–121.

Federazione Nazionale Fascista degli Esercenti le Industrie Estrattive, 1939. *I Marmi Italiani*. **8**, Roma, a cura di G. Peverelli e F. Squartina.

FIENGO, G. 1983. *Organizzazione e produzione edilizia a Napoli all'avvento di Carlo di Borbone*. Edizioni Scientifiche Italiane, Napoli.

FOLK, R. L. 1962. Spectral subdivision of limestone types. *In*: HAM, W. E. (ed.) *Classification of Carbonate Rocks*. AAPG. Memoir, **1**, 62–84.

ISRM International Society for Rock Mechanics, 1978. Suggested methods for the quantitative description of discontinuities in rock masses. *International Journal of Rock Mechanics, Mining Science & Geomechanical Abstracts*, **15**, 319–368.

NorMaL 7/81, 1981. Water absorption by total immersion and imbibition capacity (in Italian). CNR-ICR, Rome

NorMaL 11/85, 1985. Water absorption by capillarity (in Italian). CNR-ICR, Rome.

PALMSTRØM, A. 1996. Characterising rock masses by the RMi for use in practical rock engineering. Part 2: some practical applications of Rock Mass index (RMi). *Tunneling and Underground Space Technology*, **11** (3), 287–303.

PENTA, F. 1933. Marmi e pietre della provincia di Benevento. *Lo Scultore e il Marmo*, **42**, 1–2.

PENTA, F. 1935. *I materiali da costruzione dell'Italia Meridionale*. Fondazione Politecnica del Mezzogiorno, Napoli.

PENTA, F. 1937. Marmi, graniti e pietre dell'Italia meridionale. *Marmi, Pietre e Graniti*, **15** (4), 3–22.

PIERI, M. 1950. *I Marmi d'Italia. Graniti e pietre ornamentali*. Hoepli, Milano.

Regio Decreto (Royal Decree) no. 2232, 1939. Rules for acceptance of building natural stones (in Italian). Supplemento alla Gazzetta Ufficiale, 18 aprile 1940, no 92.

SALMOIRAGHI, F. 1892. *Materiali Naturali da Costruzione*. Hoepli, Milano.

UNI 9724/5, 1990. Stones. Determination of the modulus of rupture of natural building stone (in Italian). Milan.

UNI 9724/8, 1992. Stones. Determination of the elastic modulus (monoaxial) (in Italian). Milan.

UNI EN 1926, 2000. Natural stone test methods - Determination of uniaxial compressive strength.

UNI EN 12670, 2003. Natural stone – Terminology (In Italian).

UNI EN 14157, 2005. Natural stone test methods - Determination of the abrasion resistance.

UNI EN 14158, 2005. Natural stone test methods - Determination of rupture energy.

UNI EN 14581, 2005. Natural stone test methods - Determination of linear thermal expansion coefficient.

ZACCAGNA, D. 1890. *Appraisal on the civil suite Finance Administration vs Izzo*. Unpublished report, Naples Court of Appeal. Manuscript hold by the Italian Geological Survey, Rome.

Index

Note: Page numbers in *italics* denote figures. Page numbers in **bold** denote tables.